Lecture Notes in Computer Science

Lecture Notes in Artificial Intelligence   14376

Founding Editor

Jörg Siekmann

Series Editors

Randy Goebel, *University of Alberta, Edmonton, Canada*
Wolfgang Wahlster, *DFKI, Berlin, Germany*
Zhi-Hua Zhou, *Nanjing University, Nanjing, China*

The series Lecture Notes in Artificial Intelligence (LNAI) was established in 1988 as a topical subseries of LNCS devoted to artificial intelligence.

The series publishes state-of-the-art research results at a high level. As with the LNCS mother series, the mission of the series is to serve the international R & D community by providing an invaluable service, mainly focused on the publication of conference and workshop proceedings and postproceedings.

Katsuhiro Honda · Bac Le · Van-Nam Huynh ·
Masahiro Inuiguchi · Youji Kohda
Editors

Integrated Uncertainty in Knowledge Modelling and Decision Making

10th International Symposium, IUKM 2023
Kanazawa, Japan, November 2–4, 2023
Proceedings, Part II

 Springer

Editors
Katsuhiro Honda 🄳
Osaka Metropolitan University
Sakai, Osaka, Japan

Bac Le 🄳
Vietnam National University
Ho Chi Minh City, Vietnam

Van-Nam Huynh 🄳
Japan Advanced Institute of Science
and Technology
Nomi, Ishikawa, Japan

Masahiro Inuiguchi 🄳
Osaka University
Toyonaka, Osaka, Japan

Youji Kohda 🄳
Japan Advanced Institute of Science
and Technology
Nomi, Ishikawa, Japan

ISSN 0302-9743   ISSN 1611-3349 (electronic)
Lecture Notes in Artificial Intelligence
ISBN 978-3-031-46780-6   ISBN 978-3-031-46781-3 (eBook)
https://doi.org/10.1007/978-3-031-46781-3

LNCS Sublibrary: SL7 – Artificial Intelligence

This Springer imprint is published by the registered company Springer Nature Switzerland AG
The registered company address is: Gewerbestrasse 11, 6330 Cham, Switzerland

Paper in this product is recyclable.

Preface

This volume contains the papers that were accepted for presentation at the 10th International Symposium on Integrated Uncertainty in Knowledge Modelling and Decision Making (IUKM 2023), held in Ishikawa, Japan, November 2–4, 2023.

The IUKM conference aims to provide a forum for exchanges of research results and ideas and practical experiences among researchers and practitioners involved with all aspects of uncertainty modelling and management. Previous editions of the conference were held in Ishikawa, Japan (2010), Hangzhou, China (2011), Beijing, China (2013), Nha Trang, Vietnam (2015), Da Nang, Vietnam (2016), Hanoi, Vietnam (2018), Nara, Japan (2019), Phuket, Thailand (2020), Ishikawa (online), Japan (2022) and their proceedings were published by Springer in AISC 68, LNAI 7027, LNAI 8032, LNAI 9376, LNAI 9978, LNAI 10758, LNAI 11471, LNAI 12482, and LNAI 13199, respectively.

The IUKM 2023 was jointly organized by Japan Advanced Institute of Science and Technology (JAIST), University of Science – Vietnam National University Ho Chi Minh City, Vietnam, and Osaka University, Japan. During IUKM 2023, a special event was also organized for honoring Hung T. Nguyen and Sadaaki Miyamoto for their contributions to the field of uncertainty theories.

This year, the conference received 107 submissions from 16 different countries. Each submission was peer single-blind review by at least two members of the Program Committee. After a thorough review process, 58 papers (54.2%) were accepted for presentation and inclusion in the LNAI proceedings. In addition to the accepted papers, the conference program also included 10 short presentations and featured the following four keynote talks:

- Hung T. Nguyen (New Mexico State University, USA; Chiang Mai University, Thailand), On Uncertainty in Partially Identified Models
- Sadaaki Miyamoto (University of Tsukuba, Japan), Two Classes of Fuzzy Clustering: Their Theoretical Contributions
- Hyun Oh Song (Seoul National University, Korea), Contrastive Discovery of Hierarchical Achievements in Reinforcement Learning
- Thierry Denoeux (Université de Technologie de Compiègne, France), Random Fuzzy Sets and Belief Functions: Application to Machine Learning

The conference proceedings are split into two volumes (LNAI 14375 and LNAI 14376). This volume contains 30 papers related to Machine Learning, Pattern Classification and Data Analysis, and Security and Privacy in Machine Learning.

As a follow-up of IUKM 2023, a special issue of the journal *Annals of Operations Research* is anticipated to include a small number of extended papers selected from the conference as well as other relevant contributions received in response to subsequent open calls. These journal submissions will go through a fresh round of reviews in accordance with the journal's guidelines.

IUKM 2023 was partially supported by JAIST Research Fund and the U.S. Office of Naval Research Global (Award No. N62909-23-1-2105). We are very thankful to the local organizing team from Japan Advanced Institute of Science and Technology for their hard working, efficient services, and wonderful local arrangements.

We would like to express our appreciation to the members of the Program Committee for their support and cooperation in this publication. We are also thankful to the staff of Springer for providing a meticulous service for the timely production of this volume. Last, but certainly not the least, our special thanks go to all the authors who submitted papers and all the attendees for their contributions and fruitful discussions that made this conference a great success.

November 2023

Katsuhiro Honda
Bac Le
Van-Nam Huynh
Masahiro Inuiguchi
Youji Kohda

Organization

General Chairs

Masahiro Inuiguchi	Osaka University, Japan
Youji Kohda	Japan Advanced Institute of Science and Technology (JAIST), Japan

Advisory Board

Michio Sugeno	European Center for Soft Computing, Spain
Hung T. Nguyen	New Mexico State University, USA; Chiang Mai University, Thailand
Sadaaki Miyamoto	University of Tsukuba, Japan
Akira Namatame	AOARD/AFRL and National Defense Academy of Japan, Japan

Program Chairs

Van-Nam Huynh	JAIST, Japan
Bac Le	University of Science, VNU-Ho Chi Minh City, Vietnam
Katsuhiro Honda	Osaka Metropolitan University, Japan

Special Sessions and Workshop Chair

The Dung Luong	Academy of Cryptography Techniques, Vietnam

Publicity Chair

Toan Nguyen-Mau	University of Information Technology, VNU-Ho Chi Minh City, Vietnam

Program Committee

Yaxin Bi	University of Ulster, UK
Matteo Brunelli	University of Trento, Italy
Tru Cao	University of Texas Health Science Center at Houston, USA
Tien-Tuan Dao	Centrale Lille Institut, France
Yong Deng	University of Electronic Science and Technology of China, China
Thierry Denoeux	University of Technology of Compiègne, France
Sebastien Destercke	University of Technology of Compiègne, France
Zied Elouedi	LARODEC, ISG de Tunis, Tunisia
Tomoe Entani	University of Hyogo, Japan
Katsushige Fujimoto	Fukushima University, Japan
Lluis Godo	IIIA - CSIC, Spain
Yukio Hamasuna	Kindai University, Japan
Katsuhiro Honda	Osaka Metropolitan University, Japan
Tzung-Pei Hong	National University of Kaohsiung, Taiwan
Jih Cheng Huang	Soochow University, Taiwan
Van-Nam Huynh	JAIST, Japan
Hiroyuki Inoue	University of Fukui, Japan
Atsuchi Inoue	Eastern Washington University, USA
Masahiro Inuiguchi	Osaka University, Japan
Radim Jirousek	Prague University of Economics and Business, Czech Republic
Jung Sik Jeong	Mokpo National Maritime University, South Korea
Yuchi Kanzawa	Shibaura Institute of Technology, Japan
Yasuo Kudo	Muroran Institute of Technology, Japan
Vladik Kreinovich	University of Texas at El Paso, USA
Yoshifumi Kusunoki	Osaka Metropolitan University, Japan
Bac Le	University of Science, VNU-Ho Chi Minh City, Vietnam
Churn-Jung Liau	Academia Sinica, Taiwan
Marimin Marimin	Bogor Agricultural University, Indonesia
Luis Martinez	University of Jaen, Spain
Radko Mesiar	Slovak Univ. of Tech. in Bratislava, Slovakia
Tetsuya Murai	Chitose Institute of Science and Technology, Japan
Michinori Nakata	Josai International University, Japan
Canh Hao Nguyen	Kyoto University, Japan

Duy Hung Nguyen   Sirindhorn International Institute of Technology,
                           Thailand
Vu-Linh Nguyen   University of Technology of Compiègne, France
Akira Notsu   Osaka Metropolitan University, Japan
Vilem Novak   Ostrava University, Czech Republic
Sa-Aat Niwitpong   King Mongkut's University of Technology North
                           Bangkok, Thailand
Warut Pannakkong   Sirindhorn International Institute of Technology,
                           Thailand
Irina Perfilieva   Ostrava University, Czech Republic
Frédéric Pichon   Artois University, France
Zengchang Qin   Beihang University, China
Jaroslav Ramik   Silesian University in Opava, Czech Republic
Hiroshi Sakai   Kyushu Institute of Technology, Japan
Hirosato Seki   Osaka University, Japan
Kao-Yi Shen   Chinese Culture University, Taiwan
Dominik Slezak   University of Warsaw, Poland
Roman Slowinski   Poznan University of Technology, Poland
Kazuhiro Takeuchi   Osaka Electro-Communication University, Japan
Yongchuan Tang   Zhejiang University, China
Roengchai Tansuchat   Chiang Mai University, Thailand
Phantipa Thipwiwatpotjana Chulalongkorn University, Thailand
Vicenc Torra   University of Skövde, Sweden
Seiki Ubukata   Osaka Metropolitan University, Japan
Guoyin Wang   Chongqing Univ. of Posts and Telecom., China
Woraphon Yamaka   Chiang Mai University, Thailand
Xiaodong Yue   Shanghai University, China
Chunlai Zhou   Renmin University of China

Local Arrangement Team

Van-Nam Huynh   JAIST, Japan
Yang Yu   JAIST, Japan
Trong-Hiep Hoang   JAIST, Japan
Leelertkij Thanapat   JAIST, Japan
Aswanuwath Lalitpat   JAIST, Japan
Supsermpol Pornpawee   JAIST, Japan
Dang-Man Nguyen   JAIST, Japan
Chemkomnerd Nittaya   JAIST, Japan
Xuan-Thang Tran   JAIST, Japan
Jiaying Ni   JAIST, Japan
Xuan-Truong Hoang   JAIST, Japan

Contents – Part II

Security and Privacy in Machine Learning

Contents – Part I

Economic Applications

Machine Learning

Inference Problem in Probabilistic Multi-label Classification

Vu-Linh Nguyen[1]([✉]), Xuan-Truong Hoang[2], and Van-Nam Huynh[2]

[1] Heudiasyc Laboratory, University of Technology of Compiègne, Compiègne, France
vu-linh.nguyen@hds.utc.fr
[2] Japan Advanced Institute of Science and Technology, Nomi, Japan
{hxtruong,huynh}@jaist.ac.jp

Abstract. In multi-label classification, each instance can belong to multiple labels simultaneously. Different evaluation criteria have been proposed for comparing ground-truth label sets and predictions. Probabilistic multi-label classifiers offer a unique advantage by allowing optimization of different criteria at prediction time, but they have been relatively underexplored due to a shortage of insights into inference complexity and evaluation criteria. To shrink this gap, we present a generic approach for developing polynomial-time inference algorithms for a family of criteria and discuss the potential (dis)advantages of some commonly used criteria. Finally, we envision future work aimed at providing a comprehensive understanding of inference complexity and criteria selection.

Keywords: Probabilistic multi-label classification · Inference Problem · Evaluation criteria · Bayes-optimal prediction

1 Introduction

In (multi-class) classification, a predictive system perceives a training data set (consisting of input-output pairs which specify individuals of a population) and a hypothesis space (consisting of the possible classifiers), and seeks a classifier that optimizes its chance of making accurate predictions with respect to some given evaluation criterion (which is typically a loss function or an accuracy metric) which reflects how good/bad the predictive system is. Numerous classification methods in the literature are constructed based on the statistical learning theory (SLT) [17]: Once the evaluation criterion is specified, empirical risk minimization principle is adopted to find an optimal classifier that optimizes the empirical risk functional constructed on the basis of the training set.

Probabilistic classification can be seen as a special SLT based method, which estimates, for each observation of the input space, a univariate probability distribution over the output space. This characteristic of probabilistic classification makes it remarkably different from its SLT based siblings. Concisely, the intention of probabilistic classification is to provide the end user with not only all necessary information about the optimal predictions of different loss functions

K. Honda et al. (Eds.): IUKM 2023, LNAI 14376, pp. 3–14, 2024.
https://doi.org/10.1007/978-3-031-46781-3_1

[6,10], but also information about the uncertainty associated with the possible predictions. Such a unique characteristic has been making probabilistic classification one important machine learning topic.

The unique advantage of allowing optimization of different criteria of probabilistic classification has been explored in the more general setting of multi-label classification (MLC), in which each instance can belong to multiple labels simultaneously (See [5,11,12,18,19] and references therein). Yet, existing studies on inference complexity and criteria selection are arguably limited to either a small number of criteria, such as subset 0/1 accuracy, Hamming accuracy, rank loss, and F-measure [5,18] or the use of strong assumptions, such as label independence [11,12,19]. These studies clearly confirm that inference complexity can greatly depend on the nature of the chosen criterion, and the mismatch between the criterion optimized during the inference time and the criterion used to assess the classifier can negatively impact predictive performance. Therefore, there is a clear motivation to extend such studies to include a larger number of criteria and classifiers. However, such extended studies can not be done meaningfully without the availability of scalable learning and inference algorithms.

Alongside recent advances in learning probabilistic classifiers [2,7,13,16], which can in principle help to enlarge the number of classifiers, this paper provides considerable opportunities for extending existing comparative studies [5,12,18] by presenting a generic approach for developing polynomial-time inference algorithms for a family of criteria. Along the way, we also provide readers with suggestions on extending the existing set of evaluation criteria in a meaningful way without facing intractable inference problems. Moreover, insights into structures of optimal predictions of different criteria themselves, which are necessary for devising efficient inference algorithms, already suggest potential (dis)advantages of some commonly used criteria (without the need of conducting further theoretical and empirical comparisons).

We provide in Sect. 2 a minimal description of probabilistic MLC. Our main contribution, which is a Partition-then-Sort (PtS) approach for developing polynomial-time inference algorithms for a family of criteria, is presented in Sect. 3. Section 4 illustrates the use of the PtS approach when tackling a few commonly used criteria, followed by a discussion on their potential (dis)advantages. Section 5 concludes this work and sketches out future work.

2   Preliminary

We shall recall basics of probabilistic MLC and notations.

2.1   Probabilistic MLC

Let $\mathcal{X}$ denote an instance space, and let $\mathcal{L} = \{\lambda_1, \ldots, \lambda_K\}$ be a finite set of class labels. We assume that an instance $x \in \mathcal{X}$ is (probabilistically) associated with a subset of labels $\Lambda = \Lambda(x) \in 2^{\mathcal{L}}$; this subset is often called the set of relevant labels, while the complement $\mathcal{L} \setminus \Lambda$ is considered as irrelevant for x. We identify a set Λ of relevant labels with a binary vector $\boldsymbol{y} = (y^1, \ldots, y^K)$, where $y^k = [\![\lambda_k \in \Lambda]\!]$.[1] By $\mathcal{Y} = \{0,1\}^K$ we denote the set of possible labelings. We assume that an instance $\boldsymbol{x} \in \mathcal{X}$ is (probabilistically) associated with members of $\mathcal{Y}$. We denote by $\boldsymbol{p}(Y \mid \boldsymbol{x})$ the conditional distribution of Y given $\mathbf{X} = \boldsymbol{x}$. Given training data $\mathcal{D} = \{(\boldsymbol{x}_n, \boldsymbol{y}_n) \mid n = 1, \ldots, N\}$ drawn independently from $\boldsymbol{p}(\mathbf{X}, Y)$, the goal in MLC is to learn a classifier $\boldsymbol{h}$, which is a mapping $\mathcal{X} \longrightarrow \mathcal{Y}$ that assigns to each instance $\boldsymbol{x} \in \mathcal{X}$ a class $\hat{\boldsymbol{y}} := \boldsymbol{h}(\boldsymbol{x}) \in \mathcal{Y}$.

To evaluate the performance of a classifier $\boldsymbol{h}$, a criterion $u : \mathcal{Y} \times \mathcal{Y} \longrightarrow \mathbb{R}_+$ is needed, which compares a prediction $\hat{\boldsymbol{y}}$ with a ground-truth $\boldsymbol{y}$. Each classifier $\boldsymbol{h}$ is evaluated using its expected value

$$R(\boldsymbol{h}) := \mathbf{E}\big[u(\mathbf{Y}, \boldsymbol{h}(\mathbf{X}))\big] = \int u(\boldsymbol{y}, \boldsymbol{h}(\boldsymbol{x})) \, d\,\mathbf{P}(\boldsymbol{x}, \boldsymbol{y}),$$

where $\mathbf{P}$ is the joint probability measure on $\mathcal{X} \times \mathcal{Y}$ characterizing the underlying data-generating process. Therefore, the Bayes-optimal classifier of any "the higher the better" u is given by

$$\boldsymbol{h}^* := \operatorname*{argmax}_{\boldsymbol{h} \in \mathcal{H}} R(\boldsymbol{h}), \tag{1}$$

where $\mathcal{H} \subseteq \mathcal{Y}^{\mathcal{X}}$ is the hypothesis space. When $\mathcal{H}$ is probabilistic, we can follow maximum likelihood estimation and define the Bayes-optimal classifier as:

$$\hat{\boldsymbol{h}} := \hat{\boldsymbol{p}} := \operatorname*{argmax}_{\boldsymbol{p} \in \mathcal{H}} \mathrm{CLL}(\boldsymbol{p} \mid \mathcal{D}) := \operatorname*{argmax}_{\boldsymbol{p} \in \mathcal{H}} \frac{1}{N} \sum_{n=1}^{N} \log \boldsymbol{p}(\boldsymbol{y}_n \mid \boldsymbol{x}_n). \tag{2}$$

To avoid overfitting, the CLL is often augmented by a regularization term [13].

Once the classifier (2) is learned from $\mathcal{D}$, we can in principle find an optimal prediction of any criterion ℓ at the prediction time [5,11,13,18]. More precisely, assume the classifier (2) is made available, and predicts for each query instance $\boldsymbol{x}$ a probability distribution $\boldsymbol{p}(\cdot \mid \boldsymbol{x})$ on the set of labelings $\mathcal{Y}$. The Bayes-optimal prediction (BOP) of any "the higher the better" u is then given by

$$\hat{\boldsymbol{y}} = \hat{\boldsymbol{y}}(\boldsymbol{x}) \in \operatorname*{argmax}_{\bar{\boldsymbol{y}} \in \mathcal{Y}} \mathbf{E}\big(u(\boldsymbol{y}, \bar{\boldsymbol{y}})\big) = \operatorname*{argmax}_{\bar{\boldsymbol{y}} \in \mathcal{Y}} \sum_{\boldsymbol{y} \in \mathcal{Y}} u(\boldsymbol{y}, \bar{\boldsymbol{y}}) \, \boldsymbol{p}(\boldsymbol{y} \mid \boldsymbol{x}). \tag{3}$$

2.2   Inference Problem

Yet, different evaluation criteria may call for different BOPs (3) [5,13,18]. Knowledge about the probability distribution $\boldsymbol{p}(\mathcal{Y}|\boldsymbol{x})$ is necessary for finding BOP (3) of any criterion. It is known that the inference complexity, i.e., the complexity of finding BOP can greatly depend on the nature of the chosen loss function. Moreover, the inference complexity is often reduced if one either takes into account the nature of the probabilistic classifiers [8,13] or accepts the use of strong assumptions, such as label independence [9,11,14,19].

---

[1] $[\![\cdot]\!]$ is the indicator function, i.e., $[\![A]\!] = 1$ if the predicate A is true and $= 0$ otherwise.

To make the contribution generic, we shall focus on the most general setting of probabilistic MLC where the classifier (2) is made available, and predicts for each query instance $\boldsymbol{x}$ a probability distribution $\boldsymbol{p}(\cdot \,|\, \boldsymbol{x})$ on the set of labelings $\mathcal{Y}$. Apart from that, one has no further access to additional information of the probabilistic classifiers and is not allowed to make additional assumptions, such as label dependence assumption, to simplify the inference problem. To the best of our knowledge, under this setting, the inference complexity has been studied for a few criteria [5,18].

While finding a BOP of the subset 0/1 accuracy, i.e., the most probable $\boldsymbol{y} \in \mathcal{Y}$, is indeed intractable, finding a BOP of the Hamming accuracy is done in $O(K)$ given the marginal probabilities

$$\boldsymbol{p}_k(b \,|\, \boldsymbol{x}) = \sum_{\boldsymbol{y} \in \mathcal{Y}:y_k=b} p(\boldsymbol{y} \,|\, \boldsymbol{x}), b \in \{0,1\}, k \in [K]. \tag{4}$$

A BOP of the rank loss can be found in $O(K \log(K))$ by sorting the labels according to the decreasing order of their marginal probabilities $p_k := \boldsymbol{p}_k(1 \,|\, \boldsymbol{x})$.

Finding a BOP of the F-measure

$$f_\beta(\boldsymbol{y}, \hat{\boldsymbol{y}}) = \frac{(1 + \beta^2) \sum_{k=1}^{K} \hat{y}_k \, y_k}{\beta^2 \sum_{k=1}^{K} y_k + \sum_{k=1}^{K} \hat{y}_k} \tag{5}$$

requires more attention [18]. Assume probabilities on pairwise label combinations

$$\boldsymbol{p}_{k,s} := \boldsymbol{p}(y_k = 1, s_{\boldsymbol{y}} = s \,|\, \boldsymbol{x}), \text{ where } s_{\boldsymbol{y}} = \sum_{k=1}^{K} y_k, k, s \in [K], \tag{6}$$

are made available [4,18]. For any $l \in [K]_0 := \{0\} \cup [K]$, denote by

$$\mathcal{Y}^l := \left\{ \boldsymbol{y} \in \mathcal{Y} \,|\, s_{\boldsymbol{y}} := \sum_{k=1}^{K} y_k = l \right\}, \tag{7}$$

the problem of finding BOP of f_β is decomposed into an inner and an outer maximization (8)–(9) as follows:

$$\hat{\boldsymbol{y}}^l \in \underset{\bar{\boldsymbol{y}} \in \mathcal{Y}^l}{\operatorname{argmax}} \mathbf{E} \left(f_\beta(\boldsymbol{y}, \bar{\boldsymbol{y}}) \right), \tag{8}$$

$$\hat{\boldsymbol{y}} \in \underset{\bar{\boldsymbol{y}} \in \{\hat{\boldsymbol{y}}^l | l \in [K]_0\}}{\operatorname{argmax}} \mathbf{E} \left(f_\beta(\boldsymbol{y}, \hat{\boldsymbol{y}}^l) \right). \tag{9}$$

For any $l \in [K]$, the optimal prediction of the inner optimization problem (8) is

$$\hat{\boldsymbol{y}}^l = \underset{\bar{\boldsymbol{y}} \in \mathcal{Y}^l}{\operatorname{argmax}} \sum_{k=1}^{K} \bar{y}_k q_k^\beta = \underset{\bar{\boldsymbol{y}} \in \mathcal{Y}^l}{\operatorname{argmax}} \sum_{k=1}^{K} \bar{y}_{(k)} q_{(k)}^\beta, \tag{10}$$

where (k) is the index of the label with k-th highest score

$$q_k^\beta = \sum_{s=1}^{K} \frac{p_{k,s}}{\beta^2 l + s}, k \in [k].$$

Thus, to find $\hat{y}^l$ (10), it is sufficient to set the l top ranked labels $\bar{y}_{(k)} = 1$, and $\bar{y}_{(k)} = 0$ for other labels. A BOP (9) of f_β is simply the local optimal prediction $\hat{y}^l$ (10) with the highest expected f_β.

Regarding the inference complexity, finding a BOP of f_β can be done in time $O(K^3)$ given K^2 probability estimates (6) and $p(\mathbf{0}_K \mid x)$. As acknowledged in [18], faster algorithms working in $O(K^{2.376})$ are known [3].

3   A Partition-then-Sort (PtS) Approach

A closer look at the steps for finding a BOP of the f_β is enough to see that it can be conceptualized as steps of a PtS approach:

– partition the solution space $\mathcal{Y}$ into $K + 1$ "regions" (7),
– for each "region", a local optimal BOP can be found by basically sorting the labels according the decreasing order of appropriate scores,
– and pick up the local optimal BOP with the highest expected value.

In Sect. 3.1, we introduce a family of criteria which covers commonly used criteria as its members. We then present how to use the PtS approach to develop polynomial-time inference algorithms for its members and analyze the upper bound complexity. In Sect. 3.2, we show that the PtS approach can be adapted to develop polynomial-time inference algorithms for criteria which are linear combinations of single criteria which satisfy certain conditions.

3.1   The Cases of Single Criteria

Different evaluation criteria found in the literature (See [11,15] and references therein) can be seen as members of the following family

$$f(\mathbf{y}, \bar{\mathbf{y}}) = \frac{S_n + \sum_{k=1}^{K} \alpha_k g(y_k)\bar{g}(\bar{y}_k)}{S_d + \beta(s_{\mathbf{y}}) + \gamma(s_{\bar{\mathbf{y}}})}, \tag{11}$$

where S_n, S_d, α_k are constants, $\beta(\cdot)$ and $\gamma(\cdot)$ are any functions, and $(g(y_k), \bar{g}(\bar{y}_k)) \in \{(y_k, \bar{y}_k), (1 - y_k, \bar{y}_k), (y_k, 1 - \bar{y}_k), (1 - y_k, 1 - \bar{y}_k)\}\}$. Concrete examples are f_β, precision, recall, specificity and negative predictive value [1,15,20]. Moreover, there would be a plenty of room for developing new criteria. For example, by choosing concrete formulae of $\beta(\cdot)$, $\gamma(\cdot)$, $g(\cdot)$ and $\bar{g}(\cdot)$ and injecting cost sensitivity by adjusting α_k, $k \in [K]$.

In the following, we show how to use the PtS approach to develop polynomial-time inference algorithms for any criterion of the form (11).

Lemma 1. *Let f be of the form* (11). *For any $l \in [K]_0$ and $\bar{y} \in \mathcal{Y}^l$* (7), *we have*

$$\mathbf{E}(f(\boldsymbol{y}, \bar{\boldsymbol{y}})) = \sum_{\boldsymbol{y} \in \mathcal{Y}} f(\boldsymbol{y}, \bar{\boldsymbol{y}}) \cdot p(\boldsymbol{y} \mid \boldsymbol{x}) = \sum_{s=1}^{K} \left(\sum_{\boldsymbol{y} \in \mathcal{Y}^s} f(\boldsymbol{y}, \bar{\boldsymbol{y}}) \cdot p(\boldsymbol{y} \mid \boldsymbol{x}) \right) \quad (12)$$

$$= \sum_{s=1}^{K} \frac{S_n}{\nu_{s,l}^f} \sum_{\boldsymbol{y} \in \mathcal{Y}^s} p(\boldsymbol{y} \mid \boldsymbol{x}) + \sum_{k=1}^{K} \bar{g}(\bar{y}_k) q_k^f, \quad (13)$$

$$\text{where,} \qquad \nu_{s,l}^f = S_d + \beta(s) + \gamma(l), \quad (14)$$

$$q_k^f = \sum_{s=1}^{K} \frac{\alpha_k}{\nu_{s,l}^f} \left(\sum_{\boldsymbol{y} \in \mathcal{Y}^s : g(y_k)=1} p(\boldsymbol{y} \mid \boldsymbol{x}) \right), \ k \in [K]. \quad (15)$$

Therefore, to find a local optimal prediction

$$\hat{\boldsymbol{y}}^l \in \underset{\bar{\boldsymbol{y}} \in \mathcal{Y}^l}{\mathrm{argmax}}\, \mathbf{E}\left(f(\boldsymbol{y}, \bar{\boldsymbol{y}})\right) = \underset{\bar{\boldsymbol{y}} \in \mathcal{Y}^l}{\mathrm{argmax}} \sum_{k=1}^{K} \bar{g}(\bar{y}_k) q_k^f, \quad (16)$$

it is sufficient to rank the labels in the decreasing order of q_k^f (15), $k \in [K]$, and set the l top ranked labels $\bar{y}_{(k)}$ such that $\bar{g}(\bar{y}_{(k)}) = 1$ and set $\bar{y}_{(k)}$ such that $\bar{g}(\bar{y}_{(k)}) = 0$ for other labels.

Proof. (Sketch) The transition from (12) to (13) is done basically by grouping $p(\boldsymbol{y} \mid \boldsymbol{x})$, $\boldsymbol{y} \in \mathcal{Y}$, in a suitable manner. Once q_k^f (15), $k \in [K]$, are made available, finding $\hat{\boldsymbol{y}}^l$ (16) is rather straightforward. □

Proposition 1. *A BOP of any criterion of the form* (11) *can be found in at most $O(K^3)$ given K^2 probability estimates* (6), *$p(\boldsymbol{0}_K \mid \boldsymbol{x})$, and K probability estimates*

$$\boldsymbol{p}_s := \sum_{\boldsymbol{y} \in \mathcal{Y}^s} p(\boldsymbol{y} \mid \boldsymbol{x}), s \in [K]. \quad (17)$$

Proof. (Sketch) For any $l \in [K]_0$ and any $k \in [K]$, we can easily get the scores

$$q_k^f = \sum_{s=1}^{K} \frac{\alpha_k}{\nu_{s,l}^f} (\boldsymbol{p}_{k,s})^{[\![g(y_k)=y_k]\!]} (\boldsymbol{p}_s - \boldsymbol{p}_{k,s})^{1-[\![g(y_k)=y_k]\!]}. \quad (18)$$

Therefore, for any $l \in [K]_0$, computing the K score q_k^f takes time $O(K^2)$. Once these cores are computed, Lemma 1 tells us that, for any $l \in [K]_0$, a local optimal prediction $\hat{\boldsymbol{y}}^l$ (16) can be found in $O(K \log(K))$. So in total, finding $\hat{\boldsymbol{y}}^l$ for any fixed l takes time $O(K^2)$. Computing $\mathbf{E}(f(\boldsymbol{y}, \hat{\boldsymbol{y}}^l))$ can be done in $O(K)$ because

$$\sum_{s=1}^{K} \frac{S_n}{\nu_{s,l}^f} \sum_{\boldsymbol{y} \in \mathcal{Y}^s} p(\boldsymbol{y} \mid \boldsymbol{x}) = \sum_{s=1}^{K} \frac{S_n}{\nu_{s,l}^f} \boldsymbol{p}_s.$$

Therefore, finding all the local optimal prediction $\hat{\boldsymbol{y}}^l$ for the possible values of l takes time $O(K^3)$. Finding the optimal value of l is done in $O(K\log(K))$. The whole process can be done in $O(K^3)$.

This is an upper bound complexity because, as pointed out in the Sect. 4, one can see that taking into account the concrete properties of f can further reduce the complexity to $O(K^2)$ and even $O(K)$ for some specific f. □

3.2 The Cases of Combined Criteria

We shall show that the PtS approach can be adapted to develop polynomial-time inference algorithms for criteria which are of the form

$$F(\boldsymbol{y}, \bar{\boldsymbol{y}}) := \sum_{m=1}^{M} \eta_m f^m(\boldsymbol{y}, \bar{\boldsymbol{y}}), \tag{19}$$

where, for any $m \in [M]$, f^m is of the form (11) with $\alpha_k^m = 1$, $k \in [K]$. Examples are the Hamming accuracy, the markedness and the informedness [15].

Lemma 2. *Let f^m be any criterion of the form (11), which satisfies the condition $\alpha_k^m = 1$, $k \in [K]$. We can always rewrite f^m as*

$$f^m(\boldsymbol{y}, \bar{\boldsymbol{y}}) = \frac{S_n^m + \sum_{k=1}^{K} g^m(y_k)\bar{g}^m(\bar{y}_k)}{S_d^m + \beta^m(s_{\boldsymbol{y}}) + \gamma^m(s_{\bar{\boldsymbol{y}}})} \tag{20}$$

$$= \frac{S_n^m + \gamma_1^m K + \gamma_2^m s_{\boldsymbol{y}}^m + \gamma_3^m s_{\bar{\boldsymbol{y}}}^m + \sum_{k=1}^{K} \gamma_{4,k}^m y_k \bar{y}_k}{S_d^m + \beta^m(s_{\boldsymbol{y}}) + \gamma^m(s_{\bar{\boldsymbol{y}}})}, \tag{21}$$

with $\gamma_1^m \in \{0,1\}$, $\gamma_2^m \in \{-1,0,1\}$, $\gamma_3^m \in \{-1,0,1\}$, and $\gamma_{4,k}^m \in \{-1,1\}$, $k \in [K]$. For any $\boldsymbol{y} \in \mathcal{Y}^s$ and any $\bar{\boldsymbol{y}} \in \mathcal{Y}^l$, we have

$$F(\boldsymbol{y}, \bar{\boldsymbol{y}}) = \sum_{m=1}^{M} \eta_m f^m(\boldsymbol{y}, \bar{\boldsymbol{y}}) = \frac{S_n^F + \sum_{k=1}^{K} \alpha_k^F y_k \bar{y}_k}{\nu_{s,l}^F}, \tag{22}$$

where $\nu_{s,l}^F$, α_k^F and S_n^F are constants and are defined as

$$\nu_{s,l}^F := \prod_{m=1}^{M} \nu_{s,l}^m := \prod_{m=1}^{M} \left(S_d^m + \beta^m(s_{\boldsymbol{y}}) + \gamma^m(s_{\bar{\boldsymbol{y}}})\right), \tag{23}$$

$$\alpha_k^F := \sum_{m=1}^{M} \eta_m \gamma_{4,k}^m \prod_{m'\neq m} \nu_{s,l}^{m'}, \tag{24}$$

$$S_n^F := \sum_{m=1}^{M} \eta_m \left(S_n^m + \gamma_1^m K + \gamma_2^m s_{\boldsymbol{y}}^m + \gamma_3^m s_{\bar{\boldsymbol{y}}}^m\right) \prod_{m'\neq m} \nu_{s,l}^{m'}. \tag{25}$$

Proof. (Sketch) The transition from (20) to (21) is valid because by definition, we have $(g(y_k), \bar{g}(\bar{y}_k)) \in \{(y_k, \bar{y}_k), (1-y_k, \bar{y}_k), (y_k, 1-\bar{y}_k), (1-y_k, 1-\bar{y}_k)\}$. The relation (22) is valid because, for any $l \in [K]_0$ and for any $k \in [K]$, $S_n^m + \gamma_1^m K + \gamma_2^m s_{\boldsymbol{y}} + \gamma_3^m s_{\bar{\boldsymbol{y}}}$ is a constant as long as $\boldsymbol{y} \in \mathcal{Y}^s$ and $\bar{\boldsymbol{y}} \in \mathcal{Y}^l$. □

Lemma 3. *For any $m \in [M]$, let f^m be any criterion of the form (11), which satisfies the condition $\alpha_k^m = 1$, $k \in [K]$. Let F be any criterion of the form (19). For any $l \in [K]_0$ and any $\bar{\boldsymbol{y}} \in \mathcal{Y}^l$ (7), we have*

$$\mathbf{E}(F(\boldsymbol{y}, \bar{\boldsymbol{y}})) = \sum_{\boldsymbol{y} \in \mathcal{Y}} F(\boldsymbol{y}, \bar{\boldsymbol{y}}) \cdot \boldsymbol{p}(\boldsymbol{y} \,|\, \boldsymbol{x}) = \sum_{s=1}^{K} \frac{S_n^F}{\nu_{s,l}^F} \boldsymbol{p}_s + \sum_{k=1}^{K} \bar{y}_k q_k^F , \tag{26}$$

$$where \qquad q_k^F = \sum_{s=1}^{K} \frac{\alpha_k^F}{\nu_{s,l}^F} \boldsymbol{p}_{k,s}, \; k \in [K]. \tag{27}$$

Therefore, to find a local optimal prediction

$$\hat{\boldsymbol{y}}^l \in \operatorname*{argmax}_{\bar{\boldsymbol{y}} \in \mathcal{Y}^l} \mathbf{E}\left(F(\boldsymbol{y}, \bar{\boldsymbol{y}})\right) = \operatorname*{argmax}_{\bar{\boldsymbol{y}} \in \mathcal{Y}^l} \sum_{k=1}^{K} \bar{y}_k q_k^F , \tag{28}$$

it is sufficient to rank the labels in the decreasing order of q_k^F (27), $k \in [K]$, and set the l top ranked labels $\bar{y}_{(k)} = 1$ and set $\bar{y}_{(k)} = 0$ for other labels.

Proof. (Sketch) Again, the transition (26) is done basically by grouping $\boldsymbol{p}(\boldsymbol{y} \,|\, \boldsymbol{x})$, $\boldsymbol{y} \in \mathcal{Y}$, in a suitable manner. Once q_k^F (27), $k \in [K]$, are made available, finding $\hat{\boldsymbol{y}}^l$ (28) is rather straightforward. □

Proposition 2. *A BOP of any criterion of the form (19) can be found in at most $O(K^3)$ given K^2 probability estimates (6), K probability estimates (17), and $\boldsymbol{p}(\boldsymbol{0}_K \,|\, \boldsymbol{x})$.*

Proof. (Sketch) For any $s \in [K]_0$ and any $k \in [K]$, we can easily get the scores q_k^F (27) in $O(K)$. Therefore, computing the K score q_k^F takes time $O(K^2)$. Once these cores are computed, Lemma 3 tells us that, for any $l \in [K]_0$, a local optimal prediction $\hat{\boldsymbol{y}}^l$ (16) can be found in $O(K \log(K))$. So in total, finding $\hat{\boldsymbol{y}}^l$ for any fixed l takes time $O(K^2)$. Computing $\mathbf{E}(f(\boldsymbol{y}, \hat{\boldsymbol{y}}^l)$ (26) can be done in $O(K)$. Therefore, finding all the local optimal prediction $\hat{\boldsymbol{y}}^l$ for the possible values of l takes time $O(K^3)$. Finding the optimal value of l is done in $O(K \log(K))$. The whole process can be done in $O(K^3)$.

Again, this is an upper bound complexity because, as pointed out in the Sect. 4, one can see that taking into account the concrete properties of F can further reduce the complexity to $O(K \log(K))$ for some specific F. □

3.3 A Naive (and Generic) Inference Algorithm

We first present a naive inference algorithm for the cases of single criteria (11). We then show how to adapt it to have another naive and generic inference algorithm for the cases of combined criteria (19).

To simplify the notation, we encode each criterion f by using its configuration

$$\operatorname{conf}(f) = (S_n, S_d, \alpha_1, \ldots, \alpha_K, \beta(\cdot), \gamma(\cdot), g(\cdot), \bar{g}(\cdot)) \tag{29}$$

Algorithm 1. Determining a BOP of any criterion f of the form (11)

1: **Input:** The configuration conf(f), and K^2 probabilities (6) and $p(\mathbf{0}_K \mid \boldsymbol{x})$
2: **for** $l = 0$ **to** K **do**
3:     **for** $k = 1$ **to** K **do**
4:         Compute q_k^f using (18)
5:     **end for**
6:     Determine $\hat{\boldsymbol{y}}^l$ using Lemma 1; Compute $\mathbf{E}(f(\boldsymbol{y}, \hat{\boldsymbol{y}}^l))$
7: **end for**
8: Determine $\hat{\boldsymbol{y}}$ which is $\hat{\boldsymbol{y}}^l$ with the highest $\mathbf{E}(f(\boldsymbol{y}, \hat{\boldsymbol{y}}^l))$
9: **Output:** a BOP $\hat{\boldsymbol{y}}$ of f

Let f be any criterion of the form (11). Without taking into account any further detail about conf(f), we can lazily use the Algorithm 1, which summarizes our analysis presented in Sect. 3.1, to find its BOP.

To deal with combined criteria F (19), we need to replace conf(f), q_k^f (18) and $\mathbf{E}(f(\boldsymbol{y}, \hat{\boldsymbol{y}}^l))$ respectively by conf(F), q_k^F (27) and $\mathbf{E}(F(\boldsymbol{y}, \hat{\boldsymbol{y}}^l))$. The conf($F$) is

$$\text{conf}(F) := \big(\text{conf}(f^1), \ldots, \text{conf}(f^M)\big) . \tag{30}$$

4 Application and Discussion

We have pointed out that finding a BOP of a given criterion, which can be either of the form (11) or of the form (19), can be done in polynomial time. We believe that the family of criteria which has been examined in our analysis is broad enough to cover different criteria researchers/practitioners working on probabilistic MLC may experience (e.g., commonly used criteria mentioned in [1,15,20] and elsewhere). Moreover, our examination has provided one with conditions under which one can (comfortably) customize their evaluation criteria to better reflect how they wish to assess the quality of their probabilistic classifiers, while still enjoying polynomial-time inference algorithms.

We now go one step further and guide practitioners with experience in developing (inference) algorithms how to further reduce the inference complexity by taking into account the configuration of the given criterion.

4.1 Detailing the Inference Algorithms

Let us pick up the precision (31) to illustrate how to detail the inference algorithm 1. We first specify its configuration by rewriting it in the form of (11)

$$f_{\text{Pre}}(\boldsymbol{y}, \bar{\boldsymbol{y}}) := \frac{\sum_{k=1}^{K} \hat{y}_k \, y_k}{\sum_{k=1}^{K} \bar{y}_k} = \frac{0 + \sum_{k=1}^{K} 1 y_k \bar{y}_k}{0 + 0 + s_{\bar{y}}} . \tag{31}$$

Hence, we have the configuration

$$\text{conf}(f_{\text{Pre}}) = (S_n, S_d, \alpha_1, \ldots, \alpha_K, \beta(\cdot), \gamma(\cdot), g(\cdot), \bar{g}(\cdot)) \tag{32}$$

$$= (0, \quad 0, \quad 1, \ldots, \quad 1, \quad 0, \quad s_{\bar{y}}, \quad y_k, \quad \hat{y}_k) . \tag{33}$$

It is clear that f_{Pre} is a very simple criterion and we would have a (much) simpler inference algorithm than the inference algorithm 1.

Proposition 3. *Given a query instance x, assume marginal probabilities p_k, $k \in [K]$, are made available. If $\mathbf{0}_K$ is a valid prediction, it is a BOP of the precision f_{Pre}. If otherwise, a BOP of f_{Pre} is $\hat{y}^1$ which can be found in $O(K)$.*

Proof. For any $l \in [K]$, the local optimal prediction of f_{Pre} is

$$
\hat{y}^l \in= \operatorname*{argmax}_{\bar{y} \in \mathcal{Y}^l} \sum_{k=1}^{K} \bar{g}(\bar{y}_k) \left(\sum_{s=1}^{K} \frac{\alpha_k}{\nu_{s,l}^f} p_{k,s} \right) = \operatorname*{argmax}_{\bar{y} \in \mathcal{Y}^l} \sum_{k=1}^{K} \bar{y}_k \left(\sum_{s=1}^{K} \frac{1}{l} p_{k,s} \right)
$$

$$
= \operatorname*{argmax}_{\bar{y} \in \mathcal{Y}^l} \sum_{k=1}^{K} \bar{y}_k \left(\frac{1}{l} \left(\sum_{s=1}^{K} p_{k,s} \right) \right) = \operatorname*{argmax}_{\bar{y} \in \mathcal{Y}^l} \frac{1}{l} \sum_{k=1}^{K} \bar{y}_k p_k .
$$

For any $y \in \mathcal{Y}$, $f_{\mathrm{Pre}}(y, \mathbf{0}_K)$ is undefined. One common practice is to assign $f_{\mathrm{Pre}}(y, \mathbf{0}_K) = 1$. Thus, for any $l \in [K]$, the expectation

$$
\max_{\bar{y} \in \mathcal{Y}^l} \mathbf{E}\left(f_{\mathrm{Pre}}(y, \bar{y}) \right) = \max_{\bar{y} \in \mathcal{Y}^l} \frac{1}{l} \sum_{k=1}^{K} \bar{y}_k p_k \leq 1 = \mathbf{E}\left(f_{\mathrm{Pre}}(y, \mathbf{0}_K) \right) = \sum_{y \in \mathcal{Y}} p(y \mid x).
$$

Thus, if $\mathbf{0}_K$ is a valid prediction, it is a BOP of f_{Pre}.

Assume $\mathbf{0}_K$ is a non-valid prediction. For any $l \in [K]$, we already know that a local optimal prediction contains the l labels with the highest marginal probabilities p_k as relevant, and the other $K - l$ labels as irrelevant. Let (k) be the index of the label with the k-th highest marginal probability p_k, $k \in [K]$. For any $k \in [K-1]$, it is clear that

$$
\mathbf{E}\left(f_{\mathrm{Pre}}(y, \hat{y}^l) \right) = \frac{\sum_{k=1}^{l} p_{(k)}}{l} \geq \frac{\sum_{k=1}^{l+1} p_{(k)}}{l+1} = \mathbf{E}\left(f_{\mathrm{Pre}}(y, \hat{y}^{l+1}) \right) .
$$

The inequality holds since $\sum_{k=1}^{l} p_{(k)} \geq l \cdot p_{(l+1)}$. Thus, a BOP of f_{Pre} is

$$
\hat{y} \in \operatorname*{argmax}_{\bar{y} \in \{\hat{y}^l \mid l = 1, \ldots, K\}} \mathbf{E}\left(f_{\mathrm{Pre}}(y, \hat{y}^l) \right) = \operatorname*{argmax}_{\bar{y} \in \{\hat{y}^l \mid l = 1\}} \mathbf{E}\left(f_{\mathrm{Pre}}(y, \hat{y}^l) \right) ,
$$

which is found in $O(K)$ by searching for the label with highest p_k, $k \in [K]$ □

Due to the page length limit, we skip the proofs of the next propositions. Proposition 4 summarizes the complexity of finding a BOP of the recall f_{Rec} [15], which is a single criterion (11). Proposition 5 summarizes the complexity of finding a BOP of the markedness f_{Mar} [15], which is a combined criterion (19).

Proposition 4. *Given a query instance x, assume K^2 probability estimates (6) and $p(\mathbf{0}_K \mid x)$ are made available, a BOP of the recall f_{Rec} can be constructed in time $O(K^2)$. Furthermore, this BOP is either $\hat{y}^0 = \mathbf{0}_K$ or $\hat{y}^K = \mathbf{1}_K$.*

Proposition 5. *Given a query instance x, assume marginal probabilities p_k, $k \in [K]$, are made available, a BOP of the markedness f_{Mar} is constructed in time $O(K \log(K))$.*

A a BOP of the markedness f_{Mar} can be found by ranking the labels using some scores and predict top ranked labels as relevant and others as irrelevant.

4.2   Potential (Dis)advantages of Some Criteria

One might see that the BOPs of some criteria, such as the precision and recall, are trivial. Hence, in practice, it might be risky to embrace probabilistic classifiers which are specifically designed to optimize such criteria. The BOP of f_β has a more flexible structural constraint (as discussed in [18] and recalled in Sect. 2.2), which basically allows the learner to return predictions with any number of relevant labels, and might be more interesting in practice.

This might be another strong motivation to seek insights into BOPs of the criteria one wishes to use. We however leave an extensive discussion on this aspect as a future work (mainly due to the page length limit).

5   Conclusion

We present a generic approach for developing polynomial-time inference algorithms for a family of criteria and discuss the potential (dis)advantages of some commonly used criteria. Alongside recent advances in learning probabilistic classifiers, which would help to enlarge the number of classifiers, this paper provides considerable opportunities for including more criteria in future comparative studies focusing on inference complexity and criteria selection. Along the way, we provide practitioners with suggestions on extending the existing set of evaluation criteria in a meaningful way without facing intractable inference problems.

To seek a comprehensive understanding of inference complexity and criteria selection, we envision the following future work: extend our application and discussion part (presented in Sect. 4) by including more commonly used criteria, conduct empirical studies in which recent advances in learning probabilistic classifiers are employed to produce probabilistic predictions, and investigate whether the PtS approach can be used to develop polynomial-time inference algorithms for other families of criteria (especially those that provide more flexibility on customizing/injecting cost sensitivity).

Acknowledgments. This work was funded/supported by the Junior Professor Chair in Trustworthy AI (Ref. ANR-R311CHD), and the US Office of Naval Research Global under Grant N62909-23-1-2058.

References

1. Bogatinovski, J., Todorovski, L., Džeroski, S., Kocev, D.: Comprehensive comparative study of multi-label classification methods. Expert Syst. Appl. **203**, 117215 (2022)
2. Cheng, W., Hüllermeier, E., Dembczynski, K.J.: Bayes optimal multilabel classification via probabilistic classifier chains. In: Proceedings of the 27th International Conference on Machine Learning (ICML), pp. 279–286 (2010)
3. Coppersmith, D., Winograd, S.: Matrix multiplication via arithmetic progressions. J. Symb. Comput. **9**(3), 251–280 (1990)

4. Decubber, S., Mortier, T., Dembczyński, K., Waegeman, W.: Deep f-measure maximization in multi-label classification: a comparative study. In: Berlingerio, M., Bonchi, F., Gärtner, T., Hurley, N., Ifrim, G. (eds.) ECML PKDD 2018. LNCS (LNAI), vol. 11051, pp. 290–305. Springer, Cham (2019). https://doi.org/10.1007/978-3-030-10925-7_18

5. Dembczyński, K., Waegeman, W., Cheng, W., Hüllermeier, E.: On label dependence and loss minimization in multi-label classification. Mach. Learn. **88**, 5–45 (2012)

6. Elkan, C.: The foundations of cost-sensitive learning. In: Proceedings of the 17th international joint conference on Artificial intelligence (IJCAI), pp. 973–978 (2001)

7. Gerych, W., Hartvigsen, T., Buquicchio, L., Agu, E., Rundensteiner, E.A.: Recurrent bayesian classifier chains for exact multi-label classification. In: Proceedings of the 34th International Conference on Neural Information Processing Systems (NeurIPS), vol. 34, pp. 15981–15992 (2021)

8. Gil-Begue, S., Bielza, C., Larrañaga, P.: Multi-dimensional Bayesian network classifiers: a survey. Artif. Intell. Rev. **54**(1), 519–559 (2021)

9. Lewis, D.D.: Evaluating and optimizing autonomous text classification systems. In: Proceedings of the 18th Annual International ACM SIGIR Conference on Research and Development in Information Retrieval (SIGIR), pp. 246–254. ACM (1995)

10. Mortier, T., Wydmuch, M., Dembczyński, K., Hüllermeier, E., Waegeman, W.: Efficient set-valued prediction in multi-class classification. Data Min. Knowl. Disc. **35**(4), 1435–1469 (2021)

11. Nguyen, V.L., Hüllermeier, E.: Multilabel classification with partial abstention: Bayes-optimal prediction under label independence. J. Artif. Intell. Res. **72**, 613–665 (2021)

12. Nguyen, V.L., Hüllermeier, E., Rapp, M., Loza Mencía, E., Fürnkranz, J.: On aggregation in ensembles of multilabel classifiers. In: Proceedings of the 23rd International Conference on Discovery Science (DS), pp. 533–547 (2020)

13. Nguyen, V.L., Yang, Y., de Campos, C.P.: Probabilistic multi-dimensional classification. In: Proceedings of the 39th Conference on Uncertainty in Artificial Intelligence (UAI), pp. 1522–1533 (2023)

14. Pillai, I., Fumera, G., Roli, F.: Designing multi-label classifiers that maximize f measures: state of the art. Pattern Recogn. **61**, 394–404 (2017)

15. Powers, D.: Evaluation: from precision, recall and f-factor to ROC, informedness, markedness & correlation. J. Mach. Learn. Technol. **2**(1), 37–63 (2011)

16. Read, J., Pfahringer, B., Holmes, G., Frank, E.: Classifier chains for multi-label classification. Mach. Learn. **85**, 333–359 (2011)

17. Vapnik, V.N.: An overview of statistical learning theory. IEEE Trans. Neural Netw. **10**(5), 988–999 (1999)

18. Waegeman, W., Dembczyński, K., Jachnik, A., Cheng, W., Hüllermeier, E.: On the Bayes-optimality of f-measure maximizers. J. Mach. Learn. Res. **15**, 3333–3388 (2014)

19. Ye, N., Chai, K.M.A., Lee, W.S., Chieu, H.L.: Optimizing f-measures: a tale of two approaches. In: Proceedings of the 29th International Conference on International Conference on Machine Learning (ICML), pp. 1555–1562 (2012)

20. Zhang, M.L., Zhou, Z.H.: A review on multi-label learning algorithms. IEEE Trans. Knowl. Data Eng. **26**(8), 1819–1837 (2014)

A Federated Learning Model for Linear Fuzzy Clustering with Least Square Criterion

Katsuhiro Honda[✉] and Ryosuke Amejima

Osaka Metropolitan University, Sakai, Osaka 599-8531, Japan
khonda@omu.ac.jp

Abstract. Federated learning is a hot topic on privacy preserving data analysis and has also been applied to fuzzy c-means clustering. In this paper, a federated learning scheme is proposed for linear fuzzy clustering with horizontally distributed data, where each cluster is represented by a linear-shape prototype. In order to merge the client-wise independent clustering results without violating personal privacy, gradient information of each prototype instead of original observation are shared at the centralized server. The objective function is defined with the least square criterion, which is useful in handling component-wise errors and makes it possible to find cluster basis vectors without solving an Eigen problem. Therefore, attribute-wise gradient decent learning can be realized by utilizing only gradient information of prototype parameters at the central server. The global prototypes are securely updated then distributed to clients for next updating. Experimental results demonstrate that the proposed algorithm is useful for reconstructing the whole data result under privacy preservation.

Keywords: Linear fuzzy clustering · Federated learning · Horizontally distributed data

1 Introduction

Privacy preservation is a fundamental issue in handling personal data for data mining [1]. When performing collaborative data analysis utilizing multiple data sources distributed among several clients, not only cryptographic approaches [2] but also federated learning [3–5] are becoming basic techniques. Observed data can be distributed in two different forms such as vertically or horizontally distributed data [2]. In the vertically distributed data cases, multiple clients store their independent attributes on common objects while they store common attributes on their independent objects in the horizontally distributed data cases. In this paper, the horizontally distributed data cases are considered.

This work was supported in part by JSPS KAKENHI Grant Number JP18K11474 and JP22K12198.

Fuzzy c-means (FCM) clustering [6, 7] is a basic model of analyzing the intrinsic data distribution through unsupervised learning, where each cluster is characterized by its prototypical centroid. The clustering algorithm is composed of two steps of fuzzy membership estimation and cluster center updating. Pedrycz [8] proposed a federated learning model for FCM clustering, which can achieve collaborative analysis of horizontally distributed data among clients with the goal of estimating common cluster centroids. In order to merge the client-wise cluster structures in the FCM iterative algorithm without violating personal privacy, the central server shares the parameter gradient instead of the original observation.

Fuzzy c-Lines (FCL) [9] is an extension of FCM to linear fuzzy clustering, which extracts linear-shape clusters by replacing the cluster centroids with prototypical lines. Because the basis vector of each prototype is given as the principal Eigen vector of the fuzzy scatter matrix in the iterative algorithm, FCL can be identified with local principal component analysis [10, 11]. In this paper, with the goal of implementing federated learning with attribute-wise gradient information sharing, a novel model of FCL is proposed by utilizing the least square criterion under a collaborative analysis framework. Then, the main contribution of this paper is summarized as:

- The federated FCM learning model proposed by Pedrycz is extended to FCL, which achieves linear-shape cluster extraction for achieving local principal component analysis.
- A novel algorithm of realizing privacy preserving linear prototype estimation is proposed such that attribute-wise gradient information sharing is achieved by utilizing the component-wise least square criterion for intra-cluster lower-rank approximation.

The remaining parts of this paper are organized as follows: A brief review on fuzzy clustering and federated learning is presented in Sect. 2 and a novel federated learning model for FCL is proposed in Sect. 3. Section 4 shows experimental results for demonstrating the characteristics of the propose method and Sect. 5 gives the conclusions of this paper.

2   Review on Fuzzy Clustering and Federated Learning

2.1   FCM and Extension to FCL

Assume that we have m dimensional observation on n objects $\boldsymbol{x}_i = (x_{i1}, \ldots, x_{im})^\top$, $i = 1, \ldots, n$ and the goal is to partition the objects into C fuzzy clusters characterized by their prototypes. Fuzzy c-partition is represented by fuzzy memberships u_{ci}, which indicates the membership degree of object i to cluster c with $u_{ci} \in [0, 1]$ such that $\sum_{c=1}^{C} u_{ci} = 1, \forall i$.

Fuzzy c-Means (FCM) [6, 7] adopts cluster centroids $\boldsymbol{b}_c$ as the prototype of cluster c and the clustering criterion is given as the squared Euclidean distance $d_{ci} = ||\boldsymbol{x}_i - \boldsymbol{b}_c||^2$. Then, the objective function to be minimized is defined as:

$$J_{fcm} = \sum_{c=1}^{C} \sum_{i=1}^{n} (u_{ci})^\theta d_{ci}, \tag{1}$$

where θ ($\theta > 1$) is the weighting exponent for fuzzifier. A larger θ brings very fuzzy cluster boundaries while $\theta \to 1$ reduces to hard partition like k-Means [12, 13].

The clustering algorithm is the iterative process of membership estimation and centroid updating as follows:

$$u_{ci} = \frac{(d_{ci})^{\frac{1}{1-\theta}}}{\sum_{\ell=1}^{C}(d_{\ell i})^{\frac{1}{1-\theta}}}, \tag{2}$$

$$b_c = \frac{\sum_{i=1}^{n}(u_{ci})^{\theta}x_i}{\sum_{i=1}^{n}(u_{ci})^{\theta}}. \tag{3}$$

Fuzzy c-Lines (FCL) [9] is an extension of FCM to linear fuzzy clustering, where the prototype of cluster c is replaced with a line defined by the center vector b_c and the basis vector a_c with unit length as $a_c^{\top}a_c = 1$. The clustering criterion of the Euclidean distance among objects i and the prototype of cluster c is redefined as $d_{ci} = ||x_i - b_c||^2 - |a_c^{\top}(x_i - b_c)|^2$. Besides u_{ci} and b_c are updated with the same formulas with FCM using the redefined d_{ci}, the basis vector a_c is given as the principal Eigen vector of the following fuzzy scatter matrix:

$$S_c = \sum_{i=1}^{n}(u_{ci})^{\theta}(x_i - b_c)(x_i - b_c)^{\top}. \tag{4}$$

Then, the basis vectors a_c are identified with local principal component vectors [10, 11] estimated in local regions.

2.2   Federated Learning for FCM

In real world data analysis, we can expect that the collaborative analysis among multiple clients improves the quality of derived knowledge rather than the client-wise independent analysis. Here, assume that the whole dataset on n objects are horizontally distributed over T clients such that client t has n_t objects and $\sum_{t=1}^{T} n_t = n$. The goal of federated learning is to reproduce the result of whole data analysis by keeping personal data private.

In federated learning for FCM, the goal is to collaboratively estimate the global cluster centers b_c without broadcasting object-wise information such as original observations x_i and fuzzy memberships u_{ci}. When each client updates cluster centers b_c only with its own data subset, the client-wise updated centers can be distorted and they must be collaboratively merged in the central server.

Pedrycz [8] proposed a federated learning model, where each client transfer only gradient information to the central server in the centroid updating process at each iteration. Then, the gradient values are merged considering their cardinality weights and the global centroids are updated and redistributed again.

In this paper, the Pedrycz's model is further enhanced to linear fuzzy clustering with line-shape prototypes.

3 Federated Learning Model for FCL

3.1 Least Square Criterion for Collaborative FCL

In order to update the global prototype in the central server using attribute-wise gradient information, the objective function must be defined in the component-wise error function. With the goal of handling component-wise missing values [10] or noise [11], Honda and Ichihashi proposed the least square criterion for FCL. Let y_{ci} be the lower-rank approximation of object vector x_i such that $y_{ci} = b_c + f_{ci} \cdot a_c$, where f_{ci} is the principal component score of object i in cluster c. Then, the squared Euclidean distance between object i and linear prototype of cluster c is redefined as

$$d_{ci} = ||x_i - y_{ci}||^2 = \sum_{j=1}^{m} (x_{ij} - b_{cj} - f_{ci} \cdot a_{cj})^2. \tag{5}$$

In the following, the least square criterion-based FCL is enhanced to federated learning, where n objects are horizontally distributed over T clients. Here, each client t has n_t objects, where object i of client t has observation vector $x_i^t = (x_{i1}^t, \ldots, x_{im}^t)^\top$. The goal of collaborative analysis is to estimate the global linear prototypes (b_c, a_c) in conjunction with the intra-client memberships u_{ci}^t and principal component scores f_{ci}^t. The objective function is defined in the component-wise approximation principle as:

$$J_{fcl-fl} = \sum_{c=1}^{C} \sum_{t=1}^{T} \sum_{i=1}^{n_t} (u_{ci}^t)^\theta \sum_{j=1}^{m} \left(x_{ij}^t - b_{cj} - f_{ci}^t \cdot a_{cj} \right)^2. \tag{6}$$

In order to derive unique solution, a_c should be normalized as $a_c^\top a_c = 1$ in the same manner with the standard FCL.

3.2 Proposed Algorithm

The iterative algorithm of membership estimation and prototype updating is implemented with a client-server manner. In the server side, the global prototypes are initialized and updated, and then, are distributed to clients. In each client, fuzzy memberships of each object are updated and cluster prototypes are locally updated using intra-client objects only. Here, in the client-server communication, the gradient information in local prototype updating is only communicated for preserving personal privacy.

A sample algorithm is described as follows:

Algorithm: Federated Learning for Fuzzy c-Lines with Horizontally Distributed Data (FL-FCL_H)

Step 1. Initialize cluster centers b_c and basis vectors a_c, randomly, and normalize a_c so as to be $a_c^\top a_c = 1$.

Step 2. Distribute current prototypes $(\boldsymbol{b}_c, \boldsymbol{a}_c)$ from the central server to each client.

Step 3. In each client t, perform FCL implementation in several iterations using only intra-client objects.

Step 4. From each client t to the central server, feed-back gradient information of intra-client prototype updating.

Step 5. In the central server, merge the client-wise gradients and update the global prototypes.

Step 6. If the global prototypes are converged, stop. Otherwise, return to **Step 2**.

In each client, considering the optimality of the local objective function:

$$J^t_{fcl-fl} = \sum_{c=1}^{C} \sum_{i=1}^{n_t} (u^t_{ci})^\theta \sum_{j=1}^{m} \left(x^t_{ij} - b_{cj} - f^t_{ci} \cdot a_{cj} \right)^2, \tag{7}$$

the updating formulas in **Step 3** are given in the same manner with the least square criterion-based FCL as:

$$f^t_{ci} = \sum_{j=1}^{m} a_{cj} \left(x^t_{ij} - b_{cj} \right), \tag{8}$$

$$b_{cj} = \frac{\sum_{i=1}^{n_t} (u^t_{ci})^\theta \left(x^t_{ij} - f^t_{ci} \cdot a_{cj} \right)}{\sum_{i=1}^{n_t} (u^t_{ci})^\theta}, \tag{9}$$

$$a_{cj} = \frac{\sum_{i=1}^{n_t} (u^t_{ci})^\theta f^t_{ci} \left(x^t_{ij} - b_{cj} \right)}{\sum_{i=1}^{n_t} (u^t_{ci})^\theta (f^t_{ci})^2}, \tag{10}$$

where $\boldsymbol{a}_c$ is normalized so as to be $\boldsymbol{a}_c^\top \boldsymbol{a}_c = 1$. Then, after several iterations, the gradient information of prototypes is feed-backed to the central server in **Step 4** as:

$$\frac{\partial J^t_{fcl-fl}}{\partial b_{cj}} = -2 \sum_{i=1}^{n_t} (u^t_{ci})^\theta \left(x^t_{ij} - b_{cj} - f^t_{ci} \cdot a_{cj} \right), \tag{11}$$

$$\frac{\partial J^t_{fcl-fl}}{\partial a_{cj}} = -2 \sum_{i=1}^{n_t} (u^t_{ci})^\theta f^t_{ci} \left(x^t_{ij} - b_{cj} - f^t_{ci} \cdot a_{cj} \right). \tag{12}$$

On the other hand, in the central server, **Step 5** merges the client-wise gradient information for updating the global prototypes. First, in order to fairly reflect the cardinality of each client, the responsibility weight of client t is calculated as:

$$\beta_t = \frac{n_t}{\sum_{s=1}^{T} n_s}. \tag{13}$$

Then, each of prototype elements is updated with learning rate α under the gradient decent principle as:

$$b_{cj} \leftarrow b_{cj} - \alpha \sum_{t=1}^{T} \beta_t \cdot \frac{\partial J^t_{fcl-fl}}{\partial b_{cj}}, \tag{14}$$

$$a_{cj} \leftarrow a_{cj} - \alpha \sum_{t=1}^{T} \beta_t \cdot \frac{\partial J_{fcl-fl}^t}{\partial a_{cj}}, \tag{15}$$

and $\boldsymbol{a}_c$ is normalized so as to be $\boldsymbol{a}_c^\top \boldsymbol{a}_c = 1$.

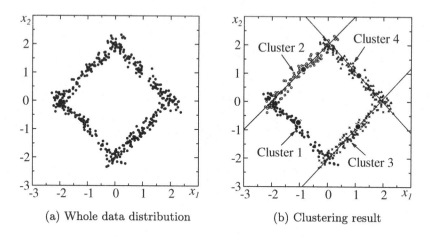

(a) Whole data distribution   (b) Clustering result

Fig. 1. Artificial dataset and clustering result in whole data analysis.

4   Experimental Results

In this section, the characteristics of the proposed method are demonstrated through numerical experiments using an artificially generated dataset.

4.1   Whole Data Analysis

An artificial dataset composed of 400 objects ($n = 400$) with 2-dimensional observations ($m = 2$) was generated as shown in Fig. 1(a), which forms 4 linear-shape substructures. First, utilizing the whole dataset, the conventional FCL was implemented with $C = 4$ and $\theta = 2.0$. Figure 1(b) shows the partition result and the derived cluster prototypes, which implies we could successfully extract the 4 linear clusters indicated by the lines with their center points.

In the following, the goal is to reconstruct similar results to the whole data analysis from the horizontally distributed datasets under federated learning.

4.2   Client-Wise Independent Analysis

In this experiment, 400 objects were horizontally distributed over 3 clients ($T = 3$) as shown in Fig. 2(a). Here, clients 1 and 2 are composed of only 3 of the 4 linear-shape substructures while client 3 is composed of only 2 of them. Before performing federated learning, client-wise independent analysis was performed with $\theta = 2$ and the clustering results were given as shown in Figs. 2(b)-(c).

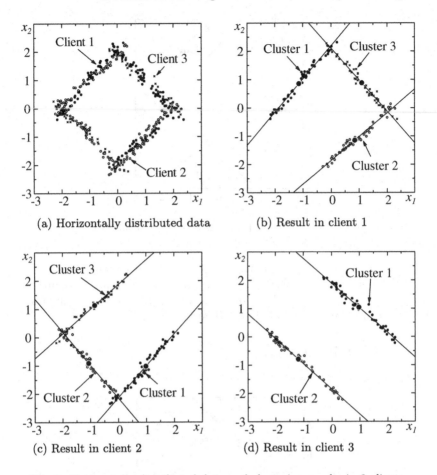

Fig. 2. Horizontally distributed data and clustering results in 3 clients.

In general clustering problems, we often do not have a priori knowledge on the number of clusters and must estimate itself. In this paper, how to estimate the number of clusters is not discussed in detail but the cluster number is assumed to be optimally selected considering some cluster validity measures [14]. For example, Fig. 2(b)-(c) clearly imply that each of clients 1 and 2 have 3 linear clusters while client 3 has 2 linear clusters only.

Next, in the collaborative analysis with horizontally distributed data, we also have another problem of how to select the total cluster number which can only be found in the whole data analysis. In this experiment, the total cluster number is estimated through cluster merging using the results of client-wise independent analysis. Figure 3 compares the 8 cluster prototypes derived in 3 clients independently, and we can find 4 groups by merging similar ones, i.e., the total cluster number should be $C = 4$.

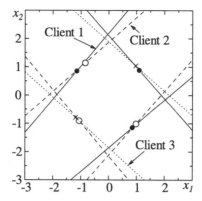

Fig. 3. Comparison of cluster prototypes given in client-wise analysis.

By the way, since each client has somewhat biased distributions as shown in Fig. 2(a), client-wise prototypical lines are distorted and are not suitable for revealing the result of whole data analysis. Then, in order to reconstruct the whole data result, we need a collaborative analysis.

4.3   Proposed Federated Learning

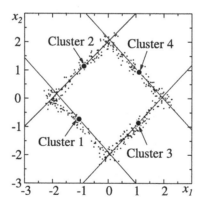

Fig. 4. Cluster prototypes given by proposed federated learning algorithm.

Finally, the proposed federated learning model was applied to the 3 data subsets, which are horizontally distributed to 3 clients. Model parameters were given as $C = 4$, $\theta = 2$ and $\alpha = 0.00001$. In client-wise FCL implementation of **Step 3**, intra-client updating was iterated in 10 times.

Figure 4 shows the estimated prototypes, which implies that the proposed federated learning model could successfully reconstruct the result of whole data analysis shown in Fig. 1(b).

5   Conclusions

In this paper, a novel federated learning algorithm was proposed for linear fuzzy clustering. Besides the conventional FCL is performed by solving an Eigen problem, the attribute-wise gradient decent learning of the proposed model was realized by using the least square criterion. The characteristics of the proposed method were demonstrated through numerical experiments such that the global prototypes derived in the central server fairly reconstructed the whole data result even when client-wise data subsets have somewhat biased distributions.

In the experiment of this paper, the cluster number was assumed to be heuristically found under the cluster merging principle, but some objective criteria will be needed for applying to much larger datasets. Development of a validity criterion for federated linear fuzzy clustering is remained in possible future works. Other future directions are to extend the proposed model to Fuzzy c-Varieties (FCV) [15] with 2 or more dimensional prototypes cases or to vertically distributed datasets. Comparison with cryptographic approaches [2, 16] can be also a promising future work.

References

1. Aggarwal, C.C., Yu, P.S.: Privacy-Preserving Data Mining: Models and Algorithms. Springer-Verlag, New York (2008)
2. Chen, T.-C.T., Honda, K.: Fuzzy Collaborative Forecasting and Clustering. SAST, Springer, Cham (2020). https://doi.org/10.1007/978-3-030-22574-2
3. McMahan, B., Ramage, D.: Federated learning: collaborative machine learning without centralized training data, Google AI Blog, April 06 (2017)
4. Yang, Q., Liu, Y., Cheng, Y., Kang, Y., Chen, T., Yu, H.: Federated Learning. Morgan & Claypool Pub, New York (2019)
5. Yang, Q., Fan, L., Yu, H.: Federated Learning. Privacy and Incentive. Springer, Berlin, Germany (2020). https://doi.org/10.1007/978-3-030-63076-8
6. Bezdek, J.C.: Pattern Recognition with Fuzzy Objective Function Algorithms. Plenum Press, New York (1981)
7. Miyamoto, S., Ichihashi, H., Honda, K.: Algorithms for Fuzzy Clustering. Springer, Heidelberg (2008). https://doi.org/10.1007/978-3-540-78737-2
8. Pedrycz, W.: Federated FCM: clustering under privacy requirements. IEEE Trans. Fuzzy Syst. **30**(8), 3384–3388 (2022)
9. Bezdek, J.C., Coray, C., Gunderson, R., Watson, J.: Detection and characterization of cluster substructure 1. Linear structure: Fuzzy c-lines. SIAM J. Appl. Math. **40**, 339–357 (1981)
10. Honda, K., Ichihashi, H.: Linear fuzzy clustering techniques with missing values and their application to local principal component analysis. IEEE Trans. Fuzzy Syst. **12**(2), 183–193 (2004)
11. Honda, K., Ichihashi, H.: Component-wise robust linear fuzzy clustering for collaborative filtering. Int. J. Approx. Reason. **37**(2), 127–144 (2004)
12. MacQueen, J. B.: Some methods of classification and analysis of multivariate observations. In: Proceedings of 5th Berkeley Symposium on Mathematical Statistics and Probability, pp. 281–297 (1967)

13. Wu, J.B.: Advances in K-means Clustering. A Data Mining Thinking. Springer, Berlin, Heidelberg (2012). https://doi.org/10.1007/978-3-642-29807-3
14. Wang, W., Zhanga, Y.: On fuzzy cluster validity indices. Fuzzy Sets Syst. **158**, 2095–2117 (2007)
15. Bezdek, J.C., Coray, C., Gunderson, R., Watson, J.: Detection and characterization of cluster substructure 2. Fuzzy c-varieties and convex combinations thereof. SIAM J. Appl. Math. **40**, 358–372 (1981)
16. Honda, K., Kunisawa, K., Ubukata, S., Notsu, A.: Fuzzy c-varieties clustering for vertically distributed datasets. Proc. Comput. Sci. **192**, 457–466 (2021)

Joint Multiple Efficient Neighbors and Graph Learning for Multi-view Clustering

Fatemeh Sadjadi[1,2(✉)] [iD] and Vicenç Torra[2] [iD]

[1] Department of Applied Mathematics, Faculty of Sciences and Modern Technologies, Graduate University of Advanced Technology, 117-76315 Kerman, Iran
sajjadifateme207@gmail.com
[2] Department of Computing Science, MIT Building, Umeå University, Västerbotten, 901 87 Umeå, Sweden
vtorra@cs.umu.se

Abstract. Graph-based multi-view clustering is a popular method for identifying informative graphs for e.g. computer vision applications. Nevertheless, optimizing sparsity and connectivity simultaneously is challenging. Multi-view clustering (MC) integrates complementary information from different views. However, most existing methods introduce noise or ignore relevant data structures. This paper introduces a Joint multiple efficient neighbors and Graph (JMEG) learning method for MC. Our approach includes a post-processing technique to optimize sparsity and connectivity by means of identifying neighbors efficiently. JMEG also uses partition space and consensus graph learning to uncover data structures efficiently. Experiments show that JMEG outperforms state-of-the-art methods with negligible additional computation cost.

Keywords: Multi-view clustering · Efficient neighborhood · Connectivity and Sparsity · Fusion graph

1 Introductions

Recent advancements in unsupervised learning have sparked renewed interest in clustering. Traditional clustering methods detect clusters solely in single-view data. However, in today's world where data is generated by multiple sources, the need for multi-view data clustering has increased. For instance, a document can be expressed in multiple languages, a picture can be taken from different angles, and a person can be identified through photographs, signatures, and fingerprints. Because of that, there is an increasing interest in multi-view data which has fostered the research on multi-view clustering.

Multi-view clustering algorithms commonly use graph-based models [7]. They build a graph representing the information of each view, and then create either a fused graph or a low-dimensional spectral embedding. In this way they summarise the information of all views in a single structure. Clustering algorithms

K. Honda et al. (Eds.): IUKM 2023, LNAI 14376, pp. 25–36, 2024.
https://doi.org/10.1007/978-3-031-46781-3_3

are then applied (e.g., k-means or spectral clustering) to obtain a set of clusters from this fused information. The use of separate graphs for each view permits to address the heterogeneity of the data.

As explained above, in multi-view graph learning [4, 7, 10–12], data in each view is represented as a graph or network. In this graph, each node represents an entity, and each edge represents a relationship between two entities. Therefore, a key point is the construction of the (weighted) graph as well obtaining a common similarity matrix, also known as affinity matrix. Naturally, different procedures to build these graphs from the data exist. They differ on e.g. what means to be similar. Sparsity and connectivity are key properties of graphs that affect learning. Different procedures can lead to graphs with different sparsity levels. It is of relevance to find a good trade-off between sparsity and connectivity. Weighted multi-view graph is when multiple views have different importance.

As a summary, multi-view graph learning has the following challenges:

1. Construction of a graph for each view, based on the corresponding similarity matrix, with good trade-off between connectivity and sparsity.
2. Effective combination of the information from different views to achieve a single clustering structure.

Our study addresses these challenges by means of

(i) Obtaining neighbourhoods for each data based on second-order connections,
(ii) Defining similarity matrices based on the neighbourhoods in each view,
(iii) Combining all similarity matrices of the multi-view data, and
(iv) Conducting extensive experiments on six real-world data sets to validate the effectiveness of our idea.

The paper starts in the next section with the preliminaries and principal concepts. In Sect. 3, we explain the proposed method, specific multi-view data, and other ways that we want to compare with them. In the next section, we summarise the principal findings of these experiments. The final two sections describe the parameter selection and explain the conclusions.

2   Preliminaries and Related Works

In this paper we will follow the following notation. Let $\{X^1, X^2, \ldots, X^V\}$ denote a multi-view data set containing V views of n samples (i.e., each sample appears in each view). Here, $X_i^\nu \in \mathbb{R}^{n \times d_\nu}$ $(1 \leqslant \nu \leqslant V)$ represent the i-th data point (i.e., the i-th row) in the ν-th view. Here, d_ν is the corresponding dimension of the ν-th view and, naturally, $1 \leqslant i \leqslant n$. In addition, let $\{W^1, W^2, \ldots, W^V\}$ denote the weight matrices for the multi-view data set.

Then, we will consider graphs defined, as usual, as a pair $G = (V, E)$ where V is the set of vertices and E is the set of edges (a subset of pairs of vertices). Then, given a data set X, we can construct graphs where the vertices are the rows of the data set, and two vertices are connected by an edge if the two records are sufficiently similar. Then, a similarity matrix will provide the information about the similarity of each pair. We will use W or A to denote these matrices. For example, for an edge between nodes i and j in E, we have a similarity $w_{ij} > 0$ in the matrix X.

Different ways exist to build a graph and a similarity matrix from a data set X. For example, considering a complete graph and W defined using the Euclidean distance. An alternative is to define connected vertices in terms of the k-nearest neighbor (knn). Then, for connected graph use also the Euclidean distance.

Let $S = [s_{ij}]$ be a matrix such that its i-th row is s_i. Its transpose, trace, inverse and Frobenius norm are S^T, $\text{Tr}(S)$, S^{-1}, and $\| S \|_F$, respectively. If D ($d_{ii} = \sum_{j=1}^{n} s_{ij}$) is the degree matrix corresponding to matrix S, then $L = D - S$ is the so-called Laplacian matrix. Also, if v is a vector, we define the l_2-norm as $\| v \|_2$, and I and 1 are the identity matrix and a column vector with all elements as one, respectively.

In this paper we extensively use results from graph-based multi-view clustering, and from multi-view spectral clustering. Due to space limitations, we refer to [17–19] for details.

2.1 Construction of Similarity Matrices

We use the following expression to compute the similarity W between vertices (to obtain matrices W^ν), and then apply k-nearest neighborhood and the γ-efficient neighborhood (to obtain matrices A^ν).

$$
w_{ij}^\nu = \frac{2e^{-\frac{\|x_i - x_j\|^2}{\lambda}}}{\sum_{h \neq i}^{n} e^{-\frac{\|x_i - x_h\|^2}{\lambda}} + \sum_{h \neq j}^{n} e^{-\frac{\|x_h - x_j\|^2}{\lambda}}}, \tag{1}
$$

where, λ is the regularization parameter. This definition was inspired by a similarity matrix obtained through subspace clustering with an entropy norm [15]. We use γ-efficient neighborhood defined as follows.

Definition 1. *Let γ and μ be natural numbers. Then, x_j is considered a γ-efficient neighbor of x_i if the following conditions are met: x_j belongs to the k-nearest neighborhood of x_i; x_j and x_i have at least μ common neighbors.*

So, x_i is a γ-efficient neighbor of k-nearest neighborhood x_j if they have at least μ common neighbors ($\gamma < k$).

2.2 Sparsity and Connectivity

The sparsity and connectivity of the similarity matrix are crucial factors in spectral clustering, as they greatly affect the performance of the clustering algorithm. There are numerous methods proposed in the literature to enhance the connectivity of clusters, such as utilizing different norms and regularization terms in subspace problems. In our proposed algorithm, we adopt the γ-efficient neighborhood approach outlined in Eq. (1), as well as utilizing k to identify k-nearest neighborhood, where k is less than a number of data sets. These two parameters, k and γ, play vital roles in controlling the sparsity and connectivity of the similarity graph. To assess the connectivity of the similarity graph, we utilize the second smallest eigenvalue of the normalized graph Laplacian, which is a widely used and effective method for calculating algebraic connectivity. In the context of clustering, we compute the second smallest eigenvalue for each of the V subgraphs of the affinity graph with K clusters, denoted as $\lambda(i)_2$ for the i-th cluster, and compare the connectivities across all clusters. We also evaluate the overall connectivity of our method by computing the average of all clusters' connectivities, which is denoted as the "connectivity" metric. By incorporating these methods, we can enhance the connectivity of clusters and improve the performance of spectral clustering. Additionally, to evaluate the overall connectivity of our method, we use the average of all clusters' connectivities:

$$\text{connectivity} := \frac{1}{K} \sum_{i=1}^{K} \lambda(i)_2. \tag{2}$$

3   Proposed Algorithm

In this section we introduce our novel learning method for multi-view clustering. We call it joint multiple efficient neighbors and graph (JMEG) learning method for MC. JMEG comprises two key components: (1) a post-processing technique to optimize sparsity and connectivity by identifying efficient neighbors, and (2) unified partition space learning and consensus graph learning. Component (1) is to establish critical connections among samples within a cluster so that samples have both large affinity coefficients and strong connections to each other. Given a coefficient matrix, we reassign the coefficients of the efficient neighbors and eliminate other entries to produce a new coefficient matrix. To bootstrap the process, we start with a random matrix (size $n \times n$), each element is a number between -1 and 1.

We demonstrate that a small number of efficient neighbors can effectively recover the cluster, and the proposed post-processing step of identifying efficient neighbors complements most existing multi-view clustering algorithms.

In our approach, the partition is more robust to noise, and graph learning helps uncover data structures. Specifically, JMEG iteratively constructs local graph matrices, generates base partition matrices, stretches them to produce a unified partition matrix, and employs it to learn a consensus graph matrix. For efficiency, JMEG adaptively allocates a large weight to the stretched base partition that is close to the unified partition, determines parameters, and imposes a low-rank constraint on graphs. Finally, clusters can be obtained directly from the consensus graph.

Experiments on five benchmark datasets demonstrate that the proposed algorithm outperforms state-of-the-art methods with negligible additional computation cost.

3.1   Finding k-Nearest Neighborhood and γ-Efficient Neighborhood

Utilizing the symmetric nonnegative weight matrix W^ν, as obtained from Eq. (1), we derive a set of new similarity matrices $\{A^1, A^2, \ldots, A^\nu\}$. To obtain the new similarity matrix for each view, we initially calculate the k-nearest neighbors for each dataset and subsequently select the γ-efficient neighborhood from them.

In order to ensure sparsity, the pruning process of selecting samples with top k coefficients in w_{ij} for x_i is commonly used. This is because larger w_{ij} values indicate stronger similarity between x_i and x_j. However, this approach does not take into account the connectivity property, which may result in intra-cluster samples not forming a connected component in the affinity graph. The limitation of simply considering the maximum k edges for each sample is that it may fail to handle noise-corrupted data, as wrong connections are preserved due to the sensitivity of max edges to noise, outliers, and samples near the intersection of two clusters. Therefore, in order to ensure both sparsity and connectivity properties in graph multi-view learning, we propose a method in this paper that defines "efficient neighbors" for each sample in each view to establish crucial latent connections within a graph.

In this paper, we propose a definition of a γ-efficient neighborhood for every data in each view, which is a neighbor that has at least μ common neighbors within the local neighborhood. This definition ensures a maximally sparse and connected neighborhood relationship, which is analyzed in Sect. 4 for sparsity and connectivity, respectively. By exploring efficient neighbors, we obtain complementary information that enhances the robustness of the method against noise-corrupted data, resulting in clustering with both sparsity and connectivity properties. Throughout the paper, we use $\aleph \in \mathbb{R}^{n \times \gamma}$ to determine the collection of efficient neighbors, where n is the number of data samples, and k and γ are used to control the sparsity and connectivity, respectively.

For this scenario, we can obtain the adjacency matrix of the k-neighborhoods based on the weight matrix Eq. (1). By taking the power of 2 of this matrix, we can determine which data points have at least μ common neighbors. This is because if there is one path of length 2 between x_i and x_j, it means that they share one common neighbor. In this paper, we consider different μ for every data, and k is related to a number of data. Now, we can obtain multiple similarity graph $\{A^1, A^2, \ldots, A^V\}$ based on extracting efficient neighbors of weight matrices $\{W^1, W^2, \ldots, W^V\}$, i.e.,

$$\min_{A^\nu} \sum_{i,j=1, i,j \in \aleph}^{n} w_{ij}^\nu a_{ij}^\nu + \alpha \parallel A^\nu \parallel_F^2 \tag{3}$$

$$s.t. A^\nu \geq 0, \nu = 1, \cdots, V.$$

After obtaining multiple similarity graphs $\{A^1, A^2, \ldots, A^\nu\}$, the next step is a combination of these graphs.

3.2   Consensus Graph Construction

Our aim is to obtain a consensus matrix of eigenvectors by fusion of all segmentation and coefficient matrices of all views, simultaneously. So, in our method we propose to use the following objective function:

$$\min_{A^\nu,A,F^\nu,\omega_\nu,\Gamma_\nu,F} \sum_{\nu=1}^{V}(\sum_{i,j=1}^{n} w_{ij}^\nu a_{ij}^\nu + \alpha \parallel A^\nu \parallel_F^2 +2\lambda_1 Tr(F^{\nu^T} L_{A^\nu} F^\nu) \tag{4}$$

$$+ \Gamma_\nu \parallel A^\nu - A \parallel_F^2 +\omega_\nu \parallel FF^T - F^\nu F^{\nu^T} \parallel_F^2) + 2\lambda_2 Tr(F^T L_A F)$$

$$s.t. A^\nu \geq 0, F^{\nu^T} F^\nu = I, F^T F = I.$$

The same as [20], construction of a consensus graph $\bar{S}$ is possible with the unified partition representation F. Joint learning of the unified partition matrix and the consensus graph can be accomplished in this process. Since graph $\bar{S}$ is shared across all views, clustering results can be obtained directly without additional clustering techniques. However, the unified partition Q obtained from Eq. (6) alone is insufficient for this purpose. To capture the local manifold structure, an adaptive neighbor strategy can be applied while learning $\bar{S}$. Hence, we derive $\bar{S}$ from F using the following equation:

$$\min_{\bar{S},Q} \sum_{i,j=1}^{n} \parallel F_i - F_j \parallel_2^2 s_{ij} + \beta \parallel \bar{S} \parallel_F^2 +2\lambda_3 Tr(Q^T L_{\bar{S}} Q) \tag{5}$$

$$s.t. F^T F = I, Q^T Q = I, \bar{S} \geq 0, \bar{s}_{ii} = 0$$

with combination of Eq. (4) and Eq. (5) we have the following optimization problem:

$$\min_{A^\nu,A,F^\nu,\omega_\nu,\Gamma_\nu,F,\bar{S},Q} \sum_{\nu=1}^{V}(\sum_{i,j=1}^{n} w_{ij}^\nu a_{ij}^\nu + \alpha \parallel A^\nu \parallel_F^2 +2\lambda_1 Tr(F^{\nu^T} L_{A^\nu} F^\nu) \tag{6}$$

$$+ \Gamma_\nu \parallel A^\nu - A \parallel_F^2 +2\lambda_2 Tr(F^T L_A F) + \omega_\nu \parallel FF^T - F^\nu F^{\nu^T} \parallel_F^2)$$

$$+ \sum_{i,j=1}^{n} \parallel F_i - F_j \parallel_2^2 s_{ij} + \beta \parallel \bar{S} \parallel_F^2 +2\lambda_3 Tr(Q^T L_{\bar{S}} Q)$$

$$s.t. a_{ii}^\nu = 0, A^\nu \geq 0, F^{\nu^T} F = I, F^T F = I, Q^T Q = I, \bar{S} \geq 0, \bar{s}_{ii} = 0, \alpha \geq 0, \beta \geq 0.$$

Proposition 1. *The optimal solution of problem (6) leads to the following six expressions:*

1) $a_{ij}^\nu = \dfrac{-1}{2(\alpha+1)} I(w_{ij} + \lambda_1 h_{ij}^\nu)$ *where* $h_{ij}^{\nu^T} =\parallel f_i^\nu - f_j^\nu \parallel_2^2.$

2) $a_{ij} = \dfrac{\lambda_2 f_{ij} - 2\sum_{\nu=1}^{V} a_{ij}^\nu}{V}$ *where* $f_{ij}^T =\parallel f_i - f_j \parallel_2^2.$

3) F^ν *is the eigenvector matrix of* $M = 2\lambda_1 L_{A^\nu} - 2\omega_\nu FF^T + \omega_\nu I$ *corresponding to the K-smallest eigenvalue.*

4) F is the eigenvector matrix of $G = \sum_{\nu=1}^{V} \omega_\nu (I_n - 2F^\nu F^{\nu^T}) + 2L_{\bar{S}} + L_A$ corresponding to the K smallest eigenvalue.

5) $\bar{s}_{ij} = \dfrac{-(f_{ij} + \lambda_2 d_{ij})}{2\beta}$ where $d_{ij}^T = \| q_i - q_j \|_2^2$.

6) The optimal solution of Q is the original spectral clustering of the laplacian matrix $L_{\bar{S}}$.

7) $\omega_\nu = \dfrac{1}{\| F^T F - F^{\nu^T} F^\nu \|_F}$.

8) $\Gamma_\nu = \dfrac{1}{\| A^\nu - A \|_F}$.

The function presented in Eq. (6) is not jointly convex and obtaining a globally optimal solution for it is a challenging task. However, by using an alternating strategy, as described in Proposition 1, we can transform Eq. (6) into several subproblems, each of which is convex. Neither the proof of the proposition, neither how these problems are solved is described in detail here because of page limit constraints. Code will be made available if the paper is accepted, proofs will be provided in an extended version of this paper.

4   Experimental Setting

We have conducted our experiments comparing the results of our method on different datasets with other methods in the literature. We describe the experiments and results in this section.

Data Sets. Table 1 describes the data sets used. They are common in the multi-view literature.

Table 1. Detail of the six multi-view data sets

Data sets	n	V	classes	$d_\nu (\nu = 1, 2, \ldots, V)$
Reuters 1200	1200	5	6	(2000, 2000, 2000, 2000, 2000)
HW	2000	6	10	(216, 76, 64, 6, 240, 47)
100Leaves	1600	3	100	(64, 64, 64)
ORL	400	4	40	(256, 256, 256, 256)
Caltech101-7	1474	6	7	(48, 40, 254, 1984, 512, 928)
Caltech101-20	2386	6	20	(48, 40, 254, 1984, 512, 928)

Methods Compared. We have compared our approach with the following methods: Auto-weighted multiple graph learning (AMGL) [22], Multi-view Low-Rank Sparse Subspace Clustering (MLRSSC) [23], Multi-view Subspace Clustering with Intactness-aware Similarity (MSCIAS) [24], Large-scale Multi-view Subspace Clustering in Linear Time (LMVSC) [25], Partition Level Multi-view Subspace Clustering (PMSC) [19], Multi-graph Fusion for Multi-view Spectral

Clustering (FGSC) [4], Incomplete Multi-view Clustering with Joint Partition and Graph Learning (JPG) [20]. We used the implementations as provided in their respective websites with default parameters. We reported best results. Our method was implemented in Python.

Parameters. For JMEG, we set k empirically to 15. μ is 2 and $\gamma < k$. The initial values of all entries of parameter α, β is 1, and λ_1, λ_2, and λ_3 were set in the range [0.1, 0.001, 1, 10, 20] and reported their best clustering results. ω_ν and Γ_ν values were adaptively tuned in the optimization procedure of the objective function for each view. Three common metrics were used to evaluate the clustering performance: Accuracy (ACC), Normalized Mutual Information (NMI), and Purity (PUR). To randomize the experiments, we ran each method ten times and reported the means for the metrics.

4.1   Experimental Results and Analysis

The multi-view datasets are used to conduct experiments and evaluate the performance of various methods, and Tables 2, 3, 4, and 5 present the results obtained. The results are indicated as average metric values. The best results have been highlighted in bold. Based on the table, several observations can be made.

– Compared to the multi-view methods, our proposed JMEG method exhibits superior performance. Across all datasets, JMEG outperforms the other methods in terms of ACC, NMI, and PUR. These results provide strong evidence that our JMEG method has significant potential as a multi-view clustering approach.
– The performance of the multi-view methods reveals that JMEG outperforms other multi-view methods (PMSC, GFSC, JPG). The primary reason behind this result is that our JMEG methods consider the segmentation matrix F based on all of the segmentation matrices of all views and all of the similarity matrices and construct a graph based on F, and can better exploit the complementary information provided by multiple views. It is worth noting that JPG consistently performs better than other multi-view clustering. However, all of these methods are inferior to our proposed JMEG method.
– The experimental results demonstrate that connectivity in JMEG outperforms other multi-view clustering methods. The superior performance of JMEG can be attributed to its ability to consider the γ-efficient neighbors, which is constructed based on k-nearest neighborhood extracted from our similarity matrix. By constructing new similarity matrices that of them are sparse, JMEG is better equipped to leverage the complementary information provided by multiple views.

Computational Efficiency. Our proposed JMEG method is evaluated for computational efficiency in this section using running time as a metric. To measure the running time, we conducted experiments on two datasets. Each method

Table 2. ACC values of different methods on the six multi-view data sets

Data sets	AMGL	MLRSSC	MSCIAS	LMVSC	PMSC	FGSC	JPG	JMEG
Reuters1200	0.16	0.45	0.49	0.58	0.40	0.44	0.59	**0.61**
Hw	0.84	0.76	0.80	0.91	0.83	0.75	0.92	**0.94**
100Leaves	0.18	0.24	0.28	0.32	0.31	0.34	0.78	**0.80**
ORL	0.26	0.28	0.33	0.36	0.40	0.38	0.81	**0.85**
Caltech101-7	0.45	0.37	0.38	0.72	0.75	0.77	0.76	**0.78**
Caltech101-20	0.30	0.28	0.31	0.53	0.54	0.39	0.55	**0.57**

Table 3. NMI values of different methods on the six multi-view data sets

Data sets	AMGL	MLRSSC	MSCIAS	LMVSC	PMSC	FGSC	JPG	JMEG
Reuters1200	-	0.22	0.27	0.33	0.21	0.25	0.44	**0.45**
Hw	0.87	0.74	0.77	0.84	0.82	0.82	0.86	**0.89**
100Leaves	-	0.19	0.23	0.27	0.26	0.30	0.86	**0.90**
ORL	0.18	0.31	0.45	0.38	0.56	0.44	0.85	**0.89**
Caltech101-7	0.42	0.21	0.23	0.51	0.55	0.58	0.61	**0.65**
Caltech101-20	0.40	0.26	0.31	0.52	0.57	0.58	0.62	**0.65**

Table 4. PUR values of different methods on the six multi-view data sets

Data sets	AMGL	MLRSSC	MSCIAS	LMVSC	PMSC	FGSC	JPG	JMEG
Reuters1200	-	0.55	0.66	0.61	0.60	0.59	0.72	**0.75**
Hw	0.87	0.87	0.86	0.91	0.87	0.68	0.91	**0.94**
100Leaves	0.53	0.47	0.61	0.68	0.73	0.79	0.80	**0.82**
ORL	0.61	0.63	0.58	0.65	0.70	0.78	0.75	**0.78**
Caltech101-7	0.75	0.41	0.44	0.75	0.84	0.89	0.90	**0.93**
Caltech101-20	0.31	0.30	0.33	0.58	0.57	0.58	0.79	**0.81**

Table 5. Conn values of different methods on the six multi-view data sets

Data sets	AMGL	MLRSSC	MSCIAS	LMVSC	PMSC	FGSC	JPG	JMEG
Reuters1200	-	0.07	0.09	0.08	0.17	0.16	0.0.18	**0.22**
Hw	-	0.07	0.09	0.11	0.19	0.17	0.20	**0.25**
100Leaves	-	0.04	0.07	0.06	0.09	0.07	0.15	**0.18**
ORL	-	0.06	0.05	0.06	0.07	0.13	0.09	**0.15**
Caltech101-7	-	0.07	0.15	0.08	0.11	0.16	0.19	**0.21**
Caltech101-20	-	0.09	0.15	0.21	0.18	0.58	0.22	**0.27**

was tested for 5 runs under the same computing environment, and the average running time was recorded. Figure 1 presents the results, indicating that our method performs the best among the multi-view methods. Multi-view clustering algorithms need to handle multiple views simultaneously, which can make them slower when dealing with cases where only a subset of the data is chosen as an efficient neighborhood instead of considering all the data. For this reason, among all the multi-view clustering algorithms, our JMEG method outperforms the others. Although the JMEG method needs to optimize multiple variables alternately, the algorithm converges quickly, as discussed in the next section.

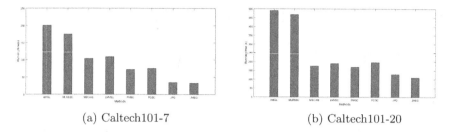

(a) Caltech101-7   (b) Caltech101-20

Fig. 1. Running time evaluation of the proposed JMEG method and other comparison methods on the two data sets

Convergence Analysis. In order to demonstrate the efficacy of the optimization strategy utilized for the objective function of our proposed JMEG method, we present the convergence curves of JMEG across two data sets, as illustrated in Fig. 2. The x-axis of each subfigure denotes the number of iterations, while the y-axis denotes the value of the objective function, as defined in Eq. (6) above. It is evident from the graphs that JMEG converges rapidly for all two datasets, achieving convergence within only 10 iterations for the Caltech101-7 and Caltech101-20 data sets. This suggests that the proposed JMEG method provides an optimized solution that is highly efficient.

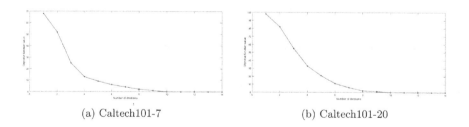

(a) Caltech101-7   (b) Caltech101-20

Fig. 2. Convergence curves on the two data sets

5   Conclusion and Future Work

This research paper introduces a novel approach, JMEG, for multi-view graph learning that aims to address the poor connectivity problem in sparse multi-view clustering. JMEG utilizes a pre-processing technique that generates multiple similarity matrices by considering γ-efficient neighbors for each data point.

Also, these matrices are then combined with segmentation matrices to create a higher-quality graph structure. Our experimental results on real-world data demonstrate the effectiveness of JMEG in improving clustering performance. Future research can explore the potential of combining JMEG with other multi-view methods to enhance clustering quality, as well as investigate its application in incomplete multi-view clustering. JMEG provides a promising approach to addressing the poor connectivity issue in multi-view clustering.

In addition, we aim to enhance the clustering performance of mixed data that contains various data types, including texts, images, and videos, in the original data space. To achieve this, we plan to extend the current JMEG method into a deep neural network in the future. Furthermore, we aim to develop a more efficient approach that can handle large-scale applications by adaptively determining the number of clusters and minimizing the impact of the k and γ values through the utilization of bipartite graph learning.

References

1. Chao, G., Sun, S., Bi, J.: A survey on multi-view clustering. arXiv preprint arXiv:1712.06246 (2017)
2. Chen, X., Ye, Y., Xu, X., Huang, J.Z.: A feature group weighting method for subspace clustering of high-dimensional data. Pattern Recogn. **45**(1), 434–446 (2017)
3. Huang, S., Kang, Z., Xu, Z.: Auto-weighted multi-view clustering via deep matrix decomposition. Pattern Recogn., 107015 (2017)
4. Kang, Z., et al.: Multi-graph fusion for multi-view spectral clustering. Knowl. Based Syst. **189**, 102–105 (2019)
5. Kang, Z., Peng, C., Cheng, Q.: Kernel-driven similarity learning. Neurocomputing **267**, 210–219 (2017)
6. Vidal, R.: Subspace clustering. IEEE Signal Process. Mag. **28**(2), 52–68 (2011)
7. Wang, H., Yang, Y., Liu, B., Fujita, H.: A study of graph-based system for multi-view clustering. Knowl.-Based Syst. **163**, 1009–1019 (2019)
8. You, C.Z., Fan, H., Shu, Z.Q.: Non-negative sparse laplacian regularized latent multi-view subspace clustering. In: Proceedings of the 19th International Symposium on Distributed Computing and Applications for Business Engineering and Science (DCABES), pp. 210–213 (2020)
9. Zheng, Q., Zhu, J., Li, Z., Pang, S., Wang, J., Chen, Li.: Consistent and complementary graph regularized multi-view subspace clustering. arXiv, 2004.03106 (2020)
10. Pan, E., Kang, Z.: Multi-view contrastive graph clustering. Adv. Neural. Inf. Process. Syst. **34**, 2148–2159 (2021)
11. Wang, H., Yang, Y., Liu, B.: Graph-based multi-view clustering. IEEE Trans. Knowl. Data Eng. **32**(6), 1116–1129 (2019)
12. Zhan, K., Nie, F., Wang, J., Yang, Y.: Multiview consensus graph clustering. IEEE Trans. Image Process. **28**(3), 1261–1270 (2018)
13. Yin, H., Hu, W., Zhang, Z., Lou, J., Miao, M.: Incremental multi-view spectral clustering with sparse and connected graph learning. Neural Netw. **144**, 260–270 (2021)
14. Lin, Y., Lai, Z., Zhou, J., Wen, J., Kong, H.: Multiview jointly sparse discriminant common subspace learning. Pattern Recogn., 109342 (2023)

15. Bai, L., Liang, J.: Sparse subspace clustering with entropy-norm. In: International Conference on Machine Learning, pp. 561–568 (2020)
16. Nie, F., Cai, G., Li, X.: Multi-view clustering and semi-supervised classification with adaptive neighbors. In: Proceedings of the AAAI Conference on Artificial Intelligence, vol. 31, no. 1 (2017)
17. Jiang, Z., Liu, X.: Adaptive KNN and graph-based auto-weighted multi-view consensus spectral learning. Inf. Sci. **609**, 1132–1146 (2022)
18. Wu, D., Lu, J., Nie, F., Wang, R., Yuan, Y.: Efficient multi-view graph clustering with comprehensive fusion, pp. 3566–3572 (2022). Embedding regularization. In: Proceedings of the 24th ACM International Conference on Multimedia, pp. 475–479 (2016)
19. Kang, Z., et al.: Partition level multiview subspace clustering. Neural Netw. **122**, 279–288 (2020)
20. Li, L., Wan, Z., He, H.: Incomplete multi-view clustering with joint partition and graph learning. IEEE Trans. Knowl. Data Eng. **35**(1), 589–602 (2021)
21. Von Luxburg, U.: A tutorial on spectral clustering. Stat. Comput. **17**, 395–416 (2007)
22. Nie, F., Li, J., Li, X.: Parameter-free auto-weighted multiple graph learning: a framework for multiview clustering and semi-supervised classification, pp. 1881–1887 (2016)
23. Brbić, M., Kopriva, I.: Multi-view low-rank sparse subspace clustering. Pattern Recogn. **73**, 247–258 (2018)
24. Wang, X., Lei, Z., Guo, X., Zhang, C., Shi, H., Li, S.Z.: Multi-view subspace clustering with intactness-aware similarity. Pattern Recogn. **88**, 50–63 (2019)
25. Kang, Z., Zhou, W., Zhao, Z., Shao, J., Han, M., Xu, Z.: Large-scale multi-view subspace clustering in linear time. In: Proceedings of the AAAI Conference on Artificial Intelligence, vol. 34, no. 4, pp. 4412–4419 (2020)

Maximum-Margin Nearest Prototype Classifiers with the Sum-over-Others Loss Function and a Performance Evaluation

Yoshifumi Kusunoki[(⊠)][iD]

Graduate School of Informatics, Osaka Metropolitan University, Gakuen-cho 1-1,
Naka-ku, Sakai, Osaka 599-8531, Japan
`yoshifumi.kusunoki@omu.ac.jp`

Abstract. This paper investigates margin-maximization models for
nearest prototype classifiers. These models are formulated through a
minimization problem, which is a weighted sum of the inverted margin and a loss function. It is reduced a difference-of-convex optimization
problem, and solved using the convex-concave procedure. In our latest
study, to overcome limitations of the previous model, we have revised
the model in both of the optimization problem and the training algorithm. In this paper, we propose another revised margin-maximization
model by replacing the max-over-others loss function used in the latest
study with the sum-over-others loss function. We provide a derivation
of the training algorithm of the proposed model. Moreover, we evaluate classification performance of the revised margin-maximization models through a numerical experiment using benchmark data sets of UCI
Machine Learning Repository. We compare the performance of our models not only with the previous model but also with baseline methods
that are the generalized learning quantization, the class-wise k-means,
and the support vector machine.

Keywords: Nearest prototype classifier · Margin maximization ·
sum-over-others and max-over-others aggregations · Convex-concave
procedure

1   Introduction

A nearest prototype classifier (NPC) is a classification model that utilizes a
(small) number of labeled prototypes arranged in the input space. In this model,
each input vector is assigned the label of its nearest prototype. This model is
attractive because the inference process for an input vector is transparent as
it is based on the selected prototype. Consequently, an end-user can comprehend the classification rationale by examining the prototype, which shares the
same attributes as the input vector. This provides a solution for the explainability problem commonly associated with machine learning methods. This study
addresses supervised learning methods for NPCs.

K. Honda et al. (Eds.): IUKM 2023, LNAI 14376, pp. 37–46, 2024.
https://doi.org/10.1007/978-3-031-46781-3_4

There are several approaches available for learning NPCs. Learning vector quantizations (LVQs) [5] are well-known NPC learning methods, in which prototypes are iteratively updated in response to classification results of training data. Sato and Yamada [10] proposed a variation of LVQ known as the generalized LVQ (GLVQ). They derived this algorithm by applying the gradient descent to an appropriate cost function. The introduction of the cost function enhances stability and controllability of the learning algorithm. Furthermore, GLVQ is extended to the generalized relevance LVQ (GRLVQ) [4] and the generalized matrix LVQ (GMLVQ) [11] to overcome the disadvantage of GLVQ that the distance is isometrically measured in the input space, i.e., attributes are equally treated.

Recently, the author has also studied optimization-based NPC learning, and proposed an application of the margin-maximization principle [6,7], which is used in the support vector machine (SVM). We define a class-wise score function for input vectors, which is the negated (Euclidean) distance between an input vector and its nearest prototypes involved in the class. For each labeled input vector, the score of the true class is required to exceed those of the other classes by a certain value called a margin. The proposed methods minimize the inverted margin along with a hinge loss function, which quantifies the extent to which the constraint is violated for each input vector. The margin can be associated with the geometric distance between the classification boundary and input vectors.

This margin-maximization problem is formulated as a difference-of-convex (DC) optimization problem, and addressed using a convex-concave procedure (CCP) [8]. CCP is an siterative method that updates a tentative solution by solving a convex approximation of the original problem, linearizing the concave parts of DC functions. This procedure can be likened to a k-means-like algorithm. For the reduction to the DC optimization problem, we replace the true-class score function by its square. This replacement introduces an additional hyperparameter and causes difficulty in controlling the resulting prototype locations. Hence, in [6], we have revised both of the optimization problem and the training algorithm to remove the replacement.

When considering our margin-maximization models as a multi-class learning algorithm, there are alternative loss functions, as discussed in the paper [2]. In the model of [6], we use the max-over-others loss function, in which the loss of each labeled input vector is defined by the maximum among the hinge losses of the true class over the others. In this paper, we propose the margin-maximization model that employs the sum-over-others loss function. The loss of each vector is defined by the sum of the hinge losses of the true class over the others. Moreover, we evaluate classification performance of our margin-maximization models using benchmark data sets in UCI Machine Learning Repository [3]. We compare the margin-maximization models with the two different loss functions with the previous version described in [7] and three baseline methods, which are GLVQ, k-means, and SVM.

This paper is organized as follows. In Sect. 2, we explain the maximum-margin NPCs including the proposed model as well as those of the previous studies. In

Sect. 3, we derive the training algorithm of the proposed model, and graphically show the classification boundary and the prototypes obtained by our models. In Sect. 4, we conduct the numerical experiment to evaluate classification performance of our models including proposed one. In Sect. 5, we provide concluding remarks.

2 Margin-Maximization Models for Nearest Prototype Classifiers

We consider a setting of supervised learning. An input space is the n-dimensional real space $\mathbf{R}^n$, and a set of class labels is defined by $C = \{1, 2, \ldots, c\}$. A training data set is composed of labeled input vectors $(x_1, y_1), (x_2, y_2), \ldots, (x_m, y_m)$, where $x_i \in \mathbf{R}^n$ and $y_i \in C$. The aim of learning is to construct a classifier $\mathscr{C} : \mathbf{R}^n \to C$ using the training data set. Let $M = \{1, 2, \ldots, m\}$ be the index set of training instances.

In this paper, we focus on classifiers based on prototypes. We arrange p labeled prototypes $(\hat{w}_1, v_1), (\hat{w}_2, v_2), \ldots, (\hat{w}_p, v_p)$, where $\hat{w}_j \in \mathbf{R}^{n+1}$ is a prototype and $v_j \in C$ is its label. Each prototype consists of $\hat{v}_j = (w_j, b_j)$, where $w_j \in \mathbf{R}^n$ is an input vector and $b_j \in \mathbf{R}_+$ is a nonnegative bias parameter. b_j controls the effect of this prototype for the classification by the prototypes. Let $P = \{1, 2, \ldots, p\}$ be the index set of prototypes. The index set associating label $k \in C$ is defined by $P_k = \{j \in P \mid v_j = k\}$.

For each $k \in C$, we define a score function $f_k : \mathbf{R}^n \to \mathbf{R}$.

$$f_k(x) = -\min_{j \in P_k} \left(\|(x - w_j)/s\|^2 + (b_j)^2 \right)^{1/2}, \tag{1}$$

where s is a positive parameter rescaling the input space. Using the score functions, we consider the following classifier $\mathscr{C}$:

$$\mathscr{C}(x) = \operatorname*{argmax}_{k \in C} f_k(x). \tag{2}$$

Combining Eqs. (1) and (2), an input vector x is classified into class k if the nearest prototype belongs to the class. In the case of a tie, it is classified into an arbitrary class with the maximum score.

The author proposed the margin-maximization model [7] for nearest prototype classifiers. We formulated a margin maximization problem as follows.

$$\begin{aligned} &\underset{w,b,s>0,\xi}{\text{minimize}} \quad 1/s^2 + \mu/c \sum_{i \in M} \sum_{k \in C} \xi_{ik} \\ &\text{subject to} \quad f_{y_i}(x_i) - f_k(x_i) + \xi_{ik} \geq 1, \ \xi_{ik} \geq 0, \ i \in M, k \in C, \end{aligned} \tag{3}$$

where μ is a positive weight for the loss term. The first term is regarded as a margin, since, under a certain assumption, s can be the minimum difference between distances from each instance x_i to the nearest prototype having the true label y_i and that having the other labels $k \neq y_i$. We remark that the constrains corresponding to $i \in M$ and $k = y_i$ are redundant and $\xi_{iy_i} = 1$ always holds, but it simplifies the formulation. Since ξ_{ik} is minimized, it is equal to the hinge loss of instance (x_i, y_i) in class pair (y_i, k).

$$\xi_{ik} = \max\{0, 1 - (f_{y_i}(x_i) - f_k(x_i))\}. \tag{4}$$

Hence, this margin maximization problem minimizes the sum of hinge losses over the other classes.

The training algorithm in [7] includes a hyperparameter to replace the true-score function $f_{y_i}(x_i)$ with its square. However, the performance of the obtained classifier is sensitive to that hyperparameter. Hence, to reduce it, we have revised its training algorithm in [6]. As same as the case of multi-class SVM discussed in [2], there are variations in the formulation of the margin-maximization model considering how to aggregate losses. In [6], we have considered the following formulation.

$$\underset{w,b,s>0,\xi}{\text{minimize}} \quad 1/s^2 + \mu \sum_{i \in M} \xi_i - \eta \sum_{i \in M} f_{y_i}(x_i)$$
$$\text{subject to} \quad f_{y_i}(x_i) - f_k(x_i) + \xi_i \geq 1 - \delta_{ik}, \ i \in M, k \in C, \tag{5}$$

where δ_{ik} is the one-hot representation of y_i, namely $\delta_{ik} = 1$ iff $y_i = k$. First, we add the third term that maximize the sum of the true scores of instances. The term, introduced from k-means, has an effect that the distance to the nearest prototype of each instance is minimized. In addition, η is a positive weight for this term. Each ξ_i in the second term is equal to the maximum of hinge losses of instance (x_i, y_i) over the classes other than y_i.

$$\xi_i = \max_{k \in C}\{1 - \delta_{ik} - (f_{y_i}(x_i) - f_k(x_i))\} \tag{6}$$

Note that $\xi_i \geq 0$ because $1 - \delta_{ik} - (f_{y_i}(x_i) - f_k(x_i)) = 0$ when $k = y_i$. This max-over-others aggregation is simpler than the above sum-over-others aggregation.

In this paper, we also consider the counter-part of (5) reflecting the sum-of-losses minimization.

$$\underset{w,b,s>0,\xi}{\text{minimize}} \quad 1/s^2 + \mu/c \sum_{i \in M} \sum_{k \in C} \xi_{ik} - \eta \sum_{i \in M} f_{y_i}(x_i)$$
$$\text{subject to} \quad f_{y_i}(x_i) - f_k(x_i) + \xi_{ik} \geq 1, \ \xi_{ik} \geq 0, \ i \in M, k \in C. \tag{7}$$

The difference between (3) and (7) is the third term in the objective function. We solve it by the same way as [6], which is shown in the next section.

3 Training Algorithm

The optimization problems (3), (5), and (7) are not convex. Hence, we reformulate them as difference-of-convex (DC) optimization problems, and apply the convex-concave procedure (CCP) [8], which is an algorithm to solve a DC optimization problem.

In [7], to reduce (3) to a DC optimization problem, we approximate the true-score function $f_{y_i}(x_i)$ by the squared one. It has a drawback that an additional hyperparameter is needed. On the other hand, in [6], we find that (5) can be reduced to a DC optimization problem without the approximation. Here, we only discuss the derivation in the case of (7).

3.1   Reduction to a DC Optimization Problem

First, by replacing ξ_{ik} with (4), we obtain the following unconstrained formulation of (7).

$$\underset{w,b,s>0}{\text{minimize}} \quad 1/s^2 + \mu/c \sum_{i \in M} \sum_{k \in C} \max\{f_k(x_i) + 1, f_{y_i}(x_i)\} - (\mu + \eta) \sum_{i \in M} f_{y_i}(x_i) \tag{8}$$

Introducing intermediate variables t_{ij} for $i \in M$, $j \in P$, and we defined

$$t_{ij} = \|(x_i - w_j)/s\|^2 + (b_j)^2. \tag{9}$$

Additionally, for a vector $t = (t_1, \ldots, t_l)$, we introduce the function of max-of negated: $h_{\max}(t) = \max\{-t_1, \ldots, -t_l\}$. Then, the k-th score function can be expressed as follows.

$$f_k(x_i) = -(-h_{\max}(t_{ik}))^{1/2}, \tag{10}$$

where $t_{ik} = (t_{ij})_{j \in P_k}$.

$f_k(x_i)$ in the second term is minimized. We introduce intermediate variable f_{ik}, and add a constraint $f_{ik} \geq -(-h_{\max}(t_{ik}))^{1/2}$. Moreover, this constraint is expressed as follows.

$$\min_{j \in P_k}\{t_{ij}\} \geq (-f_{ik})^2. \tag{11}$$

We remark that $-f_{ik}$ is nonnegative, because it is minimized.

Summarizing the above discussion, the maximum-margin model (7) can be expressed as follows.

$$\underset{w,\zeta,r,t}{\text{minimize}} \quad r + \mu/c \sum_{i \in M} \sum_{k \in C} \max\{f_{ik} + 1, f_{iy_i}\} + (\mu + \eta) \sum_{i \in M} (-h_{\max}(t_{iy_i}))^{1/2}$$

$$\text{subject to} \quad \min_{j \in P_k}\{t_{ij}\} \geq (f_{ik})^2, \ i \in M, \ k \in C$$

$$t_{ij} - \|x_i\|^2 r + 2x_i^\top w_j - \zeta_j = 0, \ i \in M, \ j \in P,$$

$$r\zeta_j \geq \|w_j\|^2, \ r \geq 0, \ j \in P, \tag{12}$$

where $1/s^2$ is replaced by r. The last two constraints are obtained by introducing a variable $\zeta_j = (b_j)^2 + \|w_j\|^2 r$ and replacing rw_j with w_j, removing $(b_j)^2$ and multiplying r in both sides. All of the nonlinear constraints except for the one of the second constraint are convex. It is an inequality including a DC function: $\phi_i + (-h_{\max}(t_{iy_i}))^{1/2}$, because $-(-h_{\max}(t_{iy_i}))^{1/2}$ is a convex function.

3.2 Convex-Concave Procedure

We solve the DC optimization problem (12) by CCP. CCP is an iterative algorithm. In each of iteration, the concave part of each DC function is linearized, and the optimization problem including the linearization is solved. For problem (12), supposing $\tilde{t}_{iy_i}$ is a current solution for variable t_{iy_i}, the concave part $(-h_{\max}(t_{iy_i}))^{1/2}$ in the objective function is replaced by

$$(-h_{\max}(\tilde{t}_{iy_i}))^{1/2} - v_i^\top (t_{iy_i} - \tilde{t}_{iy_i}), \tag{13}$$

where v_i is a subgradient of $-(-h_{\max}(t_{iy_i}))^{1/2}$ at $\tilde{t}_{iy_i}$. One of subgradients $v_i = (v_{ij})_{j \in P_{y_i}}$ is obtained as follows.

$$v_{ij} = \begin{cases} -(4t_{ij_i})^{-1/2} & j = j_i, \\ 0 & \text{otherwise,} \end{cases} \tag{14}$$

where $j_i = \operatorname{argmax}\{-\tilde{t}_{ij} \mid j \in P_{y_i}\}$. Using $-h_{\max}(\tilde{t}_{iy_i}) = \tilde{t}_{ij_i}$, (13) is expressed as follows.

$$(-h_{\max}(\tilde{t}_{iy_i}))^{1/2} - v_i^\top (t_{iy_i} - \tilde{t}_{iy_i}) = (4\tilde{t}_{ij_i})^{-1/2}(t_{ij_i} + \tilde{t}_{ij_i}). \tag{15}$$

Hence, the convex approximation of the second constraint of (12) at the current values is obtained as follows.

$$-(4\tilde{t}_{ij_i})^{-1/2}(t_{ij_i} + \tilde{t}_{ij_i}) \geq \phi_i. \tag{16}$$

However, the linearization (16) is undefined if $\tilde{t}_{ij_i} = 0$. To avoid it, we assume b_j is not less than a small positive constant δ. We replace b_j^2 with $(b_j)^2 + \delta$. As a result, the third constraint of (12) is replace with $t_{ij} - \|x_i\|^2 r + 2x_i^\top w_j - \zeta_j = \delta$.

Consequently, we solve the following optimization problem in each iteration.

$$\begin{aligned}
\underset{w,\zeta,r,t}{\text{minimize}} \quad & r + \mu/c \sum_{i \in M} \sum_{k \in C} \max\{f_{ik} + 1, f_{iy_i}\} + (\mu + \eta) \sum_{i \in M} (4\tilde{t}_{ij_i})^{-1/2} t_{ij_i} \\
\text{subject to} \quad & \min_{j \in P_k}\{t_{ij}\} \geq (f_{ik})^2, \ \xi_{ik} \geq 0, \ i \in M, \ k \in C, \\
& t_{ij} - \|x_i\|^2 r + 2x_i^\top w_j - \zeta_j = \delta, \ i \in M, \ j \in P, \\
& r\zeta_j \geq \|w_j\|^2, \ r \geq 0, \ j \in P.
\end{aligned}$$

$$(17)$$

This optimization problem can be solved by several existing softwares. The algorithm is terminated if a decrement of the objective function is small. To retrieve the original parameter (w, b, s) of (1) from a solution (w^*, ζ^*, r^*), we compute $s = 1/\sqrt{r^*}$, $w_j = w_j^*/r^*$, $b_j = (\zeta_j^* - \|w_j^*\|^2/r^* + \delta)^{1/2}$.

3.3 Initialization

A solution obtained by the convex-concave procedure depends on an initial solution. We have proposed a method [7] to initialize prototypes w_j for the problem (3) using k-means++ [1]. Furthermore, we have proposed an initialization method for the problem (5), in which prototypes w_j are obtained by k-means++ and the scale parameter s is also computed by solving (5) with fixed variables. In this paper, we apply the method of [6] to the problem (7).

After computing prototypes $w_1, \ldots, w_p$ by class-wise k-means++, we consider the following optimization problem, which is a simplified problem of (8).

$$\underset{s>0}{\text{minimize}} \quad 1/s^2 + (\mu/c) \sum_{i \in M} \sum_{k \in C} \max\{f_{ik}/s + 1, f_{iy_i}/s\} - (\mu + \eta) \sum_{i \in M} f_{iy_i}/s \tag{18}$$

where, abusing notation, we define

$$f_{ik} = -(-h_{\max}(t_{ik}))^{1/2},$$

and $t_{ik} = (t_{ij})_{j \in P_k}$, $t_{ij} = \|x_i - w_j\|^2$. All of f_{ik} are constant. Let $\bar{f} = \sum_{i \in M} f_{iy_i}$, and moreover let $r = 1/s$. Then, the above optimization problem can be expressed as follows.

$$\underset{r \geq 0}{\text{minimize}} \quad r^2 + (\mu/c) \sum_{i \in M} \sum_{k \in C} \max\{f_{ik}r + 1, f_{iy_i}r\} - (\mu + \eta)\bar{f}r \tag{19}$$

Considering the point at which the subgradient is 0, the optimal solution can be solved analytically.

Let $\bar{w}_1, \ldots, \bar{w}_p$ and $\bar{s}$ be the values obtained by the above initialization method. The initial solution (w, ζ, r) of (12) is computed as $w_j = \bar{w}_j/\bar{s}^2$, $\zeta_j = \|\bar{w}_j\|^2 \bar{s}^2$, $r = 1/\bar{s}^2$. The initial values of t_{ij} can be computed from w_j, ζ_j, and r.

3.4   Example

We graphically show prototypes and classification boundaries of three maximum-margin nearest prototype classifiers (MM-NPCs), listed below, trained for an artificial data set.

– MM-NPC(P): Solve (3) by the algorithm in [7].
– MM-NPC(M) (MM-NPC with the max-over-others loss function): Solve (5) by the algorithm in Sect. 3 (see [6]).
– MM-NPC(S) (MM-NPC with the sum-over-others loss function): Solve (7) by the algorithm in Sect. 3.

Additionally, we show those of a k-means-based nearest prototype classifier (KM-NPC), in which prototypes are obtained by applying k-means++ to the training instances for each class.

The result is shown in Fig. 1. The data has three classes identified in different colors. MM-NPC(M) and MM-NPC(S) successfully classify the data, and moreover their prototypes are placed in dense regions as expected. However, some of prototypes of MM-NPC(P) move to unexpected places far from initial places given by k-means++.

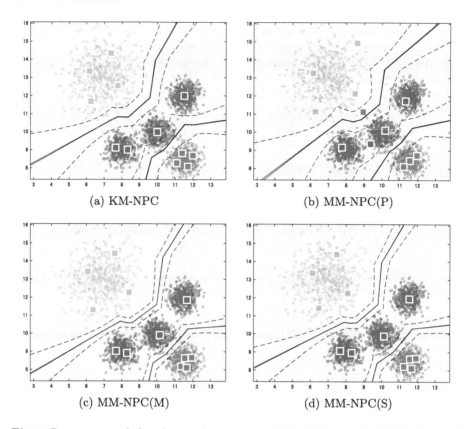

Fig. 1. Prototypes and classification boundaries of KM-NPC and MM-NPCs displayed on an artificial three-class data set.

4 Numerical Experiment

We compare three MM-NPCs (MM-NPC(S), MM-NPC(M), M-NPC(P)) with baseline methods. As baseline methods, we use two naive methods of nearest prototype classifiers (NCPs), that are GLVQ [10] and KM-NPC. Additionally, we show the performance of a support vector machine (SVM), that is one-versus-the-rest SVM with the RBF kernel.

All of the training algorithms are implemented in MATLAB. The optimization problems in MM-NPC(S), MM-NPC(M), MM-NPC(P), and SVM are solved by MOSEK [9]. An implementation of GLVQ is found in https://www.cs.rug.nl/~biehl/mcode.html.

Here, we explain hyperparameter settings in this experiment. For the five NPC methods, we need to set the number d of prototypes, which is varied in $\{1, 2, \ldots, 5\}$. For MM-NPC(S), MM-NPC(M), MM-NPC(P), and SVM, there is the weight μ for loss functions, which is varied in $\{1, 10, 100\}$. The approximation parameter τ for the true-score function used in MM-NPC(P) (see [7]) is varied in $\{0.1, 1, 10\}$. The weight η for the true-score function used in MM-NPC(S) and

MM-NPC(M) is fixed to $0.001 \times \mu$. The scale parameter σ of the RBF kernel is varied in $\{1, 2, 10, 20, 100, 200, 1000, 2000\}$.

We use four benchmark data sets obtained form UCI Machine Learning Repository [3]: "Iris" (iri), "Wine" (win), "Breast Cancer Wisconsin (Original)" (bre), and "Ionosphere" (ion).

For each method and each data set, we measure the average of classification errors by 5 times of 10-fold cross-validations. A classification error is a ratio of the number of misclassified instances to the total number in a validation data set. That is, if m is the total number of instances and $e_1, \ldots, e_{10}$ are the numbers of misclassified validation instances for 10 trials of 10-fold cross-validation, then the classification error is $(e_1 + \cdots + e_{10})/m \times 100$.

Table 1 shows results of this experiment. Results for each data set is summarized in each two rows starting with the name. Each value in the first row is an average classification error, which is the lowest value among hyperparameters examined in this experiments. The selected hyperparameter is shown in the second line.

In the view of classification performance, MM-NPCs are better than GLVQ and KM-NPC. However, we should remark that GLVQ is not completely tuned, so there is a possibility that the performance of GLVQ is improved more. On the other hand, SVM outperforms MM-NPCs in win and ion, while MM-NPCs is better in bre. Hence, we can say that MM-NPCs have classification performance comparable with kernel SVMs. We cannot say which methods among MM-NPCs is the best. We prefer MM-NPC(M) in the view of computational cost, however there is a possibility that MM-NPC(S) is better for other data sets.

Table 1. Classification errors and selected parameters

Data	MM-NPC(S)	MM-NPC(M)	MM-NPC(P)	GLVQ	KM-NPC	SVM
	(d, μ)	(d, μ)	(d, μ, τ)	(d)	(d)	(σ, μ)
iri	2.53	2.67	**2.40**	3.73	2.67	2.67
	(4,1)	(1,1)	(3,10,1)	(3)	(4)	(2,1)
win	2.14	1.69	2.02	2.92	2.92	**1.35**
	(1,1)	(1,1)	(2,1,10)	(2)	(1)	(2,1)
bre	**2.69**	2.90	2.78	3.63	3.51	3.13
	(5,10)	(1,1)	(3,100,10)	(5)	(1)	(200,10)
ion	11.05	10.94	11.51	13.45	14.25	**5.13**
	(5,10)	(5,1)	(4,1,1)	(5)	(5)	(2,1)

5   Conclusion

In this study, we present the margin-maximization model for NPCs using the sum-over-others loss function. The training algorithm is derived by transforming the model into a DC optimization problem, which is then solved using CCP.

Moreover, we discuss the procedure to provide an initial solution for CCP. This derivation is largely based on our recent study [6], in which we made revisions to the margin-maximization model to eliminate the approximation of the true-class score function. Moreover, we graphically display the obtained NPCs for the two-dimensional artificial data set. It show that the (revised) margin-maximization models successfully classify the data, while locate the prototypes within the dense region of the data. In the numerical experiment, we compare the margin-maximization models (MM-NPC(S) and MM-NPC(M)) with our previous model (MM-NPC(P)) and three baseline algorithms (GLVQ, KM-NPC, and SVM) across four benchmark data sets. The result reveals that our models outperform two naive NPC learning algorithms (GLVQ and KM-NPC) in terms of the classification error of the cross-validation.

In the future work, we plan to propose fuzzification of the score function. Justification of our models by a generalization bound for NPCs is also included in the future work.

Acknowledgements. This work was supported by JSPS KAKENHI Grant Number JP21K12062.

References

1. Arthur, D., Vassilvitskii, S.: k-means++: the advantages of careful seeding. In: Proceedings of the Eighteenth Annual ACM-SIAM Symposium on Discrete Algorithms, pp. 1027–1035 (2007)
2. Doğan, Ü., Glasmachers, T., Igel, C.: A unified view on multi-class support vector classification. J. Mach. Learn. Res. **17**(45), 1–32 (2016)
3. Dua, D., Graff, C.: UCI machine learning repository (2017). http://archive.ics.uci.edu/ml
4. Hammer, B., Villmann, T.: Generalized relevance learning vector quantization. Neural Netw. **15**(8), 1059–1068 (2002)
5. Kohonen, T.: The self-organizing map. Proc. IEEE **78**(9), 1464–1480 (1990)
6. Kusunoki, Y., Nakashima, T.: Revised optimization algorithm for maximum-margin nearest prototype classifier. In: Proceedings of IFSA 2023, pp. 276–280 (2023)
7. Kusunoki, Y., Wakou, C., Tatsumi, K.: Maximum-margin model for nearest prototype classifiers. J. Adv. Comput. Intell. Intell. Inform. **22**(4), 565–577 (2018)
8. Lipp, T., Boyd, S.: Variations and extension of the convex-concave procedure. Optim. Eng. **17**(2), 263–287 (2016)
9. MOSEK ApS: MOSEK Optimization Toolbox for MATLAB 10.0.33 (2022). https://docs.mosek.com/10.0/toolbox/index.html
10. Sato, A., Yamada, K.: Generalized learning vector quantization. In: Touretzky, D., Mozer, M., Hasselmo, M. (eds.) Advances in Neural Information Processing Systems, vol. 8. MIT Press (1995)
11. Schneider, P., Biehl, M., Hammer, B.: Adaptive relevance matrices in learning vector quantization. Neural Comput. **21**(12), 3532–3561 (2009)

Predicting Stock Price Fluctuations Considering the Sunny Effect

Kenta Nakaniwa[1], Tomoya Matsuki[1]([✉]), Makishi Iguchi[1], Akira Notsu[2], and Katsuhiro Honda[2]

[1] Osaka Prefecture University, 1-1 Gakuen-cho, Naka-ku, Sakai, Osaka 599-8531, Japan
seb01126@st.osakafu-u.ac.jp
[2] Osaka Metropolitan University, 1-1 Gakuen-cho, Naka-ku, Sakai, Osaka 599-8531, Japan
notsu@omu.ac.jp

Abstract. Predicting stock prices through machine learning is a highly anticipated research area. Previous studies have shown improved accuracy in stock price prediction using machine learning, and this study also reported a high degree of accuracy. However, the amount of past data is limited, and the psychological and subjective factors of traders are also involved, and it is still a field with many problems. The effects of weather on human psychology have already been studied and shown to have the potential to affect the psychology of stock traders. In this study, we proposed a new data generation method that takes into account the psychological effects of weather. In detail, image data combining stock price data and weather data were created and trained using a convolutional neural network (CNN). Then, by confirming the validity of the method, we confirmed that weather data can provide clues for predicting stock price fluctuations.

Keywords: Sunny Effect · Stock Price Volatility Forecast · CNN

1 Introduction

In recent years, research on stock price forecasting has been conducted not only from the traditional economics and statistics perspectives, but also by using machine learning due to advances in AI technology. Ikeda and Hayashida at the University of Kitakyushu have examined stock price prediction using deep learning [1], and Takagi and Takanobu at the Chiba Institute of Technology have studied the utility of AI for the foreign exchange market, which has strong affinity with stock prices [2]. As described above, expectations for market forecasting using machine learning are generally high, but although some research has produced results, there are still many challenges in this area, such as insufficient learning due to the limited number of historical data.

We focused on the impact of weather on stock traders and explored the possibility of using machine learning to improve the accuracy of stock price forecasts.

K. Honda et al. (Eds.): IUKM 2023, LNAI 14376, pp. 47–54, 2024.
https://doi.org/10.1007/978-3-031-46781-3_5

There have been several studies on the effects of weather on human psychology, including a study that proved a correlation between climatic conditions and the number of suicides and that lack of adaptation to climate causes negative behavior [3], and a study that proposed the consideration of external weather factors in psychiatric disorders based on the monthly variation in the number of psychiatric emergency hospitalization cases [4]. Among them, studies focusing on the psychology of stock traders in particular have attracted widespread attention, especially in the 2000 s, including a study that proved the "sunny weather effect" [5], which is well known in Japan, and a study that investigated the relationship between the phases of the moon and the market [6].

With the premise that stock price fluctuations are influenced by both weather conditions and traders' psychology, we explore the utility of employing two types of data for machine learning: historical stock price data and historical weather data. We employed a convolutional neural network to facilitate learning the spatial relationship between stock price graphs and weather data, aiming for efficient analysis.

2   Stock Prices and Psychology, Forecasting

Stock prices are one of the most important indicators for understanding economic conditions, and have therefore been the subject of research from various academic perspectives. In addition, research on stock price prediction using machine learning has been active recently, and many researchers are working on it and have produced research results.

Factors that cause stock prices to fluctuate are mainly classified into two categories: internal factors of the company (e.g., the state of corporate performance, the development, announcement, and launch of new products, mergers and acquisitions, restructuring, and corporate scandals) and external factors of the company (e.g., stock index, interest rate, currency, and price changes, wars and political changes abroad, and natural disasters). The latter, or external factors, have a particularly strong effect on short-term price movements, and the results of stock trades by traders have a large impact on stock prices.

In addition, individual traders, especially those who specialize in stock trading, predict the price movements of stocks several hours later or the next day based on the day's stock price chart, and repeat trading several times a day, and these predictions are mostly based on price movement trends rather than economic trends. The state of stock trading by these traders has a large influence on hourly and daily stock prices, and is considered one of the pivot determinants in short-term stock price prediction.

Research approaches to stock price prediction have been conducted in various fields. These include studies based on the hypothesis that stock prices will rise if there are many positive contributions in news texts [7], studies that predict stock prices using words in newspaper articles [8], and studies that explore the correlation between box-office revenues of hit movies and stock prices [9], not only from an economic perspective.

In addition to these studies that investigate the relationship between stock prices and social conditions, many studies dealing with the relationship between weather and stock prices were conducted mainly in the 2000 s. A representative example is a study by Hideaki Kato (2004) [5], who proved the "sunny weather effect" that stock prices tend to rise on sunny days. He researched the Nikkei Stock Average for the past 40 years and the cloud cover on thoes days, and showed that there was a correlation between the two. Furthermore, a study conducted at York University in Canada [10] showed that stock prices are less likely to rise when temperatures rise, and a study in the United States [6] showed that stock prices are more likely to rise during the period around the new moon than during the period around the full moon. Hence, it has been proven in several previous studies that the psychological impact of weather on traders' trading decisions is one of the important factors in stock price forecasting.

In 2017, a study was conducted by the University of Tokyo using a convolutional neural network (CNN) to learn stock price chart images and predict the rise and fall of stock prices in 30 min [11]. The output result from the CNN in this study showed a correct response rate of 52.5%, which was more accurate than random trading as a prediction for two items. A similar study [12] was conducted in 2019 using stock price candlestick images for CNNs, where the correct response rate for stock prices 5 min ahead was 63.3%, proving the usefulness of the machine learning approach for stock price prediction. Although there is a gap in accuracy between the two approaches due to differences in the number of data and the training structure, the results are both positive.

Ikeda, who conducted the latter study, has since verified further improvement in accuracy using ensemble learning and volume graphs [13], and the accuracy reached 70.7%. In a study of predicting stock price volatility by machine learning [14], he used RNN and LSTM to output results close to those obtained by conventional computation of volatility. Thus, multiple studies have shown the value of stock price prediction by machine learning.

3   Proposed Method

A neural network that predicts the magnitude of stock price fluctuations on the following day is constructed by training a single image of daily stock prices and weather data. Each image is prepared as a 40×40 pixel png file (Fig. 1,2), with the closing price of the stock for the previous 40 trading days counted from that day (the final stock price on that day) placed as a dot. The image data was created using historical data of the S&P 500, with the horizontal axis representing the date and the vertical axis representing the closing price of the stock, and the daily price movements were colored in black. In order to keep the image size uniform regardless of the size of the price movement, the image was scaled down or up so that the largest value was at the bottom and the smallest value at the top over a 40-day period. The weather data for each day were set as the background of the stock price data, and the RGB images (R - mean temperature, G - temperature difference, B - precipitation; min = 0.0, max =

1.0) are shown (Fig. 2). In order to distinguish the black color (R=0, G=0, B=0) of the stock prices, the RGB values of the weather data were adjusted to be as balanced as possible within the range of 0.5 to 1.0, so that colors close to black, indicating stock prices, are not displayed (if the value exceeds the range of 0.0 to 1.0, the value is set to 0.0 and 1.0 respectively). The formula is as follows.

R: (average temperature(°C)+20)/100+0.5
G: (maximum temperature(°C)-minimum temperature(°C))/40+0.5
B: precipitation(mm)/4+0.5

Since the weather data provided by the Japan Meteorological Agency (JMA) includes variations in observation points, the cities selected for weather data extraction in the validation of the S&P500 were selected from the daily weather data published by the JMA for 400 cities in the North American continent, with 100 cities before and 300 cities after Key West (location number 72201). In this study, the city-by-city data were converted into 2 × 2 pixel image data, which were displayed in 400 20 × 20 cells to create a background of the same size as the image of stock price fluctuations (Fig. 2).

   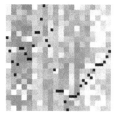

Fig. 1. Without weather data   **Fig. 2.** With weather data

We devised such a data generation method to flexibly respond to the increase/decrease and regional nature of weather data stations, while taking advantage of the effectiveness of CNNs that has been demonstrated in existing studies.

4   Experimental Setup

The purpose of this study is to measure the usefulness of weather data for predicting stock price fluctuations. We used image data generated from weather data (average temperature, temperature difference, and precipitation) for 400 cities in the North American continent and a stock price index (S&P 500) in the United States to learn from CNNs and measure the accuracy of the fluctuation prediction. The S&P500 is a stock price index derived from the stock prices of 500 companies that represent the United States in various fields, and was chosen as the subject of the study because it is less affected by stock price fluctuations due to internal effects, which can be considered as noise. In addition, since many of the traders who have a strong influence on the short-term price movements of the S&P 500 reside in the United States, historical weather data from the North American continent was used for verification.

Table 1. CNN parameters after optimization

Layer Type	Input	OutMaps	KernelShape	OutShape
MaxPooling	3,40,40		3,2	1,20,40
Convolution	1,20,40	16	4,6	16,17,35
MaxPooling	16,17,35		2,2	16,8,17
Tanh	16,8,17			16,8,17
Convolution	16,8,17	16	5,5	16,4,13
MaxPooling	16,4,13		2,2	16,2,6
Tanh	16,2,6			16,2,6
fully-connected	16,2,6			100
Tanh	100			100
fully-connected	100			1
Sigmoid	1			1

Daily image data was created from S&P 500 stock prices for the past 715 days (2018/03/01 to 2020/12/31 excluding non-trading days) and weather data for 400 cities in the North American continent for the same period published by the Japan Meteorological Agency. The data set is relatively small because the JMA has limited observation data available and it was not possible to obtain data with the same accuracy in the past. As a preliminary experiment, some forecasts were made for the next day's stock price increase or decrease, but neither the existing method nor the proposed method could be learned, probably due to the small number of data and large bias. Since the ratio of stock price fluctuation predictions for the next day differed between increases and decreases, and size adjustments were necessary, this time, in order to make a simple comparison, we used the discriminant predictions of stock price fluctuation, that is, "more than ±0.5% (161 decrease + 220 increase, 54%)" or "less than ±0.5% (334 cases, 46%)" of the large or small changes in the stock price.

In this experiment, image recognition using CNN was performed using the Neural Network Console (NNC) provided by Sony, and stock price prediction was verified. In this experiment, the network structure and other parameters of "binary_cnn.sdcproj," a template project inherent in NNC, were used as is, and the network was optimized from there. Table 1 shows the parameters after optimization for the dataset using meteorological data (listed from top to bottom in order of proximity to the input layer). The loss function is BinaryCrossEntropy and the batch size is 64. The optimization algorithm uses Adam, with parameters $\alpha = 0.001, \beta_1 = 0.9, \beta_2 = 0.999, and \varepsilon = 10^{-8}$. For training and testing, we created a dataset of S&P500 only (sp-n group) and a dataset of S&P500 and weather (sp-w group). In addition, we prepared a dataset of S&P500 and average temperature (sp-kion group), S&P500 and temperature difference (sp-kandan group), and S&P500 and precipitation (sp-kousui group) to verify the most influential weather indicators.

5   Experimental Results and Discussion

Using the data from 2018/03/01 to 2020/09/30 (651 cases (345 cases of large variation + 306 cases of small variation)) as training data and having the data from 2020/10/01 to 2020/12/31 (64 cases (36 cases of large variation + 28 cases of small variation)) predicted, the proposed method (sp-w group). The correct response rate (the percentage of data predicted to be positive or negative that actually were) was 70.31% for the proposed method (sp-w group), and 60.93% for the S&P 500-only data set (sp-n group). Incidentally, the logistic regression hardly learned anything, with 53.12% of the data. In addition, a comparison of (sp-kion group), (sp-kandan group), and (sp-kousui group) using the same network structure showed that the highest accuracy in predicting stock price fluctuations was achieved by using precipitation, at 57.81%, 56.25%, and 64.06%, respectively.

The results of learning with each of the divided weather data (average temperature, temperature difference, and precipitation) were compared, and the characteristics of each weather data in this study were clarified as described above. The learning results with the average temperature and temperature difference data in the network structure after optimization with the weather data set are less accurate than those with the stock price data alone, and these are not expected to have a direct relationship with stock price fluctuations. The only training with precipitation data showed higher accuracy than the training with stock prices alone. We believe that precipitation is a particularly important feature of the high accuracy when all weather data are used. Therefore, it can be inferred that there is a strong relationship between the amount of precipitation and the amount of fluctuation in stock prices, and that this relationship led to the large improvement in forecast accuracy.

Thus, the usefulness of weather data in predicting stock price fluctuations was mainly driven by the accuracy of precipitation data. In other words, weather conditions, such as rainfall or non-rainfall, are strongly related to stock price fluctuations. However, there are still many shortcomings to strongly support this theory, because the number of data is small, the learning is not sufficient, and seasonal effects can be expected, especially for average temperature, which means that trends (sudden rises and falls in stock prices) may have occurred during periods of low temperatures.

6   Conclusion

In this study, we devised a method for using weather data in machine learning to predict future stock price fluctuations, and investigated its usefulness. We used a stock index (S&P500), which is actively traded by traders, as the research target, and used weather data for the regions where the target stocks are located. As a result, for the S&P 500, the accuracy of the prediction of stock price fluctuations was improved by using weather data, and values that support the hypothesis were obtained. In addition, the weather data were divided and compared in terms of average temperature, temperature difference, and precipitation, and it was confirmed that the precipitation data in particular had a significant effect on improving the accuracy of the forecasts.

Future research issues include optimizing the selection of regions from which to extract weather data for specific stock price indices, increasing the number of data, and validating the results by increasing the number of data and reducing the influence of temporary trends. In addition, since seasonal changes in market conditions [15] can be considered to have a considerable influence on the experimental results, learning with time-specific data is also an item to be verified in the future.

Acknowledgements. This work was supported in part by JSPS KAKENHI Grant Number JP18K11473 and 22K12182.

References

1. Yoshikazu, I., Minoru, H.: Application of deep learning to stock price forecasting. Rev. Bus. Econ. 52, Nos. 1,2,3, and 4), The Economic Society of The University of Kitakyushu 13–26 (2017). (in Japanese)
2. Touru, T., Haruto, T.: Application for a foreign exchange trading system using by AI. In: Proceedings of the National Conference of the Association for Information Systems society of Japan 14. (Johoshisutemugakkaizennkkoku-taikaironnbunnsyu14) Information Systems society of Japan (2018). (in Japanese)
3. Yoshitaka, F.: The impact of weather and seasons on emotional disorders. (kisho.kisetsunokanjoshogaihenoeikyo), Global Environ. **8**(2), 221–228 (2003). (in Japanese)
4. Hideyuki, F.M., Shimura.: Meteorolingical factors and mental disorders - with special reference to psychiatric emergency cases. Jap. J. Biometeorol. **24**(2), 67–73 (1987). (in Japanese)
5. Hideaki, K.: The mysterious relationship between weather and stock prices. (tenki-tokabukanofushiginakankei), Toyo Keizai Shinposha (2004). (in Japanese)
6. Dichev, I.D., Janes, T.D.: Lunar cycle effects in stock returns. J. Private Equity **6**(4), 8–29 (2003)
7. Keiichi, G., Hiroshi, T., Takao, T.: Estimating news articles' negative-positive by deep learning. In: The 29th Annual Conference of the Japanese Society for Artificial Intelligence, The The Japanese Society for Artificial Intelligence (2015). (in Japanese)
8. Junya, O.: Prediction of Stock Market Price from Economical News Text Data Using Deep Learning. Bachelor's Degree Thesis of Kochi University of Technology (2014). (in Japanese)
9. Yukihiro, T., Atushi, O.: Analysis of the psychological influence on economic indicators by the relationship between the release date of hit movies and stock

prices (Economic Analysis of Technological Progress (2)) (hittoeiganokokaibitok-abukanokankeiniyorukeizaishihyohenoshinritekieikyonobunseki). In: Annual Conference Proceedings 21.2, Japan Society for Research Policy and Innovation Management (2006). (in Japanese)

10. Cao, M., Jason, W.: Stock market returns: a note on temperature anomaly. J. Banking Finan. **29**(6), 1559–1573 (2005)

11. Kunihiro, M., Yutaka, M.: Stock prediction analysis using deep learning technique. In: The 31 Annual Conference of the Japanese Society for Artifical Interigence, The Japanese Society for Artificial Intelligence (2017). (in Japanese)

12. Yoshikazu, I.: Stock movement forecasting with candlestick charts using convolutional neural networks. The review of business and economics, The University of Kitakyushu, vol. 54 no. 1, pp. 1–18 (2019). (in Japanese)

13. Yoshikazu, I.: An analysis on improving accuracy of deep learning stock price prediction with ensemble learning. The review of business and economics, The University of Kitakyushu, vol. 56, no. 1, pp. 15–34 (2021). (in Japanese)

14. Keiichi, G., Hiroshi, T., Takao, T.: Modeling the volatility clustering with recurrent neural network. Summary Collection of the National Conference on Business Information Systems 2017 Spring National Research Presentation Conference of the Japan Society of Information and Management (keieijohogakuzenkokukenkyuhap-pyotaikaiyoshishu2017nenshunkizenkokukenkyuhappyotaikai), The Japan Society for Management Information (2017). (in Japanese)

15. Katsuhiko, O. : Glabal market seasonality and investor sentiment : a text-mining approach. J. Bus. Admin. **61**(4), 110–136 (2014). (in Japanese)

Proposal of a New Classification Method Using Rule Table and Its Consideration

Yuichi Kato[1]($\boxtimes$) and Tetsuro Saeki[2]

[1] Shimane University, 1060 Nishikawatsu-cho, Matsue city, Shimane 690-8504, Japan
ykato@cis.shimane-u.ac.jp
[2] Yamaguchi University, 2-16-1 Tokiwadai, Ube city, Yamaguchi 755-8611, Japan

Abstract. This paper proposed a new classification method using a rule table and demonstrated how to derive the rule table from the Rakuten Travel dataset that represents real-world datasets and how to use it for classification problems. The usefulness of the proposed method was shown using the classification rate referring to the random forest method. The proposed rule table concept showed the expansion of the if–then rules induced by the previous statistical test rule induction method including basic rules called trunk rules behind the dataset and the usefulness for various levels of rule description for real-world datasets.

Keywords: decision table · if-then rule · classification problem · random forest · principle of incompatibility

1 Introduction

With the growth of various network societies, numerous electric datasets are generated and stored for use under different policies and/or business strategies, and such datasets are often arranged in each suitable form for the application. This paper considered a Rakuten Travel dataset (R-dataset) [1,12] which is a real-world dataset (RWD) of questionnaire surveys with the accommodation (object) rating some feature items of each object and its overall category, and a typical decision table (DT) in the field of the Rough Sets (RSs) [2]. This paper proposed a new method for arranging the R-dataset into a rule table (RT), presenting the relationships between the feature items of each object and the overall category, and applying these relationships for classifying a new object into its belonging overall category. The classification results were evaluated using the random forest (RF) [3].

In addition, as mentioned in the principle of incompatibility [4], accurate arrangement and/or summarization against the dataset as well as the comprehensive expressions for decision making are necessary to support real-world activities. However, the arrangement using the RT was too complex for human beings, as also in the case of RF, and lost comprehensibility. After showing that the RT includes if-then rule candidates (RCs) with a proper statistical significance level induced by the previous statistical test rule induction method (STRIM) [5–20] and RCs without it, the former RCs were also pointed out including the intuitively comprehensive basic rules. These three types of RCs in the dataset can organize the rule set, balancing accuracy and comprehensibility depending on the target matter. Meanwhile, RF does not provide such knowledge or information behind the dataset, although it remains a useful method for classification problems. In this way, the usefulness of the proposed method was confirmed.

2   Conventional RS and Its Rule Induction Method

The R-dataset is a typical DT in the field of the RSs [2] and the DT is formulated with an observation system S as follows: $S = (U, A = C \cup \{D\}, V, \rho)$, where $U = \{u(i)|i = 1, ..., N = |U|\}$ is a dataset, $u(i)$ denotes an object in a population, A denotes a set of given attributes of U, $C = \{C(j)| j = 1, ..., |C|\}$ denotes a set of the condition attribute $C(j)$, and D denotes a decision attribute. Meanwhile, V denotes a value set of the attribute, i.e., $V = \bigcup_{a \in A} V_a$, where V_a denotes the set of values for an attribute a and $\rho : U \times A \rightarrow V$ is called an information function. For example, let $a = C(j)$ $(j = 1, ..., |C|)$, then $V_a = \{1, 2, ..., M_{C(j)}\}$. If $a = D$, then $V_a = \{1, 2, ..., M_D\}$. Corresponding relationships with the R-dataset are given as follows: $|C| = 6$, $A = \{C(1) = Location, C(2) = Room, C(3) = Meal, C(4) = Bath\,(HotSpring), C(5) = Service, C(6) = Amenity, D = Overall\}$, and $V_a = \{1 : Dissatisfied, 2 : Slightly\,Dissatisfied, 3 \quad : \quad Neither\,Dissatisfied\,nor\,Satisfied, 4 \quad : Slightly\,Satisfied, 5 : Very\,Satisfied\}$, $a \in A$, i.e., $|V_{a=D}| = M_D = |V_{a=C(j)}| = M_{C(j)} = 5$.

The conventional RS theory finds the following subsets of U through C and D:

$$C_*(D_d) \subseteq D_d \subseteq C^*(D_d). \tag{1}$$

Here, $C_*(D_d) = \{u_i \in U|[u_i]_C \subseteq D_d\}$, $C^*(D_d) = \{u_i \in U|[u_i]_C \cap D_d \neq \emptyset\}$, $[u_i]_C = \{u(j) \in U|(u(j), u_i) \in I_C, u_i \in U\}$, and $I_C = \{(u(i), u(j)) \in U^2|\rho(u(i), a) = \rho(u(j), a), \forall a \in C\}$, where $[u_i]$ denotes the equivalence class with the representative u_i induced by the equivalence relation I_C , and $D_d = \{u(i)|(\rho(u(i), D) = d\}$. In equation (1), $C_*(D_d)$ and $C^*(D_d)$ are called the lower and upper approximation of D_d, respectively, and $(C_*(D_d), C^*(D_d))$ is the rough set for D_d. Being found $C_*(D_d) = \{u(i)| \wedge_j (\rho(u(i), C(j)) = v_{j_k})\}$ by using the DT, the following if–then rule with necessity is obtained using the inclusion relation in equation (1): if CP then $D = d$, where the condition part (CP) is specifically $CP = \wedge_j(C(j) = v_{j_k})$. Similarly, the if–then rule with possibility is induced using $C^*(D_d)$. Thus, the conventional RS theory derives relationships between $C = (C(1), ..., C(6))$ and D. The specific algorithm and RSs can be respectively referred to in the literature [2, 21].

However, in most cases, $u(i) = (u^C(i), u^D(i))$ is randomly collected from a population of interest so that the attribute values $u^C(i) = (v_{C(1)}(i), ..., v_{C(6)}(i))$ $(v_{C(j)}(i) (\in V_{C(j)}))$ or $u^D(i) = v_d(i) (\in V_D)$ follow random variations. The collection of the dataset from the same population indicates that U will variate

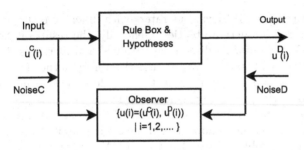

Fig. 1. Data generation model: Rule Box contains if-then rules and Hypotheses regulate how to apply rules for Input and transform Input into Output.

such that the corresponding induced rules also variate since the conventional RS theory directly uses the DT attribute values [5,8,13]. As a result, the rules induced by the conventional RSs do not fulfill the function, e.g., for the classification problem. In statistics, $C(j)$ and D are recognized as random variables, and $v_{C(j)}(i)\,(\in V_{C(j)})$ and $v_d(i)\,(\in V_D)$ are their respective outcomes. The conventional RS theory lacks these statistical views and does not have a model for collecting the DT values.

3   Outlines of Data Generation Model

The previous STRIM proposed a data generation model as shown in Fig. 1 in which Rule Box and Hypotheses transformed an input $u^C(i)$ into the output $u^D(i)$. Here, Rule Box contains pre-specified if-then rules and Hypotheses regulate how to apply those rules for the input and transform the input into the output. The model was used for generating a dataset on a simulation experiment as follows: (1) specifying some proper if-then rules in Rule Box, (2) randomly generating an input and transforming it into the output based on those rules and Hypotheses, (3) repeating (1) and (2) N times and forming $U = \{u(i) = (u^C(i),\, u^D(i))|i = 1, ..., N = |U|\}$. The generated U was used for investigating the rule induction abilities by applying it for any rule induction method (see details [5–20]). The previous STRIM [20] could induce the pre-specified rules while the rules induced by the method like RSs [15,21], CART [14,22] or association rule [19], [23] included a lot of meaningless rules and hardly corresponded with the pre-specified rules.

4   Introduction of RT and Its Application for Classification Problem

The validity and usefulness of the previous STRIM have been confirmed in simulation experiments which generate the dataset obeying pre-specified rules, that is, a well-behaved dataset. This paper newly expanded Rule Box and Hypotheses in Fig. 1 so as to adapt the R-dataset as one of RWD which includes an ill-behaved dataset caused by various raters with different rating standard, and investigated its rule induction abilities. Generally, the result of the rule induction from an RWD cannot be directly ascertained and increases the complexity of rules' description. This paper investigates the ability in the classification problem having the complimentary relation to the rule induction problem in which induced rules directly affect the classification rate.

Let there be a new relationship between $C(j)$ $(j = 1, ..., 6)$ and D by using the R-dataset. This section shows how to form the RT and apply the new relationship for a classification problem. The R-dataset of $N=$ 10,000 was first formed by randomly selecting 2,000 samples, each of $D = m$ $(m = 1, ..., 5)$ from about 400,000 surveys in the 2013–2014 dataset. The R-dataset was randomly divided into two groups. One is the R_L-dataset with $N_L=5,000$ for learning the R-dataset and the other is the R_C-dataset with $N_C=5,000$ for the classification experiment.

Regarding the learning process, let us consider a specific example of learning data: $(C(1), ..., C(6), D) = (1, 2, 3, 4, 5, 1, 3)$. This data can be derived by an if–then rule: if $CP = (C(1) = 1) \wedge (C(2) = 2) \wedge (C(3) = 3)$ (hereafter denoted with $CP = (123000)$) then $D = 3$. This rule is called the rule with rule length 3 $(RL = 3)$ as it involves three conditions. Assuming $RL = 3$, $CP = (023400)$ can be considered another RC. Thus, all the RCs with $RL = 3$ in this example can be $_6C_r|_{r=3} = 20$ different ways. Accordingly, all the possible RCs with $RL = 3$ is $_6C_r(5)^r|_{r=3} = 2,500$. The R_L-dataset was arranged in the RT with $RL = 3$ $(RT(r = 3))$ as shown in Table 1. The first row of the table $(1,2,3) = (1,1,1)$ represents the CP: $(C(1) = 1) \wedge (C(2) = 1) \wedge (C(3) = 1)$. Meanwhile, $(79,0,0,0,0)$ is the frequency distribution of D satisfying the condition in the R_L-dataset. In other words, $D = 1$ represent the maximum frequency (if there are the same frequencies, D is randomly selected between them). Most of the distributions of D in Table 1 widely fluctuate corresponding to the same CP, which was caused by different raters with varying standards. If an RC has a frequency of $(0,0,0,0,0)$, it is called the empty RC. In Table 1, each RC is called a sub-rule of the RT. In addition, by using the $RT(r)$, the rule set $\cup_{r=1}^{|C|=6} RT(r)$ includes all the rules behind the R_L-dataset. This RT is the newly expanded Rule Box in Fig. 1.

With respect to the classification process of transforming an input $u^C(i)$ into the output $u^D(i)$, sub-rules of $_6C_r|_{r=3} = 20$ that match the input pattern should be considered to adapt to different rating standards. Therefore, this paper adopted the vote of each sub-rule's output. Figure 2 provides a specific example where the input $u^C(i) = (4, 5, 1, 4, 4, 3)$ $(u^D(i) = 4)$ is classified by 20 sub-rules with $RL = 3$ using the RT in Table 1. In the first row, the input values $(C(1) = 4, C(2) = 5, C(3) = 1)$ correspond to the CP of the sub-rule: $(1,2,3) = (4,5,1)$ in the RT. It is classified as $D = 1$ by selecting the maximum frequency. Similarly, in the 19th row, $D = 2$ or 3 is randomly selected. By arranging non-empty cases (i.e., deleting empty sub-rules) and by counting the votes, the maximum frequency is 9 at $\hat{D} = 4$, which is the final result that happens to coincide with $u^D(i)$. In the case of same values, one of them is randomly selected. These processes are the new Hypotheses in Fig. 1.

Table 1. Example of RT with $RL = 3$ of the R_L-dataset.

Condition Part $(C(j1), C(j2), C(j3))(j1 < j2 < j3)$ $(j1, j2, j3) = (k1, k2.k3)$	Frequency of $D(n_1, n_2, ..., n_5)$	Decision Part D
(1,2,3) = (1,1,1)	(79, 0, 0, 0, 0)	1
(1,2,3) = (1,1,2)	(9, 2, 0, 0, 0)	1
(1,2,3) = (1,1,3)	(9, 0, 0, 0, 0)	1
...	...	...
(1,2,3) = (5,5,5)	(8, 6, 4, 28, 445)	5
(1,2,4) = (1,1,1)	(82, 1, 0, 0, 0)	1
...	...	...
(1,2,4) = (5,5,5)	(10, 9, 5, 19, 393)	5
...	...	...
(4,5,6) = (1,1,1)	(188, 17, 0, 0, 0)	1
...	...	...
(4,5,6) = (5,5,5)	(4, 5, 5, 23, 409)	5

input: $(C(1), ..., C(6), D)$	(4, 5, 1, 4, 4, 3, $D = 4$)

$\downarrow$

$(j1, j2, j3)$	$(k1, k2, k3)$	Distribution of D	D
1: (1,2,3)	(4, 5, 1)	(6, 0, 1, 2, 0)	1
2: (1,2,4)	(4, 5, 4)	(6, 1, 2, 43, 48)	5
3: (1,2,5)	(4, 5, 4)	(1, 0, 1, 37, 27)	4
4: (1,2,6)	(4, 5, 3)	(4, 1, 3, 14, 10)	4
5: (1,3,4)	(4, 1, 4)	(19, 14, 3, 4, 1)	1
...	...	...	...
19: (3,5,6)	(1, 4, 3)	(4, 5, 5, 2, 0)	2 or 3
20: (4,5,6)	(4, 4, 3)	(3, 4, 28, 54, 9)	4

$\downarrow$

arrange 20 cases output of $\hat{D}$	(1, 5, 4, 4, 1, 2, 2, 4, 4, 4, 1, 1, 2, 4, 4, 4, 1, 1, 2, 4) $\rightarrow$ (6,4,0,9,1) $\rightarrow$ 4

Fig. 2. Example of classification process by RT with $RL = 3$

5   Classification Experiments on R-Dataset Using RT and Comparison With RF

The RT accompanied with the classification method proposed in Sect. 4 was applied for the R-dataset to investigate its ability and confirm the usefulness. The experiment was executed for every $RT(r)$ $(r = 1, \ldots, 6)$ as follows: 1) composing $RT(r)$ using the R_L-dataset, 2) forming the classification dataset by randomly sampling $N_b = 500$ from the R_C-dataset, 3) classifying the N_b dataset according to the classification process (see Fig. 2), and 4) repeating the previous three procedures by $N_r = 50$ times. Table 2 summarizes one of the results classified by $RT(r)$ $(r = 1, \ldots, 6)$ with $(m_{RT(r)}, SD_{RT(r)})$ [%]. Here $m_{RT(r)}$ and $SD_{RT(r)}$ denote the N_r times mean of the classification rate and its standard deviation, respectively. The following represents the comparisons between the results by $RT(RL = r)$ $(r = 1, \ldots, 6)$:

(1) When RL is small, sub-rule accuracy tend to be low, while their coverage tends to be high. Consequently, there are rarely any empty sub-rules, and the frequency distribution bias of D is low, leading to a lower classification ability. Here, $accuracy = |U(d) \cap U(CP)|/|U(CP)| = P(D = d|CP)$, $coverage = |U(d) \cap U(CP)|/|U(d)| = P(CP|D = d)$, $U(d) = \{u(i)|u^{D=d}(i)\}$, $U(CP) = \{u(i)|u^C(i)\ satisfies\ CP\}$.

(2) When RL becomes too high, the sub-rule accuracy tend to increase, while the coverage decreases. Consequently, there are many empty sub-rules, leading to a decrease in the classification ability.

(3) Suppose that $X_{r1} \sim N(\mu_{r1}, \sigma_{r1}^2)$ and $X_{r2} \sim N(\mu_{r2}, \sigma_{r2}^2)$. Here, X_{r1} and X_{r2} denote random variables of the classification rate by $RT(r1)$ and $RT(r2)$, respectively. The N_r times mean $\overline{X_{r1}}$ and $\overline{X_{r2}}$, and their normalized difference $\overline{X_{r1}} - \overline{X_{r2}}$ is given as follows

$$Z = \frac{\overline{X_{r1}} - \overline{X_{r2}} - (\mu_{r1} - \mu_{r2})}{\left(\frac{\sigma_{r1}^2}{N_r} + \frac{\sigma_{r2}^2}{N_r}\right)^{0.5}}. \tag{2}$$

Under null hypothesis $H0$: $\mu_{r1} = \mu_{r2}$, $Z = \frac{\overline{X_{r1}} - \overline{X_{r2}}}{\left(\frac{\sigma_{r1}^2}{N_r} + \frac{\sigma_{r2}^2}{N_r}\right)^{0.5}} \sim N(0, 1)$.

For example, placing $\overline{X_{r1}} = m_{RT(4)}$, $\sigma_{r1} = SD_{RT(4)}$, $\overline{X_{r2}} = m_{RT(3)}$, $\sigma_{RT(3)} = SD_{RT(3)}$, $z = 2.83$ with p-value $= 2.31E\text{-}3$, resulting in the rejection of $H0$, i.e., statistically $\mu_{RT(4)} > \mu_{RT(3)}$.

Accordingly, $RT(r = 4)$ was used for the classification experiment in the R_C-dataset due to the highest classification rate. Table 3 presents one of the results classified by $RT(r = 4)$, arranged as the confusion matrix of all the datasets (500×50). For example, the first row shows the total data number of $D = 1$ is $(3,827 + 1,032 + 95 + 66 + 52) = 5,072$, the rate classified $D = 1$ is $3,827/5,072 = 0.755$, $D = 2$ is $1032/5,072 = 0.203$, and the class error is 0.245.

The same experiment was conducted for the same R_L-dataset and R_C-dataset by RF for comparison with the classification results by the RT method. Specifically, the same R_L-dataset was first used for the learning RF model as shown in Fig. 3: classmodel = randomForest(x, y, $mtry = 2$). Here, the function: randomForest is in the R-language library [24], $\{u^C(i) \in R_L\text{-datase}\}$ was set to x, and $\{u^D(i) \in R_L\text{-datase}\}$ was set to y after changing their data classes appropriately. The parameter $mtry = 2$ was found to be the least error rate of out of bag (OOB) in the preliminary experiment. Figure 3 shows one of the outputs of

Table 2. Summary of classification experiment for R-dataset by $RT(r)$, RF and tr-STRIM.

r	$RT(1)$	$RT(2)$	$RT(3)$	$RT(4)$	$RT(5)$	$RT(6)$	RF	$tr - STRIM$
$(m_r,$	(56.0,	(63.0,	(67.4,	(68.6,	(65.2,	(56.1,	(68.6	(60.5,
$SD_r)$	2.08)	2.07)	2.09)	2.25)	1.95)	2.16)	1.70)	2.12)

Table 3. Results of R-dataset classification experiment by $RT(r = 4)$.

D	1	2	3	4	5	class error
1	3827(0.755)	1032(0.203)	95(0.019)	66(0.013)	52(0.010)	0.245
2	1336(0.266)	2718(0.542)	770(0.153)	155(0.031)	38(0.008)	0.458
3	169(0.034)	994(0.203)	3099(0.632)	575(0.117)	67(0.014)	0.368
4	30(0.006)	86(0.017)	750(0.147)	3412(0.670)	813(0.160)	0.330
5	8(0.002)	6(0.001)	61(0.012)	739(0.150)	4102(0.834)	0.166

"classmodel" is $(729 + 201 + 32 + 15 + 11) = 988$ for the dataset of $D = 1$ and class error $= (201+32+15+11)/988 = 0.262$. Table 4 corresponding to Table 3 shows one of the results of the classification experiment by the RF implemented in Fig. 3 and each class error is similar as that in Fig. 3. The N_r times mean rate of the classification rate and its standard deviation was $(m_{RF}, SD_{RF}) = (68.6, 1.70)$ [%], as shown in Table 2.

The comparison of the classification results of the R-dataset between $RT(r = 4)$ and RF revealed no significant difference between $\mu_{RT(4)} > \mu_{RF}$ and $\mu_{RT(4)} < \mu_{RF}$ using equation (2) ($z = 0.075$), indicating that $\mu_{RT(4)} = \mu_{RF}$ for time being. Figure 4 also shows the comparison of class errors with D between Table 3 (by $RT(r = 4)$) and Table 4 (by RF) which shows the same tendency and both appears to execute the classification based on the close rules each other. RF is also one of the classification methods which uses the voting result through a large number of decision trees with randomly selected variables to decrease the correlation among those decision trees, improving the CART method by constructing a tree structure [3,22]. However, the difference between them is that the RT explicitly induces rules whereas RF cannot.

6   Proposal of Trunk Rules and Its Consideration

The classification experiments on the R-dataset by the RT method accompanied with classification procedures showed the equivalent ability to that by RF. However, the RT method arranged the R-dataset into numerous sub-rules and used the RT for voting to obtain the classified result, although the method was adaptive to the ill-behaved RWD, resulting in RT losing comprehensibility for human beings. Meanwhile, the previous STRIM [5–19] used the following principle for exploring the CP of if–then rules: $P(D = d|CP) \neq P(D = d)$, setting the null hypothesis $H0$: the CP is not a rule candidate $(P(D = d|CP) = P(D = d))$

```
Call:
randomForest(x = x, y = y, mtry = 2)
Type of random forest: classification
Number of trees: 500
No. of variables tried at each split: 2
OOB estimate of error rate: 30.9%
Confusion matrix:
          1      2      3      4      5 class.error
1       729    201     32     15     11       0.262
2       239    543    165     22      4       0.442
3        28    192    667    132      6       0.349
4         6     12    169    652    148       0.339
5         1      1      8    152    865       0.158
```

Fig. 3. Results of error rate of OOB and confusion matrix in R_L-dataset.

Table 4. Results of classification experiment for R_C-dataset by RF

D	1	2	3	4	5	class error
1	3785(0.744)	1066(0.209)	149(0.029)	56(0.011)	33(0.006)	0.256
2	1316(0.261)	2821(0.559)	752(0.149)	135(0.027)	22(0.004)	0.441
3	156(0.032)	936(0.190)	3200(0.650)	555(0.113)	77(0.016)	0.350
4	7(0.001)	98(0.019)	821(0.162)	3283(0.649)	848(0.168)	0.351
5	5(0.001)	8(0.002)	51(0.010)	760(0.156)	4060(0.831)	0.169

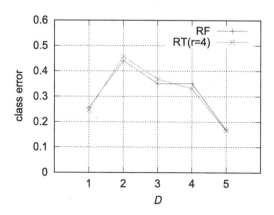

Fig. 4. Class error tendency corresponding to Tables 3 and 4

and executing the statistical test using the R_L-dataset. The frequency distribution of D, e.g., (79,0,0,0,0) at the first row in Table 1, rejects the null hypothesis, i.e., it is a rule candidate. However, $H0$ cannot be rejected by the second (9,2,0,0,0) and the third (9,0,0,0,0) as they do not satisfy the necessary test sample size n (in this specification, approximately $n \geq 25$ and the sample size of the second is $9 + 2 = 11$ (see [13])). Thus, the RT can be divided into two types of sub-rules: rejecting $H0$ and not rejecting. Table 5 shows the number of induced RCs from the whole of RT: $\cup_{r=1}^{|C|=6} RT(r)$, which satisfies the principle with $p - value < 1.0E - 5$. It arranges the induced RCs by every $RL = r$, which coincides with the result by the previous STRIM. That is, the RT expands the range of RCs by the previous STRIM, and the RT with $RL = r$ method was labeled as expanded STRIM ($ex - STRIM|RL = r$).

Although the details were omitted due to space limitations, all the induced RCs with $RL = 1$ in Table 5 are of the following form:

$$if\ C(j) = d\ then\ D = d\ (j = 1, \ldots, 6,\ d = 1, \ldots, 5). \tag{3}$$

Table 6 shows the RCs with $RL \geq 2$, having only $C(j) = d$, extracted from Table 5 and arranged in the same manner as Table 5. The number of all RCs with $RL = r$ constructed by only $C(j) = d$ is given by $_6C_r$ and presented in Table 6, except for the cases of $D = 2$ with $RL = 4$, 5, and 6. In addition, eight RCs of $D = 2$ with $RL = 4$ were discovered: $CP=$ (222002), (220202), (220022), (022220),(022202), (022022), (020222), and (002222). The values in both the condition and decision parts of the if–then rule coincide with each other, considering that the rating scale of $C(j)$ ($j = 1, \ldots, 6$) is the same ordinal scale including D. Consequently, the RT or the previous STRIM induced such understandable RCs, which were labeled trunk rules, and an inducing method trunk STRIM (tr-STRIM). As shown in Table 2, a classification result by the tr-STRIM was $(m_{tr-ST}, SD_{tr-ST}) = (60.5, 2.12)$ [%], which is positioned in the middle of $RT(r)$ with $r= 1$ and $r= 2$.

The inclusion relationship of RCs induced by the three types of STRIM is arranged as follows:

(1) $Rset(ex - STRIM|RL = r) \supset Rset(STRIM|RL = r) \supset Rset(tr - STRIM|RL = r)$,
(2) $\cup_{r=1}^{|C|=6} Rset(ex-STRIM|RL =r) \supset Rset(STRIM) \supset Rset(tr-STRIM)$,
(3) $Rset(ex - STRIM|r = 1) \supset Rset(ex - STRIM|r = 2) \supset, \ldots, Rset(ex - STRIM|r = 6)$.

In this context, $Rset(method)$ refers to the rule set induced by a particular method. The average classified results can be summarized as follows, including no-show relationships due to space limitations: $C(Rset(ex - STRIM|RL = r), data) > C(Rset(STRIM|RL = r), data) > C(Rset(tr - STRIM|RL = r), data)$, where $C(Rset(method), data)$ represents the average classification result against a dataset by $Rset(method)$. To express the classification results qualitatively, the trunk rules were improved by adding the branch rules with statistical significance level and further by the leaf rules without it. These studies can be beneficial in considering a level of "the principle of incompatibility" depending on the targeted matter. For instance, these studies can be useful in various policies or business strategies where RF cannot explicitly induce rules, indicating that the contents of the dataset and the level need to be considered.

Table 5. Number of induced rules with statistical significance level from whole of RT.

	$RL = 1$	2	3	4	5	6	Total
$D = 1$	6	81	226	78	13	2	406
2	6	61	158	49	1	0	275
3	6	43	124	134	33	3	343
4	6	45	149	141	22	2	365
5	6	50	142	151	53	5	407
Total	30	280	799	553	122	12	1796

Table 6. Number of extracted trunk rules for each RL from Table 5

	$RL = 1$	2	3	4	5	6	Total
$D = 1$	6	15	20	15	6	1	63
2	6	15	20	8	0	0	49
3	6	15	20	15	6	1	63
4	6	15	20	15	6	1	63
5	6	15	20	15	6	1	63
Total	30	75	100	68	24	4	301

7   Conclusion

The validity and usefulness of the previous rule induction method, STRIM have been confirmed using a simulation experiment. However, the examination of its usefulness in an RWD was left for future study [8,13,20]. This study specifically used the R-dataset with DT in the field of RS for experimentally addressing the issue by newly proposing the RT method for adaption to the RWD. The validity or usefulness of the method was confirmed by applying it to the classification problem, as their confirmation of the rule induction from the RWD cannot be generally ascertained, even though the result directly reflects the classification result. The usefulness of the method was confirmed using the classification result by RF. The following aspects were specifically considered:

(1) The newly proposed RT method demonstrated how to arrange the dataset into $RT(r)$ which is a set of sub-rules and all the possible RCs with $RL = r$, and how to use $RT(r)$ to classify a new object.
(2) The validity and usefulness of the RT method were experimentally examined by applying it to the classification problem after selecting a proper $RT(r)$. The result of the classification rate showed equivalence to that by RF, i.e., $\mu_{RT(r=4)} = \mu_{RF}$.
(3) However, the induced $RT(r)$ and the decision trees generated by RF are increasingly difficult to understand. Both methods offer limited insights and information regarding the dataset in the form of if–then rules. Thus,

the whole of RT: $\cup_{r=1}^{|C|=6} RT(r)$ was subsequently reviewed and then organized into trunk rules using the trunk-STRIM method. The relationships between these rules were shown as $Rset(ex - STRIM|RL = r) \supset Rset(STRIM|RL = r) \supset Rset(tr - STRIM|RL = r)$ and were useful in providing "the principle of incompatibility" depending on the targeted matter. On the other hand, RF does not provide such knowledge, although it remains a useful method for classification problems.

The investigation of the following points is recommended for future studies:

(1) To validate the findings of this study, future research should focus on applying the three types of STRIM, namely the previous, ex-STRIM, and tr-STRIM, to various other RWDs and replicate the findings to validate the findings of this study.

(2) When using ex-STRIM with $RL = r$ as a classification method, whether $\mu_{(ex-ST|r)} = \mu_{RF}$ is always equivalent or not should be studied, considering factors, such as the size of the learning dataset and changes in RWD.

Acknowledgments. This work was supported by JSPS KAKENHI Grant Number JP20K11939. We truly thank Rakuten Inc. for presenting Rakuten Travel dataset [1].

References

1. https://www.nii.ac.jp/dsc/idr/rakuten/
2. Pawlak, Z.: Rough sets. Int. J. Comput. Inf. Sci. **11**(5), 341–356 (1982)
3. Breiman, L.: Random forests. Mach. Learn. **45**(1), 5–32 (2001)
4. Zadeh, L.A.: Outline of a new approach to the analysis of complex systems and decision processes. IEEE Trans. Syst. Man Cybern. **3**, 28–44 (1973)
5. Matsubayashi, T., Kato, Y., Saeki, T.: A new rule induction method from a decision table using a statistical test. In: Li, T., Nguyen, H.S., Wang, G., Grzymala-Busse, J., Janicki, R., Hassanien, A.E., Yu, H. (eds.) RSKT 2012. LNCS (LNAI), vol. 7414, pp. 81–90. Springer, Heidelberg (2012). https://doi.org/10.1007/978-3-642-31900-6_11
6. Kato, Y., Saeki, T., Mizuno, S.: Studies on the necessary data size for rule induction by STRIM. In: Lingras, P., Wolski, M., Cornelis, C., Mitra, S., Wasilewski, P. (eds.) RSKT 2013. LNCS (LNAI), vol. 8171, pp. 213–220. Springer, Heidelberg (2013). https://doi.org/10.1007/978-3-642-41299-8_20
7. Kato, Y., Saeki, T., Mizuno, S.: Considerations on rule induction procedures by STRIM and their relationship to VPRS. In: Kryszkiewicz, M., Cornelis, C., Ciucci, D., Medina-Moreno, J., Motoda, H., Raś, Z.W. (eds.) RSEISP 2014. LNCS (LNAI), vol. 8537, pp. 198–208. Springer, Cham (2014). https://doi.org/10.1007/978-3-319-08729-0_19
8. Kato, Y., Saeki, T., Mizuno, S.: Proposal of a Statistical Test Rule Induction Method by Use of the decision Table. Appl. Soft Comput. **28**, 160–166 (2015)
9. Kato, Y., Saeki, T., Mizuno, S.: Proposal for a statistical reduct method for decision tables. In: Ciucci, D., Wang, G., Mitra, S., Wu, W.-Z. (eds.) RSKT 2015. LNCS (LNAI), vol. 9436, pp. 140–152. Springer, Cham (2015). https://doi.org/10.1007/978-3-319-25754-9_13

10. Kitazaki, Y., Saeki, T., Kato, Y.: Performance comparison to a classification problem by the second method of quantification and STRIM. In: Flores, V., et al. (eds.) IJCRS 2016. LNCS (LNAI), vol. 9920, pp. 406–415. Springer, Cham (2016). https://doi.org/10.1007/978-3-319-47160-0_37

11. Fei, J., Saeki, T., Kato, Y.: Proposal for a new reduct method for decision tables and an improved STRIM. In: Tan, Y., Takagi, H., Shi, Y. (eds.) DMBD 2017. LNCS, vol. 10387, pp. 366–378. Springer, Cham (2017). https://doi.org/10.1007/978-3-319-61845-6_37

12. Kato, Y., Itsuno, T., Saeki, T.: Proposal of dominance-based rough set approach by STRIM and its applied example. In: Polkowski, L., et al. (eds.) IJCRS 2017. LNCS (LNAI), vol. 10313, pp. 418–431. Springer, Cham (2017). https://doi.org/10.1007/978-3-319-60837-2_35

13. Kato, Y., Saeki, T., Mizuno, S.: Considerations on the principle of rule induction by STRIM and its relationship to the conventional rough sets methods. Appl. Soft Comput. **73**, 933–942 (2018)

14. Kato, Y., Kawaguchi, S., Saeki, T.: Studies on CART's performance in rule induction and comparisons by STRIM. In: Nguyen, H.S., Ha, Q.-T., Li, T., Przybyła-Kasperek, M. (eds.) IJCRS 2018. LNCS (LNAI), vol. 11103, pp. 148–161. Springer, Cham (2018). https://doi.org/10.1007/978-3-319-99368-3_12

15. Saeki, T., Fei, J., Kato, Y.: Considerations on rule induction methods by the conventional rough set theory from a view of STRIM. In: Nguyen, H.S., Ha, Q.-T., Li, T., Przybyła-Kasperek, M. (eds.) IJCRS 2018. LNCS (LNAI), vol. 11103, pp. 202–214. Springer, Cham (2018). https://doi.org/10.1007/978-3-319-99368-3_16

16. Kato, Y., Saeki, T., Mizuno, S.: Considerations on the Principle of Rule Induction by STRIM and Its Relationship to the Conventional Rough Sets Methods; Applied Soft Computing, 73 pp, pp. 933–942, Elsevier (2018)

17. Kato, Y., Saeki, T., Fei, J.: Application of STRIM to datasets generated by partial correspondence hypothesis. In: Fagan, D., Martín-Vide, C., O'Neill, M., Vega-Rodríguez, M.A. (eds.) TPNC 2018. LNCS, vol. 11324, pp. 74–86. Springer, Cham (2018). https://doi.org/10.1007/978-3-030-04070-3_6

18. Kato, Y., Saeki, T.: Studies on reducing the necessary data size for rule induction from the decision table by STRIM. In: Mihálydeák, T., Min, F., Wang, G., Banerjee, M., Düntsch, I., Suraj, Z., Ciucci, D. (eds.) IJCRS 2019. LNCS (LNAI), vol. 11499, pp. 130–143. Springer, Cham (2019). https://doi.org/10.1007/978-3-030-22815-6_11

19. Kato, Y., Saeki, T.: New rule induction method by use of a co-occurrence set from the decision table. In: Gutiérrez-Basulto, V., Kliegr, T., Soylu, A., Giese, M., Roman, D. (eds.) RuleML+RR 2020. LNCS, vol. 12173, pp. 54–69. Springer, Cham (2020). https://doi.org/10.1007/978-3-030-57977-7_4

20. Kato, Y., Saeki, T.: Application of Bayesian STRIM to datasets generated via partial correspondence hypothesis. In: 18th International Conference on Machine Learning and Data Mining, MLDM 2022, pp. 1–15 (2022)

21. Grzymala-Busse, J.W. : LERS- A System for Learning from Examples Based on Rough Sets In: Słowiński, R. (ed.) Kluwer Academic Publishers: Intelligent Decision Support; Handbook of Applications and Advances of the Rough Sets Theory, pp. 3–18 (1992)

22. Decision tree learning - Wikipedia

23. Association rule learning - Wikipedia

24. randomForest: Breiman and Cutler's Random Forests for Classification and Regression r-project.org

TouriER: Temporal Knowledge Graph Completion by Leveraging Fourier Transforms

Thanh Vu[1,2], Huy Ngo[1,2], Ngoc-Trung Nguyen[3], and Thanh Le[1,2(✉)]

[1] Faculty of Information Technology, University of Science,
Ho Chi Minh City, Vietnam
{19120374,19120242}@student.hcmus.edu.vn, lnthanh@fit.hcmus.edu.vn
[2] Vietnam National University Ho Chi Minh City, Ho Chi Minh City, Vietnam
[3] Faculty of Information Technology, University of Education,
Ho Chi Minh City, Vietnam
trungnn@hcmue.edu.vn

Abstract. In recent times, numerous studies on static knowledge graphs have achieved significant advancements. However, when extending knowledge graphs with temporal information, it poses a complex problem with larger data size, increased complexity in interactions between objects, and a potential for information overlap across time intervals. In this research, we introduce a novel model called TouriER, based on the MetaFormer architecture, to learn temporal features. We also apply a data preprocessing method to integrate temporal information in a reasonable manner. Additionally, the utilization of Fourier Transforms has proven effective in feature extraction. Through experiments on benchmark datasets, the TouriER model has demonstrated better performance compared to well-known models based on standard metrics.

Keywords: Temporal Knowledge Graph · Link Prediction · MetaFormer · Fourier Transforms

1 Introduction

A knowledge graph (KG) is a type of knowledge base that can be graphically represented that reflects the real world. Knowledge graphs can be used to answer questions, query information, and solve other problems in the high-tech industry. Because of the knowledge graph's usefulness, it is increasingly being used in applications directly related to human life. As a result, the need to improve knowledge graphs is pressing in the context of ever-growing data, but no knowledge graph is truly full; there are always missing links. Link prediction is a task that makes predictions and adds missing links to the knowledge graph, assisting in its refinement so that it can better serve applications.

This research is supported by the research funding from the Faculty of Information Technology, University of Science, Ho Chi Minh city, Vietnam.

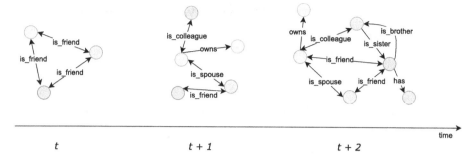

Fig. 1. A temporal knowledge graph

Temporal knowledge graphs (TKGs) have the same characteristics and applications as static knowledge graphs, but they include a new information field, time. A temporal knowledge graph can be thought of as a collection of knowledge graph *snapshots* taken at various periods in time. When doing link prediction research on temporal knowledge graphs, which is both a strength and a drawback. The advantage is that we now have a new field of knowledge to use and create more accurate forecasts. In comparison to static knowledge graphs, relationships in temporal knowledge graphs can become exceedingly complicated and unpredictable. Figure 1 shows the change of a temporal knowledge graph over time, where the knowledge graph at each point in time is called a *snapshot*.

The goal of the temporal link prediction problem is to complete the temporal knowledge graph by adding the missing relationships at each snapshot. Although other studies on temporal knowledge graph completion have made remarkable achievements, their ability to extract features is still limited, leading to poor accuracy. Therefore, we do this research to find a model that can extract features better. In this paper, we introduce TouriER, a MetaFormer-based [18] model that integrates the data preprocessing method from SpliMe [11] with token mixer utilizing Fourier Transform inspired by FNet [9] to obtain more information about knowledge graph embeddings. Our research focuses on building a model that can perform the problem on static knowledge graph to take advantage of the existing strengths from static knowledge graph completion. Along with that, thanks to the integration of data processing methods from SpliMe, we can apply the proposed model on temporal knowledge graphs. Our main contributions are as follows:

- Our model takes advantage of both of mathematics and MetaFormer architecture that is successful in the field of computer vision, promoting the development of other models in the field of temporal knowledge graph completion.
- Our model is the first to apply Fourier Transforms for efficient data extraction on temporal link prediction.
- We perform our model evaluation and analysis with various datasets, showing that our model performs well with temporal knowledge graph data.

The structure of this paper is as follows: In Sect. 2, we discuss ways to solve the link prediction problem in both static knowledge graph and temporal knowledge graph. Next, in Sect. 3, we will detail our proposed solution. In addition, our experiments and evaluations are presented in Sect. 4. Finally, in Sect. 5 we summarize the research and offer potential future directions.

2 Related Work

Although there is a lot of research done directly on the temporal knowledge graph completion problem recently, we found that the static knowledge graph completion problem has a lot more research. Referring to other studies performed on the static knowledge graph can be helpful in taking the strengths from that to solve the problem of link prediction on the temporal knowledge graph. Therefore, in this section, besides reviewing researches on link prediction in temporal knowledge graphs, we also provide a brief overview of static knowledge graph embedding models.

2.1 Static Knowledge Graphs

Many knowledge graph embedding techniques are proposed, which can be categorized as translational distance-based models, semantic matching-based models, and neural network-based models.

TransE [1], the earliest translational distance-based model, models triple as a translation operation from a head entity to a tail entity. But it does not do well in dealing with these properties: reflexive, one-to-many, many-to-one, and many-to-many relations. TransH [14], which models a relationship as a hyperplane with a translation operation on it, is proposed to improve that.

RESCAL [10] represents entities as vectors to capture latent semantics and relations as matrices to capture interactions between entities. DistMult [17] represents relations as diagonal matrices to simplify RESCAL.

R-GCN [12] is the first model to apply graph convolution networks to knowledge graphs. It learns entity embeddings by aggregating neighborhood embeddings through a message-passing framework and using a decoder such as DistMult to explore these embeddings. ConvE [3] reshapes and concates entities and relations embeddings and uses convolutional and fully connected layers to extract more feature interactions between these embeddings.

2.2 Temporal Knowledge Graphs

TTransE [8], an extension of TransE for temporal knowledge graphs, uses time information in three ways: encoding each time point with a synthetic relation; representing time as a vector in the same space as entities and relations; and representing time as real values in (0, 1]. HyTe [2], a model that also uses a scoring function based on the translational distance, represents time as a hyperplane and projects the triple into its time hyperplane to incorporate temporal knowledge into the relational and entity embeddings. Leveraging the DistMult scoring function, TA-DistMult [4] learns representations for time-augmented knowledge graph facts with a digit-level long-short-term memory. DE-SimplE [5] proposed a diachronic embedding function that can capture entity features at any given time, and they also use another scoring function from static knowledge graph embedding models. To capture more features from time information, ATiSE [16] fits the evolution process of an entity or relation as a multi-dimensional additive time series composed of a trend component, a seasonal component, and a random component. The authors of ATiSE also consider the transformation result of their model from the subject to the object to be akin to the predicate in a positive fact.

Several temporal knowledge graph embedding models are proposed that embed entities, relations, and time information into complex vector spaces. TNTComplEx [7] and TIMEPLEX [6] are two TKGE models that build upon the ComplEx [13] model in distinct ways. TNTComplEx models subject, relation, object, and time as an order 4 tensor, and afterward applies tensor decomposition for the temporal knowledge graph completion task. TIMEPLEX introduces three weighted time-dependent terms into the ComplEx score function and sums over the trinomial Hermitian products. TeRo [15] defines the functional mapping induced by each time step as an element-wise rotation from the time-independent entity embeddings to the time-specific entity embeddings in complex vector spaces and uses these embeddings in the translational-based scoring function.

3   Proposed Method

3.1   Problem Formulation and Notations

A temporal knowledge graph $\mathcal{G}$ is formalized as $\mathcal{G} = \{(s,r,o,t)|s \in \mathcal{E}, o \in \mathcal{E}, r \in \mathcal{R}, t \in \mathcal{T}\}$, which is a set of quadruples. Where $\mathcal{E}$ represents the set of entities, $\mathcal{R}$ represents the set of relations, $\mathcal{T}$ denotes the set of time steps, and s, r, o, t represent the subject, relation, object, and the corresponding time step, respectively. Entities and relations are represented by embedding, they are denoted as $\mathbf{e}_s$, $\mathbf{e}_r$, $\mathbf{e}_o$ for the embedding of subject, relation and object respectively. In the next subsection we take the approach of using synthetic temporal relation, so $\mathbf{e}_r$ is the embedding of the synthetic temporal relation corresponding to the quadruple, that's also why we don't use a separate embedding for time $\mathbf{e}_t$. A synthetic temporal relation is denoted as r_a. The number of entities, relations and timestamps is represented by n_e, n_r and n_t, while the dimension of the embedding vector representing the entity and relationship is symbolized by d_e and d_r.

In this paper, we apply the data preparation method from SpliMe, which takes a quinruple as input. Where time t is decomposed into a time value pair t_s and t_e. Therefore, the t component in the quadruples which is mentioned in this paper is actually talking about the time value pair $t = (t_s, t_e)$.

Table 1. Methods used to generate synthetic temporal relations

Dataset	Method	Parameters
WIKIDATA12K	Merge	$Shrink = 4$
YAGO11K	Split (CPD)	$Pref, \epsilon = 5$
ICEWS14	Timestamp	–

The link prediction task's goal is to predict missing links based on the facts already present in the knowledge graph. Our model takes as input a query q where $q = (s, r_a)$ and makes predictions about the missing object o for each sample (1-N scoring manner, in which the query is scored against all entities).

3.2 Data Augmentations

Synthetic Temporal Relationship. Facts are represented as quadruples in temporal knowledge graph completion, as opposed to triples in static knowledge graph completion. This makes it difficult for models developed for static knowledge graphs to be adapted to temporal knowledge graphs. By concatenating relations with timestamps, synthetic temporal relationships can bring temporal knowledge graphs into the domain space of static knowledge graphs. Time span data may have overlaps (for example, 2000–2005 and 2001–2007) resulting in more synthetic relations if we concatenate relation and time span directly, so that we use an approach by Radstok et al. [11] to find optimal timestamps to concatenate with. We must convert time points in time point format to time span format for datasets containing time in time point format, hence this method defines $t_s = t_e = t$. Table 1 shows the details of the strategy we employ to preprocess each dataset as suggested by the original paper.

Among all of the methods described in the SpliMe framework, *Timestamping* is the simplest method, it turns each temporal fact in the TKG into a set of facts, one for each timestamp when the fact was true. At each iteration, the *Splitting* approach using Change Point Detection (CPD) refines the temporal scope of the synthetic relations in the knowledge graph by making them more specific, CPD is used to determine where to place split points most efficiently. In contrast to the splitting approach, the *Merging* method begins by constructing a large number of combinations (r, t) and then combines pairs of these combinations to reduce the number of synthetic relations.

Query Embeddings. We linearly transform the embedding vectors from 1D to 2D matrices in order to help our model learn better entity-to-relation links; both should be the same size after transformation. The two embeddings from the entity and the relation are then stacked (which can be considered as transforming into two channels in the image). By transforming embeddings to an image-like form, we can adapt to using computer vision models in link prediction task and take advantage of models that perform well in computer vision problems.

Fig. 2. Visualization of transforming 1D embeddings into image-like 2D embeddings

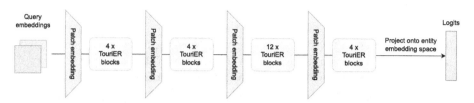

Fig. 3. Overall architecture of TouriER

Figure 2 depicts how embeddings are transformed and layered to generate a relational query. The transformation and stacking embedding procedure can be formalized as follows:

$$q = \Phi_s[\phi(\mathbf{e}_s), \phi(\mathbf{e}_r)] \tag{1}$$

where $\phi(\mathbf{e}_s) = \mathbf{e}_s \mathbf{W}_s + b_s$ and $\phi(\mathbf{e}_r) = \mathbf{e}_r \mathbf{W}_r + b_r$, and Φ_s is the stacking operator, notice that the r symbols in the formula (1) are referring to the synthetic temporal relation r_a. The parameters of the weight matrices and bias vectors are learned during the training phase to improve the fit of embeddings into the model.

3.3 Model Architecture

In this section, we delve into the architecture of the proposed model. Our approach capitalizes on the advantages of PoolFormer [18], a MetaFormer-based architecture that incorporates a straightforward Pooling layer as a token mixer layer, which has proven effective in addressing computer vision challenges. Besides, we integrated a Fast Fourier Transform (FFT) layer, drawing inspiration from FNet [9], to serve as a replacement for pooling layer. The use of FFT allows for the extraction of crucial features from knowledge graph embeddings, thereby improving the overall performance of the model. The general architecture of our proposed model is depicted in Fig. 3, while the specific components contained within the TouriER block is depicted in Fig. 4.

Our model is structured as four primary stages, with each stage's output being downsampled to retain the most significant features for missing entity prediction. Each stage have a different number of TouriER blocks, with stages 1–4 having block numbers of 4, 4, 12, and 4 respectively.

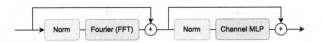

Fig. 4. A TouriER block

Discrete Fourier Transform. The Discrete Fourier Transform (DFT) is a mathematical transformation that converts a sequence of discrete data points into a frequency domain sequence of complex numbers. When DFT is applied to tokens, they are transformed into the frequency domain, where they become complex numbers representing various frequency components that can be retrieved as features. In our implementation, we employ FFT to reduce the computation time of calculating DFT. Given a sequence $\{x_n | n \in [0, N-1]\}$, the DFT is defined by:

$$X_k = \sum_{n=0}^{N-1} x_n e^{-\frac{2\pi i}{N} nk}, \ 0 \le k \le N-1 \tag{2}$$

where x_n is a 1-dimensional array input, N is the size of the input array, e is the base of the natural logarithm. We apply the 1-dimensional DFT twice to our image-like input, once to each row and once to each column, then we take the result as 2-dimensional DFT.

The FFT layer used in the proposed model serves to bring the embedding to the frequency domain, allowing us to distinguish between high and low frequencies. In this case, high frequency represents a huge intensity difference in input embedding, while low frequency represents a little intensity difference. The separation of high and low frequencies will help the model learn the pattern of embedding better. Instead of calculating the magnitude spectrum by taking the absolute of the complex values as the output of the Fourier transform, we take the real part to create a matrix of values. A FFT block is implemented as in Snippet 1.1.

Snippet 1.1. Implementation of a FFT block

```
class FFTBlock(nn.Module):
    def __init__(self):
        super().__init__()

    def forward(self, x):
        x = torch.fft.fft(torch.fft.fft(x, dim=-1), dim=-2).real
        return x
```

TouriER Block. Each of our TouriER blocks is a MetaFormer-based block, where the main components consist of an FFT layer as the token mixer to extract features from embeddings and an MLP layer which is used to capture relationships across channels (entity-relation interactions). There are two skip

Table 2. Statistics of benchmark datasets

Dataset	n_e	n_r	n_t	Train	Val	Test	Timestamp
WIKIDATA12K	12,554	24	70	32,497	4,062	4,062	Time span
YAGO11K	10,623	10	59	16,406	2,050	2,051	Time span
ICEWS14	6,869	230	365	72,826	8,941	8,963	Point

connections to preserve information that can be lost after passing through too many stages due to the proposed model's multi-stage architecture.

Scoring Function. A linear transformation is used to map the output of a forward pass through all of the TouriER blocks into the embedding space $\mathbb{R}^{d_e}$. The scoring function is as follows:

$$\psi(s, r_a, o) = f(TouriER(q))\mathbf{W} + b \tag{3}$$

where $TouriER(.)$ denotes our proposed architecture, q represents the query embeddings, f is a nonlinear activation function, for which we are currently utilizing ReLU, $\mathbf{W}$ and b are the weight matrix and translational vector for linearly projecting results into the output embedding space, respectively.

Loss Function. To train our model, we use the standard Binary Cross Entropy (BCE) loss. The logits from the scoring function are exposed to a sigmoid function before being given into the loss function along with labels. We also use label smoothing to improve the model's robustness and generalization.

4 Experiments

4.1 Datasets

WIKIDATA12K, YAGO11K [2], and ICEWS14 [4] are the three benchmark datasets we used to evaluate our proposed models. Whereas WIKIDATA12K and YAGO11K have a time span time data type, ICEWS14 has a time point time data type. As a result, when tested on many data sets from different types of time data, the model evaluation should be more objective, allowing the results to be evaluated more objectively. Table 2 shows the statistics of benchmark datasets.

4.2 Evaluation Method

We compare the findings achieved on the Mean Reciprocal Rank (MRR), Hit@1, Hits@3, and Hits@10 metrics, with the MRR metric providing the best results. We also employ filtered MRR rather than raw MRR to exclude quadruples that already existed in the dataset but are ranked higher than candidate ones.

Table 3. Results on WIKIDATA12K and YAGO11K.

Model	WIKIDATA12K				YAGO11K			
	MRR	H@1	H@3	H@10	MRR	H@1	H@3	H@10
HyTE	.253	.147	–	.483	.136	.033	–	.298
TTransE	.172	.096	.184	.329	.108	.020	.150	.251
TA-DistMult	.218	.122	.232	.447	.161	.103	.171	.292
TNTComplEx	.301	.197	–	.507	.180	.110	–	.313
DE-SimplE	.253	.147	–	.491	.151	.088	–	.267
TIMEPLEX	.334	.228	–	.532	.236	.169	–	.367
TeRo	.299	.198	.329	.507	.187	.121	.197	.319
ATiSE	.280	.175	.317	.481	.170	.110	.171	.288
SpliMe	.358	.222	.433	**.610**	.214	.065	**.299**	**.458**
TouriER (ours)	**.402**	**.314**	**.447**	.572	**.278**	**.212**	.294	.408

4.3 Results

The results presented for HyTE, DE-SimplE, and TNTComplEx are obtained from the work of Jain et al. [6], results of TTransE and TA-DistMult are adopted from TeRo [15], while the results for other models are sourced from their respective original papers. A thorough analysis of these results reveals that our proposed model consistently outperforms the baseline models when applied to time span datasets. This demonstrates the effectiveness of our strategy in addressing the temporal link prediction problem, particularly in cases involving overlapping time spans. The evaluation results of the proposed model compared with baseline models are summarized in Table 3 and Table 4 where best results are emphasized in bold, second best results are underlined.

Our method of data augmentation plays a crucial role in overcoming this challenge. The development of our model is greatly motivated by the SpliMe, which focuses on transforming temporal link prediction into static link prediction by incorporating synthetic temporal relations. In light of this inspiration, we would like to emphasize the superior performance of our model when compared to the approach employed in the SpliMe. On two data sets, WIKIDATA12K and YAGO11K, our model increases by 12.3%, 29.9% on the MRR metric, and 41.4%, 126.2% on the Hits@1 metric, respectively. In addition, the proposed model outperforms SpliMe by 3.2% on the Hits@3 metric with the WIKIDATA12K dataset. On the ICEWS14 dataset, our model shows complete superiority when compared to SpliMe, outperforming all metrics. Specifically, our model improves 97.2%, 600.0%, 59.2%, 48.0%, respectively, on the MRR, Hits@1, Hits@3 and Hits@10 scales on the ICEWS14 dataset compared to SpliMe. By building upon the foundation laid by SpliMe and incorporating our novel techniques, we successfully enhanced the model's ability to handle temporal link prediction tasks more accurately and efficiently.

Table 4. Results on ICEWS14.

Model	ICEWS14			
	MRR	H@1	H@3	H@10
HyTE	<u>.297</u>	<u>.108</u>	<u>.416</u>	**.601**
TTransE	.255	.074	–	**.601**
SpliMe	.213	.047	.294	.544
TouriER (ours)	**.435**	**.349**	**.480**	<u>.596</u>

5   Conclusions and Future Work

In this research, we proposed a model based on MetaFormer, a familiar architecture in the field of computer vision, and paired it with a Fourier Transform function and a data augmentation method to improve the quality of synthetic relations. As a result, our model beats the baseline models in benchmark datasets using time span data, demonstrating the efficacy of the suggested strategy when applied to the link prediction job.

Some of the other work we are pursuing includes analyzing model performance on different aspects, testing other token mixer layers in place of the FFT layer as well as changing the number of layers of the overall architecture. We are working to enhance the proposed model in time point data in the hopes of enabling this model in producing good results in all sorts of time data.

References

1. Bordes, A., Usunier, N., Garcia-Duran, A., Weston, J., Yakhnenko, O.: Translating embeddings for modeling multi-relational data. In: Burges, C., Bottou, L., Welling, M., Ghahramani, Z., Weinberger, K. (eds.) Advances in Neural Information Processing Systems, vol. 26. Curran Associates, Inc. (2013). https://proceedings.neurips.cc/paper_files/paper/2013/file/1cecc7a77928ca8133fa24680a88d2f9-Paper.pdf
2. Dasgupta, S.S., Ray, S.N., Talukdar, P.: HyTE: hyperplane-based temporally aware knowledge graph embedding. In: Proceedings of the 2018 Conference on Empirical Methods in Natural Language Processing, pp. 2001–2011. Association for Computational Linguistics, Brussels, Belgium, October–November 2018. https://doi.org/10.18653/v1/D18-1225. https://aclanthology.org/D18-1225
3. Dettmers, T., Minervini, P., Stenetorp, P., Riedel, S.: Convolutional 2D knowledge graph embeddings. In: Proceedings of the AAAI Conference on Artificial Intelligence, vol. 32, no. 1 (2018). https://doi.org/10.1609/aaai.v32i1.11573. https://ojs.aaai.org/index.php/AAAI/article/view/11573
4. García-Durán, A., Dumančić, S., Niepert, M.: Learning sequence encoders for temporal knowledge graph completion. In: Proceedings of the 2018 Conference on Empirical Methods in Natural Language Processing, pp. 4816–4821. Association for Computational Linguistics, Brussels, Belgium, October–November 2018. https://doi.org/10.18653/v1/D18-1516. https://aclanthology.org/D18-1516

5. Goel, R., Kazemi, S.M., Brubaker, M., Poupart, P.: Diachronic embedding for temporal knowledge graph completion. In: Proceedings of the AAAI Conference on Artificial Intelligence, vol. 34, no. 04, pp. 3988–3995, April 2020. https://doi.org/10.1609/aaai.v34i04.5815. https://ojs.aaai.org/index.php/AAAI/article/view/5815
6. Jain, P., Rathi, S., Mausam, Chakrabarti, S.: Temporal knowledge base completion: new algorithms and evaluation protocols. In: Proceedings of the 2020 Conference on Empirical Methods in Natural Language Processing (EMNLP), pp. 3733–3747. Association for Computational Linguistics, Online, November 2020. https://doi.org/10.18653/v1/2020.emnlp-main.305. https://aclanthology.org/2020.emnlp-main.305
7. Lacroix, T., Obozinski, G., Usunier, N.: Tensor decompositions for temporal knowledge base completion. In: International Conference on Learning Representations (2020). https://openreview.net/forum?id=rke2P1BFwS
8. Leblay, J., Chekol, M.W.: Deriving validity time in knowledge graph. In: Companion Proceedings of the The Web Conference 2018, WWW 2018, International World Wide Web Conferences Steering Committee, Republic and Canton of Geneva, CHE, pp. 1771–1776 (2018). https://doi.org/10.1145/3184558.3191639
9. Lee-Thorp, J., Ainslie, J., Eckstein, I., Ontanon, S.: FNet: mixing tokens with Fourier transforms. In: Proceedings of the 2022 Conference of the North American Chapter of the Association for Computational Linguistics: Human Language Technologies, pp. 4296–4313. Association for Computational Linguistics, Seattle, United States, July 2022. https://doi.org/10.18653/v1/2022.naacl-main.319. https://aclanthology.org/2022.naacl-main.319
10. Nickel, M., Tresp, V., Kriegel, H.P.: A three-way model for collective learning on multi-relational data. In: Proceedings of the 28th International Conference on International Conference on Machine Learning, ICML 2011, pp. 809–816. Omnipress, Madison, WI, USA (2011)
11. Radstok, W., Chekol, M., Velegrakis, Y.: Leveraging static models for link prediction in temporal knowledge graphs. In: 2021 IEEE 33rd International Conference on Tools with Artificial Intelligence (ICTAI), pp. 1034–1041 (2021). https://doi.org/10.1109/ICTAI52525.2021.00165
12. Schlichtkrull, M., Kipf, T.N., Bloem, P., van den Berg, R., Titov, I., Welling, M.: Modeling relational data with graph convolutional networks. In: Gangemi, A., et al. (eds.) ESWC 2018. LNCS, vol. 10843, pp. 593–607. Springer, Cham (2018). https://doi.org/10.1007/978-3-319-93417-4_38
13. Trouillon, T., Welbl, J., Riedel, S., Gaussier, E., Bouchard, G.: Complex embeddings for simple link prediction. In: Balcan, M.F., Weinberger, K.Q. (eds.) Proceedings of The 33rd International Conference on Machine Learning. Proceedings of Machine Learning Research, vol. 48, pp. 2071–2080. PMLR, New York, New York, USA, 20–22 June 2016. https://proceedings.mlr.press/v48/trouillon16.html
14. Wang, Z., Zhang, J., Feng, J., Chen, Z.: Knowledge graph embedding by translating on hyperplanes. In: Proceedings of the AAAI Conference on Artificial Intelligence, vol. 28, no. 1, June 2014. https://doi.org/10.1609/aaai.v28i1.8870. https://ojs.aaai.org/index.php/AAAI/article/view/8870
15. Xu, C., Nayyeri, M., Alkhoury, F., Shariat Yazdi, H., Lehmann, J.: TeRo: a time-aware knowledge graph embedding via temporal rotation. In: Proceedings of the 28th International Conference on Computational Linguistics, pp. 1583–1593. International Committee on Computational Linguistics, Barcelona,

Spain (Online), December 2020. https://doi.org/10.18653/v1/2020.coling-main. 139. https://aclanthology.org/2020.coling-main.139

16. Xu, C., Nayyeri, M., Alkhoury, F., Yazdi, H., Lehmann, J.: Temporal knowledge graph completion based on time series Gaussian embedding. In: Pan, J.Z., et al. (eds.) ISWC 2020. LNCS, vol. 12506, pp. 654–671. Springer, Cham (2020). https://doi.org/10.1007/978-3-030-62419-4_37

17. Yang, B., Yih, S.W.T., He, X., Gao, J., Deng, L.: Embedding entities and relations for learning and inference in knowledge bases. In: Proceedings of the International Conference on Learning Representations (ICLR) 2015, May 2015. https://www.microsoft.com/en-us/research/publication/embedding-entities-and-relations-for-learning-and-inference-in-knowledge-bases/

18. Yu, W., et al.: MetaFormer is actually what you need for vision. In: Proceedings of the IEEE/CVF Conference on Computer Vision and Pattern Recognition (CVPR), pp. 10819–10829, June 2022

Machine Learned KPI Goal Preferences for Explainable AI based Production Sequencing

Rudolf Felix[(⊠)]

PSI FLS Fuzzy Logik and Neuro Systeme GmbH, Joseph-von-Fraunhofer Straße 2,
44227 Dortmund, Germany
felix@fuzzy.de

Abstract. In this paper we show how machine learned KPI goal preference relations based on interactions between KPI goals are used to explain results of an AI algorithm for optimization of real-world production sequences. It is also shown how such algorithms can be both parameterized and reparametrized in an explainable ad-hoc and post-hoc manner. The explanations are also used to manage contradictory and counterfactual optimization effects so that uncertainty in the decision situations before releasing the sequences to production is handled better than if only the pure sequences were presented.

Keywords: Machine learned ad-hoc and post-hoc explanations · AI based optimization algorithm · Real-world production sequencing

1 Introduction

Real-world production sequencing is a complex optimization problem [6] that must handle multiple optimization goals derived from key performance indicators (KPIs) [4]. In former work [2] it was shown that a decision-making model based on interactions between goals (DMIG) helps to handle complex optimization problems better than additive approaches [3]. In contrast to such approaches DMIG self-recognizes both positive and negative interactions between optimization goals directly out of input data [5]. The interactions are used to self-adapt the behavior of the optimization process and to self-adjust user-given (initially possibly inconsistent) preferences or priorities and balance them best for the current input data [4]. The degree of inconsistencies in the input data in production sequencing is a consequence of inconsistencies between KPI goals that describe a) what is situationally required to be produced due to the demand which quantities of which products to produce compared to b) KPIs that describe what currently is possible to produce due to the technical and capacitive restrictions describing the current production line abilities. In the paper we discuss how to handle and to explain such inconsistencies based on the real-world optimization of production sequences in factories which produce passenger cars. In case of the DMIG based algorithm, the self-recognition of the inconsistencies is machine learned from the input data. The algorithm self-adapts its behavior and provides situational solutions for this complex problem that intelligent human operators would probably generate in a similar way if the complexity

K. Honda et al. (Eds.): IUKM 2023, LNAI 14376, pp. 79–90, 2024.
https://doi.org/10.1007/978-3-031-46781-3_8

were not so high. Having this in mind, we classify the algorithm as AI-based according to criteria summarized in [20].

2 Relation to Other Optimization and Explainable AI Approaches

2.1 Relation to Other Optimization Approaches

The algorithm for optimization of order sequences in car production of which AI-based explanation aspects are discussed in this paper was already put into relation to other optimization algorithms in former papers, for instance in [2] and in some papers referenced there. The basic difference is that many other approaches are based on additive optimization functions and do not explicitly reason about the interactive structure of the optimization criteria or goals in case that the separability assumption does not hold [3]. In this sense DMIG based optimization is different. For more details please see also [2] and papers referenced there. Although there is a huge variety of scheduling and sequencing methods [6] like those based on genetic algorithms, evolution strategies, tabu search as meta heuristics, MILP and other LP oriented methods [16], most of them are different kind of additive cost or utility function-based optimization methods [6, 16]. Compared to these methods DMIG-based optimization is different because of the common property of the other approaches of not reasoning explicitly about the interactive structure of the goals as discussed in [2, 3]. This holds also for car factory motivated approaches like [21]. In this paper we do not further justify qualitatively the DMIG algorithm since this was already done in former papers like [2] and [5] including a running example. In addition, several benchmarks were done in various tender processes in the last twenty years. The results lead to more than one hundred real-world installations of the DMIG-based sequence optimization. The order of magnitude of real-world benefits are mentioned in [14]. Recently, in [15] machine learning was mentioned as a method for prediction of sequence position deviations but not for sequence optimization.

2.2 Relation to Other Explainable AI Approaches

The AI-based explainability of the DMIG-based sequence optimization algorithm is different in several aspects compared to other existing approaches that are rather sophisticated mathematical black boxes [9]. Due to the use of the concept of interactions between goals, there is more insight based on which pros and cons of the KPI goals the results are obtained. So, although the black box of the algorithm is not opened, there is more explainability of balancing the optimization because balancing the goal interactions anyhow is part of the user interaction during the sequencing process. Also, since the KPIs are a kind of meta features of the sequencing, and the interactions of the KPI goals are machine learned from input data, the approach is intrinsic with respect to machine learning of explanations and different from approaches like [11]. The way machine learned explanations are provided based on the DMIG concept allows both ad-hoc and post-hoc explanations. The ad-hoc explanations are used to support human operators in an active manner like [7, 18] and [19] but in a more complex multi-dimensional context compared for instance to the explanation of image information [8] or to neighbor-based evaluation of data points. In case of post-hoc explanations compared to [10] the difference is also that instead of cost functions as prior knowledge non-additively related KPI goals are considered. From the perspective of machine learning of interpretability [1], both machine learned interactions between the goals and their preferences are an additional contribution to AI-based explainability of optimization algorithms. Machine learned preferences for conflicting goals may be directly used to generate explanations by generating contrary or inverse comparisons as in [12] and provide for predictive process monitoring like [13] and for explanation of counterfactual results compared to expected results without opening the black box similar to [17]. From the point of view of the sequence optimization process itself the machine learned DMIG-based explanation helps not only to discover deviations in the sequence as discussed in [15] but show in addition how to optimize the deviations away.

3 Sequencing of Orders as a Complex Resource Planning Problem with Interactive Optimization of KPI Goals

The optimization of production sequences in automotive factories aims to calculate the sequence with which the orders pass through the body shop, paint shop and the assembly line(s). The assembly line is built of a number s (for instance s = 50) of assembly stations arranged one after another. In each assembly station some equipment components are successively assembled into the prefabricated vehicles (orders) already painted and equipped with drive system and chassis. Many of the equipment components are individually configured depending on how each of the cars is configured. The current arrangement of the assembly line is a result of former factory planning assumptions about customer demands. These assumptions led to the design of the line as it is equipped with certain resource capacities and technical abilities. In case that customer demands develop differently compared to the previous assumptions, the capabilities of the production line and the production demands are at least partly inconsistent and to a certain extent contradictory to the previous design. The optimization algorithm calculating the sequences must be able to self-recognize the partial contradictions (inconsistencies) and to balance them in the best way. The balancing is operating on a set of initially pre-prioritized preferences of KPI optimization goals that reflect both the efficiency business goals of the production process like capacity utilization, and business goals that reflect the requirements derived from the current customer demand as combination of car types and delivery due dates. Because of the dynamics of the entire demand portfolio, the KPI-based reaction of the optimization algorithm must be dynamic, too, and cannot be coded a priori. The optimization algorithm must be able to self-recognize the inconsistencies and their degrees since the inconsistencies imply reprioritization of KPI preferences which lead to mutually exclusive sequences. Please note that as soon as one of the sequences is released to production the others become counterfactual.

3.1 Sequencing and its Complexity Aspects in Car Factories

In every car factory the complexity of the sequencing process is implied by at least three different groups of aspects. Firstly, it is characterized by the fact that configurations depend on how customers configure their vehicles individually. Since the number of equipment features and options that can be ordered is huge, the resulting variety of vehicles to be sequenced depends exponentially on this high number of equipment variants and their characteristics. As a rule, within hundreds of thousands of vehicles only a few of them each year are identical. Secondly, at the same time, the number of possible sequences itself depends factorially on the number n of vehicles to be manufactured. Since we typically have $100 < n < 1000$ and more, we have a very high numerical complexity. Overall, an exponentially high number of equipment variants combined with the factorially high number of sequences makes up the enormously complex optimization space of sequencing. Thirdly, an existing production line is subject to technical and capacitive production restrictions and restrictions of the logistics processes of the equipment features (components) to be supplied to the corresponding workstations of the assembly line. For example, the assembly time of a sunroof in the sunroof-designated workstation can vary depending on the complexity of the component "sunroof". A metal sunroof is the easiest to install and requires the least installation time, and logistically it is the simplest feed to the assembly line. A glass sunroof is already more complex in terms of both aspects and a panoramic roof requires the maximum complex feed and the maximum installation time. Other workstations on the assembly line, which are responsible for other equipment features of the vehicles, have their corresponding restrictions and assembly times in an analogous manner.

3.2   What, from a KPI-Based Point of View, is a Good Production Sequence?

A good production sequence is a sequence of orders corresponding to the vehicles to be manufactured, which: a) complies with the technical and capacitive restrictions as physically motivated KPI goals, b) meets the important efficiency KPI goals such as the quantities to be produced, and c) other efficiency KPI goals, such as evenly distributing the workload across workstations, are sufficiently fulfilled. The latter is so important because the uniformity of the workload in the workstations over the sequence has a lasting influence on the quality of the assembly results.

Since the distribution of equipment features (order components or order characteristics) is very irregular due to the high variety of orders, the capacity of the production line and its restrictions is designed with a certain reserve for an average number of orders (vehicles) and an average potential of logistical supply of the components (equipment features or order characteristics) to be assembled. Depending on the distribution of the equipment components belonging to the orders to be sequenced and the continuously changing capacities, conflicts arise between the fulfillment of some of the efficiency KPI goals, the workload KPI goals and the KPI restrictions. Accordingly, a sequencing algorithm must be able to recognize all the conflicting KPI goals and balance them appropriately. Since the distribution of conflicts is not known in advance due to the variety of variants of the equipment compositions within the orders, it is partly unknown and intrinsically uncertain from the perspective of the planning personnel (being the user of the optimization algorithm) which KPI goal conflicts are present in the current optimization situation and to which extent. Therefore, the algorithm must a) recognize the resulting conflicting goals themselves, b) correspond to the trade-offs through appropriate preferential distribution of order characteristics, and c) provide comprehensible explanatory information to the operators of the sequencing algorithm. The latter is particularly important when the trade-offs cannot be fully resolved because the distribution of equipment features in the number of orders to be sequenced is not fully compatible with the current capabilities of the assembly line.

Furthermore, the sequencing algorithm must react very quickly and ad hoc to new situations. For example, due to the high number of ad-hoc changes in production, the entire shift planning must not take longer than half an hour, with potentially several sequencing runs to be completed during this period before all relevant pros and cons have been appropriately weighed and the sequence is released for production.

3.3 Sequence Optimization based on Interactions Between KPIs as Optimization Goals

For the sequencing problem described above, a sequence optimization algorithm was designed that is organized based on the DMIG principle as described more in detail in [4] with some basic ideas given already in [5] and further developed to a more general scheduling algorithm principle as described in [2].

In DMIG decision making is modelled based on fuzzy interactions between decision goals. Decision goals are defined as positive and negative fuzzy impact sets of decision alternatives. The impact values are assumed to be estimated and provided as input data and are modelled as membership values of alternatives noted as δ in the definition given below. So, goals are modelled as fuzzy sets. Interactions between the goals are then modelled as fuzzy relations based on intersections and non-intersections of these fuzzy sets that represent the goals. The formal definition of intersections between decision goals is already described in different publications [2, 3], for instance, and is repeated here for better readability of this paper:

Def. 1a) Let A be a non-empty and finite set of decision alternatives, G a non-empty and finite set of goals, $A \cap G = \emptyset$, $a\ A$, $g\ G$, $\delta\ (0,1]$.

For each goal g we define two types of fuzzy sets S_g and D_g each from A into $[0, 1]$ for all $a \in A$ by following membership functions:

i. $S_g(a):= \delta$, if a affects g positively with degree δ, $S_g(a):=0$ else.
ii. $D_g(a):= \delta$, if a affects g negatively with degree δ, $D_g(a):=0$ else.

Def. 1b) Let S_g and D_g be defined as in Def. 1a). S_g is called the positive impact set of g and D_g the negative impact set of g. S_g and D_g are fuzzy sets over A and the positive and the negative impact functions are their corresponding membership functions.

The fuzzy set S_g contains alternatives with a positive impact on the goal g and δ is the degree of the positive impact. In other words, S_g is the fuzzy set of decision alternatives being good for reaching the goal g. The fuzzy set D_g contains alternatives with a negative impact on the goal g and δ is the degree of the negative impact. In other words, D_g is the set of decision alternatives being bad for reaching the goal g:

Def. 1c) Let X, Y be fuzzy sets. $I(X,Y)$ is defined as a fuzzy inclusion relation of X in Y and $N(X,Y): = 1-I(X,Y)$ is defined as a fuzzy non-inclusion relation of X in Y.

Based on the inclusion and non-inclusion relations between the impact sets of the goals as described above (and for instance in [2]), 8 basic fuzzy types of interactions between goals are defined. The different types of interactions describe the spectrum from a high compatibility of goals (analogy) to a high competition or contradiction (trade-off) [2] between them. For the sake of completeness, independency and unspecified dependency are defined as well.

Def. 2) Let S_{g1}, D_{g1}, S_{g2} and D_{g2} be fuzzy sets given by the corresponding membership functions as defined in *Def. 1)*. For simplicity we write S_1 instead of S_{g1} and D_1 instead of D_{g1} etc. Let $g_1, g_2 \in G$ where G is a set of goals. Let T be a t-norm.

The fuzzy types of interaction between two goals are defined as binary fuzzy relations on $G \times G$ as follows:

1. g_1 *is independent of* g_2: $<$ $=>$ $T(N(S_1, S_2), N(S_1, D_2), N(S_2, D_1), N(D_1, D_2))$
2. g_1 *assists* g_2: $<=>$ $T(I(S_1, S_2), N(S_1, D_2))$
3. g_1 *cooperates with* g_2: $<=>$ $T(I(S_1, S_2), N(S_1, D_2), N(S_2, D_1))$
4. g_1 *is analogous to* g_2: $<=>$ $T(I(S_1, S_2), N(S_1, D_2), N(S_2, D_1), I(D_1, D_2))$
5. g_1 *hinders* g_2: $<=>$ $T(N(S_1, S_2), I(S_1, D_2))$
6. g_1 *competes with* g_2: $<=>$ $T(N(S_1, S_2), I(S_1, D_2), I(S_2, D_1))$
7. g_1 *is in trade-off to* g_2: $<=>$ $T(N(S_1, S_2), I(S_1, D_2), I(S_2, D_1), N(D_1, D_2))$
8. g_1 *is unspecified dependent from* g_2: $<=>$ $T(I(S_1, S_2), I(S_1, D_2), I(S_2, D_1), I(D_1, D_2))$

The interactions between goals are used for orientation during the decision-making process. They reflect the uncertainty of the goals dependencies on each other and describe the pros and cons of the decision situation with respect to these dependencies and indicate how to aggregate them appropriately. For example, for cooperative goals a conjunctive aggregation with similarly high preferences is appropriate. If the goals are competitive, then an aggregation based on an exclusive disjunction with mutually exclusive preferences is appropriate. For more details please see [2, 3] and [4].

The goals-interaction-based sequence optimization algorithm (short: sequencing) starts with the first sequence position 1 and selects an order based on the DMIG principle and assigns this order to the position 1. The algorithm iterates using DMIG in each iteration step i and selects for each position i an order and assigns it to the position i until all n positions in the sequence are selected and got assigned an order. In each iteration step i the set of decision alternatives is the set of orders not yet assigned. So, in each iteration step, the set A (see *Def. 1a)*) of decision alternatives is the set of not yet assigned orders. The set of goals G (see *Def. 1a)*) is derived from the sequencing KPI goals described in Sects. 3.1 and 3.2.

To illustrate this let us consider a KPI that expresses the evaluation of the capacity limitations C_k of the assembly line for the decision to set the order a at position i as decision alternative a_i. The KPI is applied to all stations of the assembly line, and it is numerically qualified by the optimization algorithm if in the current step i the assembly station will be close to an overload or rather not with respect to what the previous capacity situation at all the assembly stations is. The KPI values are real numbers e_i that evaluate (therefore we write e, i is the iteration index) the capacity utilizations and the demand in working time and are mapped to the interval $[-1,1]$ as evaluation values a_{ei}. If we are close to an overload, then the evaluation value will be accordingly close to -1. If capacity is still there, then the evaluation value will be accordingly close to 1. The transformation into DMIG is done by mapping the evaluation values $[-1,0)$ to negative impact values $\delta = a_{ei}$ forming D_{KPI} and $(0,1]$ to positive impact values $\delta = a_{ei}$ forming S_{KPI} according to Def. 1a)ii. And 1a)i. obtaining the negative and the positive impact sets respectively. In this way the C_k capacity limitation values are used as input for capacity utilization KPIs as decision goals. All KPIs and sequencing restrictions are interpreted in this way with their corresponding sets *SKPI*, *DKPI* and we obtain as many *SKPIs*, and *DKPIs* sets as we have KPIs (300 up to 500 KPIs in real-world scenarios and more may be to handle):

Now DMIG is applied in every iteration step i as $DMIG_i$ $1 \leq i \leq n$, i,n N and the decision (a_i,a_{ei}): $= a_i$ is made by selecting the order a_i with the KPIs impact values $\delta = a_{ei}$ a_i A_i: $= \{A / \{$all already assigned a_k of the set of orders$\}\}$. The result for the ith iteration step is the setting of an a_i to the ith position of the entire sequence $((a_1,a_{e1}) \ldots a_i = (a_i,a_{ei}) \ldots, (a_n,a_{en}))$ $A*$. Where $A*$ is the set of all possible permutations of all n sequence positions with attached orders. The final sequence is then $(a_1, \ldots, a_n)$.

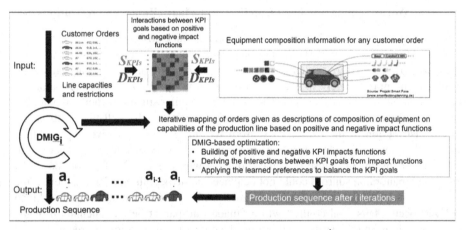

Fig. 1. The DMIG based sequencing algorithm in the ith iteration step.

The interactions between the KPI goals in Fig. 1 are represented by a matrix since for any pair of goals the type of interaction is calculated. Please note that due to the non-symmetricity of the inclusion and non-inclusion relations I and N the matrix of interactions is not necessarily symmetric. The interactions describe the pros and cons of any decision in any iteration step of the optimization algorithm. They are an important source of information for the algorithm to machine learn appropriate preference settings as adjustment parameters and both to explain its behavior in any decision situation and to modify the behavior if the results must go KPIs-wise towards a different direction compared with the previous one. Going in different preference -oriented directions is possible because the evaluation of the KPIs is done goal- interaction-based with look-ahead on the effect on global KPIs, which are calculated in terms of the sequencing goals. These goals are defined to describe the characteristics of both the sequence achieved until the iteration step i and the sequence still to achieve in the iteration steps $> i$ with the remaining $(n-i)$ orders not yet assigned. Considering such global look-back and look-ahead estimations of sequence KPIs ensures that the iteration algorithm is not simply a greedy one. This is possible, since, as already shown in [3], during the iteration process the degree of the separability condition between the KPI goals based on their interactions is continuously estimated as part of an efficient non-greedy heuristics on how to achieve the KPI goals using consistent machine learned goal preferences [4].

4   Machine Learned Explainable Parameterization of AI-Based Sequence Optimization Algorithm

Let us consider some typical parameterization scenarios and illustrate how machine learned preference relations derived from the interactions between the KPI goals are used for both the explanation of the results and the (re-)parameterization of the optimization algorithm.

4.1   Real-World Sequencing Goal Conflicts and Their (Re)parameterizations

If, for instance, for a particular sequence of 1000 sequence positions 200 orders must be positioned that contain a panorama sunroof and the panorama sunroof must be regularly distributed over the entire sequence because of the high workload the panorama roof creates, then approximately to every fifth position an order containing the panorama roof as equipment must be attached. Therefore, in a regular situation where no potential conflicts with other equipment occur, every fifth position of the KPI that controls the regular distribution of the panorama sunroof will have a high positive impact value. On all other four antecedent positions the impact values will rather be negative. If, however, for instance because of some problems with the supply of panorama roofs, orders with panorama sunroof could not be placed at the beginning of the sequence, the positive impact will increase anyhow and goal conflicts with some maximum workload KPI will occur. This goal conflict will be made transparent be the calculation of the goal interactions and the system will report the conflict and situationally recommend a reparameterization of the optimization goals since the originally intended situation is now counterfactual to the new reality.

4.2   Machine Learned Explanations Based on Interactions Between KPI Goals

The recommendations are possible because adequate parameterizations are automatically learned both based on the current conflict situation and historized data that contains the information on how the conflicts were resolved in similar situations in the past. In every iteration of the sequencing process the situation-given interactions between the decision goals together with the evaluated positive and negative impact sets are not only built for optimization but also stored and accumulated. In this way we obtain a behavioral history of the sequence optimization process and the master data situations associated with the data. As shown in [4] from this data preference relations regarding the optimization goals are learned. So, in connection with this preference learning algorithm we obtain both a machine learned exploration of the optimization space and a machine learned insight into the behavior of the optimization process. Based on this accumulated learned preference relations we therefore obtain deeper insight in the structure of the input-output behavior of the optimization algorithm and knowledge about the situational presence of goal conflicts and how to presumably balance them currently and in future if the structure of the customer orders and the production line capabilities does not substantially change. The preference learning is not running on flat raw numerical input data but on the relations that reflect the current interactions between the KPI goals derived from the raw data. Based on the interactions consistent preference relations are predicted and used for the control of the optimization algorithm and for explanation of its results.

4.3   Example of Explainable (Re-)parameterization of KPI Goal Preferences

Let us for instance assume that a set of KPI goals $G = \{g_1, g_2, g_3, g_4, g_5, g_6\}$ with the following goal interactions has been learned in a given decision situation:

> g_1 cooperates with g_2, g_2 cooperates with g_1,
> g_2 cooperates with g_3, g_3 cooperates with g_2,
> g_4 assists g_3, g_4 assists g_5, g_1 assists g_5,
> g_3 competes with g_6, g_6 competes with g_3,
> g_2 hinders g_4, g_5 hinders g_4

as shown in [4]. Based on these interactions the following goal preference relation can be derived and learned: $g_3 > g_4 > g_1 > g_5 > g_2 > g_6$ (please see also Fig. 2).

Derived from the learned interactions between the goals, which in this example situation for the two goal pairs g_3 and g_6 and g_4 and g_2 are competing, the preference learning algorithm concludes that prioritizing simultaneously KPI goals g_3 and g_6 and g_4 and g_2 is rather inconsistent and therefore not recommended. On the other hand, we know that if we set, let say g_6 to the most important goal, the decision algorithm will generate a contrary result to setting g_3 as the most important KPI goal. So, by iterating over the goals and applying the DMIG an automatically learned space of mutually exclusive results will be learned from the data the optimization algorithm generates while running on given KPI input data. From an explanation perspective the mutually exclusive results are counterfactuals of the form "If g_3 hadn´t been selected but g_6 then not the result for optimizing g_6 but for g_3 would have been obtained".

In Fig. 2 we see a part of the user interface of a DMIG-based real-world sequence optimization tool. The circles with numbers are not part of the user interface and are integrated in the figure for descriptions of the various functionalities mentioned in this paper. The grey chart (circle1) visualizes the sequence with its distributed equipment elements. The filled rectangles must be read column-wise. Each column represents an order a_i. The bordered row with the circle 5 shows that the distance between the indicated equipment component type (that is displayed in this row along the sequence) is quite equally distributed with distance of 8 to 10 orders. The equipment that corresponds to this distance KPI could be the sunroof equipment component mentioned in the above example. The table indicated with circle 2 shows the sequence row-wise and is content-wise redundant to 1 but in many cases appreciated by the users. Circle 3 indicates the KPI goal interactions matrix (positive interactions are green, negative interactions are red) and on the right-hand side of the interaction matrix some corresponding capacity limitation effects are shown by blue vertical bars. The higher the bar the higher the capacity demand of each shown station of the assembly line. Finally, circle 4 indicates in which way the parameterization along the pros and cons can span the mutual exclusive parameterization space and what the current parameterization options are. Here suggested KPI goals preference settings as learned from the interactions are $g_3 > g_4 > g_1 > g_5 > g_2 > g_6$ (see the sliders in Fig. 2).

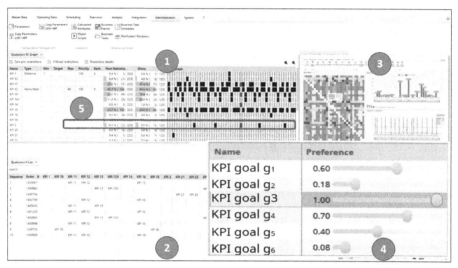

Fig. 2. User Interface of the DMIG-based Optimization Tool with its AI based Explanation Components and anonymized KPIs.

4.4   Which Kind of Machine Learned Explanations are Used Here?

With this kind of machine learned information, the DMIG-based sequence optimization algorithm generates various types of both parametrization recommendations and explanations of the behavior of the optimization algorithm. The learned preference information is situationally used in the car factory control room to justify KPI-oriented ad-hoc effects of reparameterizations, and to explain their influence on operator goals or operator´s questions to be discussed before and while releasing the sequences to production. Especially in case that, at first glance, some KPIs appear counterfactual and are not that satisfactory as required by the operators or situationally some priority changes are to be discussed, the learned parameterizations and explanations help to understand and manage the (to a certain extent uncertain) decision situation in the control room. When evaluating the sequences in the control room, ad hoc explanations like *"if more orders with a panorama sunroof have to be produced then we have to increase the capacity of two particular assembly stations and to reduce the number of right-hand drive cars for the next 100 sequence positions"* are derivable and useful. In in real-world control room applications this kind of machine learned explanations help human operators to reason on how to ad-hoc adjust or to readjust optimization parameters like minimum distances needed between panorama sunroofs or capacity parameters in the assembly stations, for example.

In addition, post-hoc explanations are possible as well: Since the DMIG-based sequence optimization runs in daily production, many real-world sequences are generated by the algorithm. A historization of all these sequences is a good source for training data. Based on these data the same machine learning of preferences, now based on the historized and accumulated interaction data, provides for further machine learned post-hoc explanation of the optimization results. Since these results accumulate many optimization situations, the machine learned interactions and preferences give a deep insight into both the accumulated structure of orders to produce and the accumulated abilities of the production lines for which the data is valid. Therefore, the operators use the learned interactions to adapt former preferences to the new situation based on decisions in the past. In addition, these machine learned post hoc explanations are also used for justification of long-term redesign of physical parameters of the assembly line and for reallocation of resource capacities needed for the production line. In this way accountable and factual arguments for instance for investment or personnel allocation decisions are provided trough the machine learned KPI goal interactions and preferences. Possible uncertainty about the interactions and preferences is machine learned and made explicit.

5 Conclusions

In this paper we have shown how machine learned KPI goal preference relations based on interactions between KPI goals are used to explain results of an AI algorithm for optimization of real-world production sequences. It turns out that machine learned accumulated interactions between the optimization goals and machine learned preference relations are useful to help to operate such an algorithm and to explain its behavior. It was also discussed how these preference relations are used to both parameterize and reparametrize in an explainable ad hoc and post hoc manner the real-world sequence optimization process. Counterfactual reasoning is supported by these explanations, too. In conflictive and therefore uncertain decision situations under time pressure the explanations help the operators in a car factory control room to release the optimized sequences to production preserving KPI based accountability and overview regarding the effects on the KPI goals.

References

1. Carvalho, D., Pereira, E., Cardoso, J.: Machine learning interpretability: a survey on methods and metrics. Electronics **8**, 832, MDPI (2019)
2. Felix, R.: Optimization of partly conflicting goals in complex resource planning. In: Proceedings of the 8th EUSFLAT Conference, Milano, Italy, pp. 669–674 (2013)
3. Felix, R.: On separability of preferences and partly consistent induced interacting goals. In: Proceedings of the IFSA-EUSFLAT International Joint Conference, Gijón, Spain (2015)

4. Felix, R.: Learning of consistent preference relations for decision making and optimization in context of interacting goals. In: Proceedings of the IFSA-EUSFLAT-AGOP International Joint Conference, Bratislava, Slovakia, pp. 266-273.Cooper, M.C. (2021)

5. Felix. R. The optimization of capacity utilization with a fuzzy decision support model. In: Proceedings EUSFLAT-ESTYLF, Palma de Mallorca, pp. 461–463 (1999)

6. Fuchigami, H.Y., Rangel, S.: A survey of case studies in production scheduling: analysis and perspectives. J. Computat. Sci. 425–436, Elsevier (2018)

7. Ghai, B., Liao, V., Zhang, Y., Bellamy, R., Mueller, K: Explainable active learning (XAL): toward AI explanations as interfaces for machine teachers. Proc. ACM Hum. Comput. Interact. 1(1), 28 (2020)

8. Gordo, A., Almazán, J., Murray, N., Perronnin, F.: LEWIS – Latent Embeddings for Word Images and their Semantics. arXiv:1509.06243

9. Guidotti, R., Monreale, S., Ruggieri, A., Turini, F., Pedreschi, D, Gianotti, F.: A survey of methods for explaining black box models. ACM Comput. Surv. 51(5), 1–42 (2018)

10. Jeyasothy, A., Laugel, T., Lesot, M., Marsala, C., Detyniecki, M.: Integrating Prior Knowledge in Post-hoc Explanations. arXiv:2204.11634

11. Kowsari, K., Bari, N., Vichr, R., Goodarzi, F.A.: FSL-BM: fuzzy supervised learning with binary meta-feature for classification. In: Arai, K., Kapoor, S., Bhatia, R. (eds.) FICC 2018. AISC, vol. 887, pp. 655–670. Springer, Cham (2019). https://doi.org/10.1007/978-3-030-03405-4_46

12. Laugel, T., Lesot, M.-J., Marsala, C., Renard, X., Detyniecki, M.: Comparison-based inverse classification for interpretability in machine learning. In: Medina, J., Ojeda-Aciego, M., Verdegay, J.L., Pelta, D.A., Cabrera, I.P., Bouchon-Meunier, B., Yager, R.R. (eds.) IPMU 2018. CCIS, vol. 853, pp. 100–111. Springer, Cham (2018). https://doi.org/10.1007/978-3-319-91473-2_9

13. Mehdiyev, N., Fettke, P.: Explainable artificial intelligence for process mining: a general overview and application of a novel local explanation approach for predictive process monitoring. In: Pedrycz, W., Chen, S.-M. (eds.) Interpretable Artificial Intelligence: A Perspective of Granular Computing. SCI, vol. 937, pp. 1–28. Springer, Cham (2021). https://doi.org/10.1007/978-3-030-64949-4_1

14. Rudolf, G., Felix, R.: KPI-orientierte Optimierung von Produktionssequenzen in der Automobilindustrie und KI-Methoden, Slide 27, VDA Forum Automobillogistik (2019)

15. Stauder, M., Kühl, N.: AI for in-line vehicle sequence controlling: development and evaluation of an adaptive machine learning artifact to predict sequence deviations in a mixed model production line. Flexible Serv. Manufact. J. 34, 709–747 (2022)

16. Tahriri, F., Dawal, S.Z., Taha, Z.: Fuzzy mixed assembly line sequencing and scheduling: optimization model using multiobjective dynamic fuzzy GA. Sci. World J. 505207, 20 (2014)

17. Wachter, S., Mittelstadt, B., Russell, C.: Counterfactual explanations without opening the black box: automated decisions and the GDPR. Harvard J. Law Technol. 31, 841–887 (2018)

18. Wang, K., et al.: Cost-effective active learning for deep image classification. IEEE Trans. Circ. Syst. Video Technol. 27(12), 2591–2600 (2016)

19. Wang, L., et al.: Active learning via query synthesis and nearest neighbour search. Neurocomputing 147, 426–434 (2014)

20. Wang, P.: On defining artificial intelligence. J. Artif. Gen. Intell. 10(2) 1–37, Sciendo (2019)

21. Zhang, R., Chang, P.C., Wu, C.: A hybrid genetic algorithm for the job shop scheduling. Int. J. Product. Econ. 145(1), 38–52 (2013). ISSN 0925-5273

Unearthing Undiscovered Interests: Knowledge Enhanced Representation Aggregation for Long-Tail Recommendation

Zhipeng Zhang[1,2,4], Yuhang Zhang[1], Tianyang Hao[2], Zuoqing Li[2],
Yao Zhang[3(✉)], and Masahiro Inuiguchi[4]

[1] School of Computer Science and Artificial Intelligence, Liaoning Normal University,
Dalian, China
[2] Dalian Gona Technology Group Co., Ltd., Dalian, China
[3] School of Mechanical Engineering and Automation, Dalian Polytechnic University,
Dalian, China
`zhangyao@dlpu.edu.cn`
[4] Department of Systems Innovation, Osaka University, Osaka, Japan

Abstract. Graph neural networks have achieved remarkable performance in the field of recommender systems. However, existing graph-based recommendation approaches predominantly focus on suggesting popular items, disregarding the significance of long-tail recommendation and consequently falling short of meeting users' personalized needs. To this end, we propose a novel approach called Knowledge-enhanced Representation Aggregation for Long-tail Recommendation (KRALR). Firstly, KRALR employs a user long-tail interests representation aggregation procedure to merge historical interaction information with rich semantic data extracted from knowledge graph (KG). By utilizing random walks on the KG and incorporating item popularity constraints, KRALR effectively captures the long-tail interests specific to the target user. Furthermore, KRALR introduces a long-tail item representation aggregation procedure by constructing a co-occurrence graph and integrating it with the KG. This integration enhances the quality of the representation for long-tail items, thereby enabling KRALR to provide more accurate recommendations. Finally, KRALR predicts rating scores for items that users have not interacted with and recommends the top N un-interacted items with the highest rating scores. Experimental results on the real-world dataset demonstrate that KRALR can improve recommendation accuracy and diversity simultaneously, and provide a wider array of satisfactory long-tail items for target users. Code is available at https://github.com/ZZP-RS/KRALR.

Keywords: Long-tail recommendation · Knowledge graph · Graph neural network · Representation aggregation

© The Author(s), under exclusive license to Springer Nature Switzerland AG 2024
K. Honda et al. (Eds.): IUKM 2023, LNAI 14376, pp. 91–103, 2024.
https://doi.org/10.1007/978-3-031-46781-3_9

1   Introduction

With the explosive growth of information on the Internet, addressing the problem of information overload and providing personalized recommendation services to users in various domains has become the norm through recommender system (RS). Typically, RSs model user preferences based on their past interaction information and predict items that users may be interested in. The ultimate objective is to provide users with a curated list of recommendations.

Graph neural networks (GNNs) have demonstrated impressive performance in the realm of RSs; however, existing GNNs-based recommendation approaches suffer from the long-tail problem. Due to the power-law distribution of data in RSs, models trained on such data tend to favor recommending popular items to users. Additionally, traditional GNNs-based recommendation approaches yield poor-quality representations for long-tail items due to their lack of interaction information, exacerbating the long-tail problem. From a market perspective, popular items face intense competition and offer limited profit margins, whereas long-tail recommendations can bring substantial marginal profits to recommendation platforms and businesses [1]. From a user perspective, a large number of repetitive popular items can lead to user fatigue, whereas long-tail recommendations can enhance recommendation diversity and increase user satisfaction [2,3]. From an item perspective, popular items are well-known to the general public, and even without recommendations, users with demand will still purchase them [4]. The key to long-tail recommendation lies in capturing users' long-tail interests. However, long-tail information in user interaction data is scarce, and long-tail items may have poor vector representations due to their limited number of interactions. Therefore, the crucial aspect of addressing the long-tail problem involves exploring users' long-tail interests and enhancing the representations of long-tail items. In the past, many long-tail recommendation approaches focused on improving recommendation diversity at the cost of sacrificing recommendation accuracy. Some approaches [1,3] even exclusively recommended long-tail items to users. However, solely emphasizing recommendation diversity while neglecting recommendation accuracy not only consumes network resources but also impacts user experience. Therefore, there is an urgent need for a long-tail recommendation method that balances both recommendation accuracy and diversity.

Inspired by the rich semantic information present in the knowledge graph (KG) and the powerful representation ability of GNNs, we propose a novel approach called Knowledge-enhanced Representation Aggregation for Long-tail Recommendation (KRALR). Specifically, on the user side, starting from the items interacted by the user, KRALR performs random walks based on their connections with entities in the KG, generating a large number of collaborative paths that contain uncertain information. We apply a constraint by calculating the average item popularity of each collaborative path and select a few paths to replace the user's first-order neighbors. On the item side, items interacted by the same user or items with similar attributes are inherently related. Based on this, we construct a co-occurrence graph to enhance item representations. Finally, we predict the ratings of items with which the target user has not interacted and recommend top N items with the highest predicted ratings to the target user. The contributions of this paper can be summarized as follows: (1) We propose a novel approach called KRALR, which effectively utilizes the uncertain information in the KG to model users' long-tail interests and enhances item representations through the construction of a co-occurrence graph, ultimately achieving long-tail recommendation. (2) We propose a user long-tail interest aggregation procedure, which captures users' deep-level long-tail interests through random walks on the KG, and overcomes the problem of diminishing propagation in these long-tail interests. (3) We propose a long-tail item representation aggregation procedure, which enhances the representation of long-tail items by constructing a co-occurrence graph and increasing information propagation connections.

2   Proposed Approach

In this section, we first give the problem formulation of this paper. Then, we present a detailed description of our proposed KRALR. Figure 1 illustrates the basic flowchart of KRALR.

2.1   Problem Formulation

In a typical RS, we have a set of M users denoted as $U = \{u_1, u_2, ..., u_M\}$ and a set of N items denoted as $I = \{i_1, i_2, ..., i_N\}$. The user-item interaction bipartite graph $\mathcal{G}_{UI} = \{(u, i)|u \in U, i \in I, y_{ui} = 1\}$ is defined based on users' implicit feedback, where $y_{ui} = 1$ indicates that a user has interacted with an item (e.g., clicked, viewed), while $y_{ui} = 0$ indicates no interaction. Additionally, we have a knowledge graph $\mathcal{G}_{KG} = \{(h, r, t)|h, t \in \epsilon, r \in \mathcal{R}\}$, where triples like (Tim Robbins, ActorOf, The Shawshank Redemption) represent that Tim Robbins is an actor of The Shawshank Redemption.

- **Input:** Given the user-item interaction bipartite graph $\mathcal{G}_{UI}$ and the knowledge graph $\mathcal{G}_{KG}$, as well as a target user u,
- **Output:** The top N items that are most relevant to the target user and contain as many long-tail items as possible.

2.2   User Long-Tail Interests Representation Aggregation over KG

The key to long-tail recommendation is aggregating long-tail information in the user representation, which cannot be achieved solely based on the user's historical interacted items. However, KG contains a substantial amount of uncertain information, it provides us with a solution to this problem. Based on this, we conduct a user long-tail interest representation aggregation on the KG, which allows users to establish associations with the KG through the connections between their interacted items and the entity nodes in the KG. We use each interacted item as a starting point and perform random walks on the KG to sample a large number of paths with a length of K. To extract the desired long-tail signals from these paths, we have designed a Long-tail collaborative signal filter to select long-tail collaborative signals. Finally, we utilize the selected long-tail collaborative signals to reconstruct the user's first-order neighbors and perform information aggregation based on the reconstructed first-order neighbors.

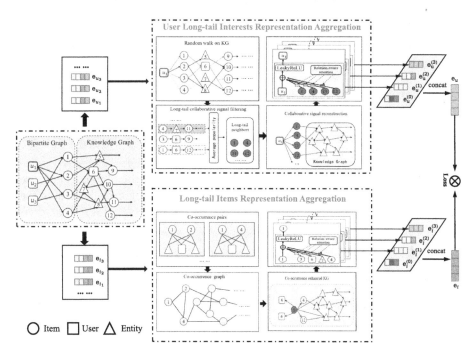

Fig. 1. The flowchart of the proposed KRALR approach. KRALR primarily consists of the following steps: 1) User long-tail interests representation aggregation, which involves exploring collaborative signals through random walks on the KG, and filters out long-tail collaborative signals by incorporating popularity constraints. 2) Long-tail items representation aggregation, which involves constructing a co-occurrence graph and enhancing item representation by integrating information from the KG. 3) Model prediction and optimization, which focuses on optimizing the model to predict rating scores for un-interacted items of the target user, and recommends the top N items with the highest predicted rating scores.

Random Walk on KG. There exists a certain connection between the items a user has interacted with in the past and the items they have not interacted with yet [5]. Following this logic, we utilize the high-order connectivity of the graph and perform random walks on the KG to sample collaborative signals. We define the set of the user's historical interacted items as $N_u = \{i | i \in I, y_{ui} = 1\}$, where the item nodes in N_u are connected to other entity nodes in the KG. These connected entity nodes or other item nodes can further serve as potential interests for the user. Firstly, for the target user u, we consider their set of historical interacted items N_u as the initial seed set S^1. From S^1, we randomly select a node and consider its neighbors as the candidate set for the second-order propagation, denoted as $S^2 = \{t|(h,r,t) \in \mathcal{G}_{KG}, h \in S^1\}$. We continue this process layer by layer, expanding until the K-th order $S^K = \{t|(h,r,t) \in \mathcal{G}_{KG}, h \in S^{k-1}\}$ is selected. By repeating this process, multiple paths can be obtained.

Long-Tail Collaborative Signal Filtering. The paths obtained through random walk in the KG contain a large amount of collaborative signal. However, for long-tail recommendation, we only need to extract the long-tail collaborative signal from these paths. To achieve this, we start from the characteristic of sparse interactions with long-tail items and design a long-tail collaborative signal filtering method to filter out other collaborative signals, which involves calculating the average item popularity for each path and selecting the top s paths with the lowest average item popularity based on this constraint. The average item popularity for each path is defined as follows:

$$\bar{P} = \frac{\sum\limits_{k=1}^{K} a_k p_k}{K'},\tag{1}$$

where, p_k represents the popularity of node k, which is defined as the number of interactions between node k and users in the training set. a_k serves as a controller that determines whether node k is accumulated or not, with a value of 0 or 1 depending on whether node k is an item node. K' denotes the number of item nodes in the path. It is worth noting that the filtered collaborative signals are not all long-tail signals. This can be observed from the fact that the starting point of each path is from N_u. Additionally, the reason why the filtered paths must contain long-tail collaborative signals is that only paths containing long-tail items can lower the average item popularity of the paths, thereby passing the filter's selection process.

User Collaborative Signal Reconstruction. Although the filtered paths contain long-tail collaborative signals, directly using these paths for information propagation can lead to unstable results. This is because we cannot determine the exact position of the long-tail collaborative signals within the paths. Long-tail collaborative signals located at the end of the paths can become extremely scarce due to the layered information propagation process. Therefore, we reconstruct the user's first-order neighbors using the item nodes from the filtered paths, denoted as $N'_u = \{i|i \in \mathcal{P}\}$, where $\mathcal{P}$ represents the filtered paths. By replacing the original first-order neighbors N_u with the reconstructed set N'_u, we ensure that the user can consistently aggregate a sufficient amount of long-tail collaborative signals. This enables effective long-tail recommendations.

User Representation Attentive Aggregation. Next, we will utilize the user embedding e_u and its set of item neighbor nodes N'_u to perform attention-based aggregation and obtain a high-order user representation. Firstly, for any user u, we define its neighborhood information as follows:

$$e_{N'_u} = \sum_{i \in N'_u} g(u, r, i) e_i, \tag{2}$$

where e_i represents the embedding of item i, $g(u, r, i)$ denotes the attention weight of user u on item i. Here, we consider the user-item interaction as a relationship and instantiate it with relation-aware attention [6], which is then normalized using the Softmax function:

$$g'(u, r, i) = (W_r e_i)^{\mathrm{T}} \tanh(W_r e_u + e_r), \tag{3}$$

$$g(u, r, i) = \mathrm{Softmax}(g'(u, r, i)), \tag{4}$$

where e_u, e_r, and e_i represent the embeddings of user, relation, and item respectively. W_r is a transformation matrix based on relation r, used to project the embeddings of user and item into the vector space of relation r. tanh represents the activation function. We utilize the user embedding e_u and its neighboring information e'_{N_u} to obtain the high-order representation of the user:

$$e_u^{l+1} = f(e_u^l, e_{N'_u}^l), \tag{5}$$

where e_u^{l+1} represents the $(l+1)$-th layer GNN representation of user u, and e_u^0 is the initial embedding of user u. We utilize the bi-interaction aggregator [6] to instantiate the function f:

$$\begin{aligned} f(e_u^l, e_{N'_u}^l) =\ & \mathrm{LeakyReLU}(W_1(e_u^l + e_{N'_u}^l)) \\ & + \mathrm{LeakyReLU}(W_2(e_u^l \odot e_{N'_u}^l)), \end{aligned} \tag{6}$$

where W_1 and W_2 are trainable transformation matrices, and $\odot$ represents the element-wise product.

2.3   Long-Tail Items Representation Aggregation over KG

In the context of recommendation systems, the probability of recommending an item to a user depends on the similarity of their embeddings [7]. In GNNs, the update of node representations is influenced by their connections with other nodes. In other words, a node can obtain richer representations if it has more connections. However, this gives rise to an unfairness issue: long-tail items have significantly fewer interactions compared to popular items, which may result in poorer quality representations for long-tail items. Inspired by co-occurrence graphs, we reconstruct the information propagation graph using item co-occurrences.

Co-occurrence Graph Construction. KG contains a large number of entity nodes and relationships between entities, where each node and edge play different roles in constructing item embeddings. Additionally, from a user's perspective, two items that have been interacted with by the same user indicate some common characteristics between them. Furthermore, two items with similar attributes also imply a connection between them. Based on these principles, we introduce connections between items in the KG to enhance item embeddings. Firstly, we compute the frequency of all item-item pairs: if item i_m and item i_n have been interacted with by the same user or share common attributes, the frequency of the pair (i_m, i_n) is increased by 1. Then, we set a threshold x, which is a hyperparameter used to filter out item-item pairs with low co-occurrence frequencies. For the connection between two items, we define it as:

$$Co(i_m, i_n) = \begin{cases} 1, \text{ if frequency } (i_m, i_n) > x; \\ 0, \text{ otherwise.} \end{cases} \quad (7)$$

Co-occurrence Enhanced KG. In this way, we obtain the item co-occurrence graph $\mathcal{G}_{II}$. Next, we combine the user-item interaction bipartite graph $\mathcal{G}_{UI}$, item co-occurrence graph $\mathcal{G}_{II}$, and KG $\mathcal{G}_{KG}$ to form the final collaborative information propagation graph $\mathcal{G}_{UIE} = \{(h, r, t) | h, t \in U \cup \epsilon, r \in \mathcal{R} \cup Interact \cup Co - occurrence\}$. We then perform high-order information propagation on $\mathcal{G}_{UIE}$ to update node representations.

Long-Tail Item Representation Attentive Aggregation. Similar to the update of user representations, we update the representations of item nodes and entity nodes based on their own representations and neighborhood information. For an item or entity node h, we define its set of neighbor nodes as $N_h = \{t | (h, r, t) \in \mathcal{G}_{UIE}\}$. The neighborhood information for h is then defined as follows:

$$e_{N_h} = \sum_{t \in N_h} g(h, r, t) e_t, \quad (8)$$

where e_t represents the embedding of the neighbor node t of h, similar to the aggregation of user representations, the attention score of h on t is:

$$g'(h, r, t) = (W_r e_t)^T \tanh(W_r e_h + e_r), \quad (9)$$

$$g(h, r, t) = \text{Softmax}(g'(h, r, t)), \quad (10)$$

The high-order representation of h is:

$$e_h^{l+1} = f(e_h^l, e_{N_h}^l), \quad (11)$$

$$f(e_h^l, e_{N_h}^l) = \text{LeakyReLU}(W_1(e_h^l + e_{N_h}^l)) \\ + \text{LeakyReLU}(W_2(e_h^l \odot e_{N_h}^l)), \quad (12)$$

After L layers of GNN, we can obtain the l-th order representation for each user and each item:

$$E_u = \{e_u^{(0)}, e_u^{(1)}, ...e_u^{(L)}\}, \tag{13}$$

$$E_i = \{e_i^{(0)}, e_i^{(1)}, ...e_i^{(L)}\}. \tag{14}$$

2.4 Model Prediction and Optimization

Model Prediction. Given the layer-wise GNN representations E_u and E_i for users and items, respectively, we concatenate the representations from each layer to obtain a single vector representation:

$$e_u^* = e_u^{(0)}||e_u^{(1)}||...||e_u^{(L)}, \tag{15}$$

$$e_i^* = e_i^{(0)}||e_i^{(1)}||...||e_i^{(L)}, \tag{16}$$

where $||$ represents the concatenation operation. Finally, we predict the level of interest of user u in item i by taking their inner product:

$$\hat{y}_{u,i} = e_u^{*\mathrm{T}} e_i^*. \tag{17}$$

Model Optimization. In KRALR, we employ two types of optimizers, namely interactive-aware learner and knowledge-aware learner. The interactive-aware learner aims to learn users' diverse preferences from their interactions. We utilize the BPR loss to implement it.

$$\mathcal{L}_{BPR} = \sum_{(u,i,j)\in O} -\ln\sigma(e_u^{*\mathrm{T}} e_i^* - e_u^{*\mathrm{T}} e_j^*), \tag{18}$$

where $O = \{(u, i, j)|i \in N_u', j \in I, j \notin N_u, j \notin N_u'\}$, i represents the positive sample of user u, while j represents a negative sample randomly drawn. We train the model by minimizing the distance between u and the positive sample i, and maximizing the distance between u and the negative sample j, using the sigmoid function σ.

The knowledge-aware learner aims to model entities based on the triplets in $\mathcal{G}_{UIE}$. For any triplet (h, r, t), we define its confidence score as follows:

$$\beta(h, r, t) = ||W_r e_h + e_r - W_r e_t||_2^2. \tag{19}$$

It means calculating the distance between the head entity h and the tail entity t, where a lower confidence score indicates a more reliable and truthful triplet. The loss function of training KG is as follow.

$$\mathcal{L}_{KG} = \sum_{(h,r,t,t')\in\mathcal{T}} -\ln\sigma(g(h, r, t') - g(h, r, t)), \tag{20}$$

where $\mathcal{T} = \{(h,r,t,t')|(h,r,t) \in \mathcal{G}_{UIE}, (h,r,t') \notin \mathcal{G}_{UIE}\}$, the negative samples are constructed by randomly replacing t.

Finally, we obtain the objective function by learning Eqs. (18) and (20) jointly:

$$\mathcal{L} = \mathcal{L}_{BPR} + \mathcal{L}_{KG} + \lambda ||\theta||_2^2, \qquad (21)$$

where $\theta = \{E, W_r, W_1^{(l)}, W_2^{(l)}, \forall l \in \{1, ..., L\}\}$ represents the set of model parameters. E denotes the embedding table for all nodes. L_2 regularization, parameterized by λ, is applied to θ to prevent overfitting.

3 Experiments

In this section, we will present our experimental setup and showcase our experimental results on real datasets.

We provide explanations for some parameters mentioned in this paper. We set the number of paths obtained through random walks to 1500, and the filtered path count, denoted as s, is set to 3. To validate the effectiveness of our KRALR, we conducted experiments on the Last-FM dataset. We prepared the same settings as in [6] to construct knowledge graphs for the Last-FM datasets by mapping items into Freebase entities. Table 1 presents the profile of the dataset for the Last-FM dataset.

Table 1. The profile of Last-FM Dataset

Dataset		Last-FM
User-item interaction	Users	23,566
	Items	48,123
	Interaction	3,034,796
Knowledge graph	Entities	58,266
	Relations	9
	Triplets	464,567

The direct criterion for measuring the quality of a recommendation system is its recommendation accuracy. Therefore, we evaluate the recommendation accuracy of KRALR using Precision and Recall. Additionally, for a long-tail recommendation model, accuracy is not the sole criterion for evaluating its recommendations. Recommendation diversity is also an important metric for assessing long-tail recommendations. We employ the Aggregate Diversity (AD) [8] measure to evaluate the recommendation diversity of KRALR. Since short-head and long-tail are relative concepts, to directly reflect the performance of KRALR in long-tail recommendations, we use the Average Recommendation Popularity (ARP) [9] as a measurement metric.

We denote $Top(u, N)$ as the top N items recommended to user u. Furthermore, we define $Rec(u)$ as the set of items that the user u has interacted with in the test set. The calculations for Precision and Recall are defined as follows:

$$Precision = \frac{1}{m} \sum_{u \in U} \frac{|Top(u, N) \cap Rec(u)|}{|Top(u, N)|}, \tag{22}$$

$$Recall = \frac{1}{m} \sum_{u \in U} \frac{|Top(u, N) \cap Rec(u)|}{|Rec(u)|}, \tag{23}$$

where m represents the number of users in the training set, the larger the values of these two metrics, the higher the recommendation accuracy.

AD calculates the number of distinct items recommended to all users, while ARP calculates the average popularity of the recommended items for users. A higher AD value indicates a greater variety of recommended items, while a lower ARP value indicates a stronger long-tail recommendation capability of the recommendation system. The calculations for AD and ARP are defined as follows:

$$AD = | \bigcup_{u \in U} \{i \in Top(u, N)\}|. \tag{24}$$

$$ARP = \frac{\sum_{i \in Top(u,N)} p_i}{N}. \tag{25}$$

where p_i represents the popularity of item i.

To validate the effectiveness of our proposed approach, we selected the following approaches for comparison:

- **BPRMF** [10]: A basic matrix factorization approach used to rank items of interest for each user according to their preferences.
- **CKE** [11]: A typical regularization-based approach that enhances matrix factorization using semantic embeddings derived from TransR.
- **CFKG** [12]: A model applies TransE to a unified graph that contains users, items, entities, and relations, transforming the recommendation task into the likelihood prediction of the $(u, interactive, i)$ triple.
- **KGAT** [6]: A novel propagation-based model that utilizes an attention mechanism to distinguish the importance of neighbors in the collaborative knowledge graph.

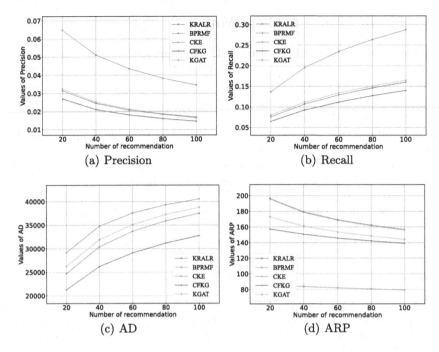

Fig. 2. Comparative experiment results of accuracy metrics (a) precision and (b) recall, and diversity metrics (c) AD and (d) ARP on Last-FM dataset.

Figure 2 presents the experimental results of KRALR compared to other approaches on the Last-FM dataset. From the figure, it can be observed that KRALR outperforms the other comparative approaches in terms of precision metrics such as Precision and Recall. Additionally, KRALR also demonstrates better performance than other comparative approaches in terms of the diversity metric AD. Moreover, KRALR shows stronger capability in recommending long-tail items compared to other approaches, as indicated by the ARP metric. Therefore, we can conclude that KRALR not only performs well in terms of recommendation accuracy but also excels in the task of long-tail recommendation.

It is because KRALR implements user long-tail interests representation aggregation over KG, which effectively leverages the rich semantic information in the KG to capture users' long-tail interest representations. Consequently, it resolves the issue faced by GNNs-based recommendation approaches in capturing users' long-tail interests. Furthermore, KRALR incorporates long-tail items representation aggregation over KG by utilizing both the KG and co-occurrence graph to enrich the available information for long-tail items. As a result, it enhances the quality of representation for long-tail items. By obtaining high-quality representations for both user long-tail interests and long-tail items, KRALR outperforms other related approaches in terms of recommendation accuracy and diversity for long-tail recommendations.

4 Conclusions and Future Work

In this paper, we introduced the novel KRALR approach that leverages the rich semantic data in KG and the powerful representation capabilities of GNNs. Our approach successfully captures user long-tail interests and enhances the embedding quality of long-tail items. The experimental results on real-world dataset demonstrate that KRALR effectively improves recommendation accuracy and diversity simultaneously. For future work, it would be valuable to apply KRALR to address the challenging cold-start recommendation problem, particularly for users and items with no prior interaction information. This presents a more difficult scenario for RSs and warrants further investigation.

Acknowledgements. This work was supported in part by the National Science Foundation of China (No. 61976109); Liaoning Province Ministry of Education (No. LJKQZ20222431); China Scholarship Council Foundation (No. 202108210173).

References

1. Park, Y.J., Tuzhilin, A.: The long tail of recommender systems and how to leverage it. In: Conference on Recommender Systems, RecSys 2008, pp. 11–18. Association for Computing Machinery, New York (2008). https://doi.org/10.1145/1454008.1454012
2. Zhang, Z., Kudo, Y., Murai, T., Ren, Y.: Improved covering-based collaborative filtering for new users' personalized recommendations. Knowl. Inf. Syst. **62**, 3133–3154 (2020). https://doi.org/10.1007/s10115-020-01455-2
3. Zhang, Z., Zhang, Y., Ren, Y.: Employing neighborhood reduction for alleviating sparsity and cold start problems in user-based collaborative filtering. Inf. Retrieval J. **23**, 449–472 (2020). https://doi.org/10.1007/s10791-020-09378-w
4. Li, J., Lu, K., Huang, Z., Shen, H.T.: On both cold-start and long-tail recommendation with social data. IEEE Trans. Knowl. Data Eng. **33**(1), 194–208 (2021). https://doi.org/10.1109/TKDE.2019.2924656
5. Zhang, Z., Dong, M., Ota, K., Kudo, Y.: Alleviating new user cold-start in user-based collaborative filtering via bipartite network. IEEE Trans. Comput. Soc. Syst. **7**(3), 672–685 (2020). https://doi.org/10.1109/TCSS.2020.2971942
6. Wang, X., He, X., Cao, Y., Liu, M., Chua, T.: KGAT: knowledge graph attention network for recommendation. In: Proceedings of the 25th ACM SIGKDD International Conference on Knowledge Discovery and Data Mining, KDD 2019, pp. 950–958. Association for Computing Machinery, New York (2019). https://doi.org/10.1145/3292500.3330989
7. Zhang, Z., Dong, M., Ota, K., Zhang, Y., Kudo, Y.: Context-enhanced probabilistic diffusion for urban point-of-interest recommendation. IEEE Trans. Serv. Comput. **15**(6), 3156–3169 (2022). https://doi.org/10.1109/TSC.2021.3085675
8. Zhang, Z., Dong, M., Ota, K., Zhang, Y., Ren, Y.: LBCF: a link-based collaborative filtering for over-fitting problem in recommender system. IEEE Trans. Comput. Soc. Syst. **8**(6), 1450–1464 (2021). https://doi.org/10.1109/TCSS.2021.3081424
9. Wan, Q., He, X., Wang, X., Wu, J., Guo, W., Tang, R.: Cross pairwise ranking for unbiased item recommendation. In: Proceedings of The Web Conference 2022, WWW 2022, pp. 2370–2378. Association for Computing Machinery, New York (2022). https://doi.org/10.1145/3485447.3512010

10. Rendle, S., Freudenthaler, C., Gantner, Z., Schmidt-Thieme, L.: BPR: Bayesian personalized ranking from implicit feedback. In: Proceedings of the Twenty-Fifth Conference on Uncertainty in Artificial Intelligence, UAI 2009, pp. 452–461. AUAI Press, Arlington, Virginia, USA (2009). https://doi.org/10.5555/1795114.1795167
11. Zhang, F., Yuan, N.J., Lian, D., Xie, X., Ma, W.Y.: Collaborative knowledge base embedding for recommender systems. In: Proceedings of the 22nd ACM SIGKDD International Conference on Knowledge Discovery and Data Mining, KDD 2016, pp. 353–362. Association for Computing Machinery, New York (2016). https://doi.org/10.1145/2939672.2939673
12. Ai, Q., Azizi, V., Chen, X., Zhang, Y.: Learning heterogeneous knowledge base embeddings for explainable recommendation. Algorithms **11**(9), 137 (2018). https://doi.org/10.3390/a11090137

A Novel Methodology for Real-Time Face Mask Detection Using PSO Optimized CNN Technique

Anand Nayyar[1], Nhu Gia Nguyen[1(✉)], Sakshi Natani[2], Ashish Sharma[2], and Sandeep Vyas[2]

[1] Duy Tan University, Da Nang 550000, Viet Nam
{anandnayyar,nguyengianhu}@duytan.edu.vn
[2] Department of Electronics and Communication Engineering, Jaipur Engineering College and Research Center Jaipur, Jaipur 302022, Rajasthan, India

Abstract. The coronavirus (covid-19) pandemic has had a global impact on human beings. Many people have been infected, and millions have died due to COVID-19. A lot of precautionary measures are suggested by the World Health Organization (WHO) to prevent the spread of COVID-19, like the use of sanitizers, social distancing, and face masks. Wearing the face-masks incorrectly makes them useless and spreads the virus. In this manuscript, a convolution neural network (CNN) based real-time face mask detection technique is proposed. However, designing the architecture of CNN is a complex task and requires a deep knowledge of its internal structure and fine tuning of hyper-parameters for better accuracy. In this manuscript, the design complexity of the convolution neural network is reduced by optimizing the hyper-parameters of CNN using the particle swarm optimization (PSO) technique known as PCNN. The proposed PCNN model is tested on an image dataset. The obtained results are compared with other machine learning models like fundamental CNN, support vector machine, decision tree, and naive Bayes algorithms. The performance of the proposed model is tested using validation measures like Accuracy, precision, Recall, and F-1 score. It is observed that proposed model has higher accuracy compared to convention machine learning algorithms.

Keywords: PSO · CNN · Machine Learning · Face mask detection

1 Introduction

Covid 19 pandemic has made an immense impact on everyone's life. The Coronavirus malady is effectuated by an acute respiratory syndrome (SARS-CoV-2). The world has suffered its repercussions and has faced significant drawbacks in every field. COVID-19 is a contagious infection caused by the coronavirus. The pandemic has brought various changes in every person's lifestyle worldwide. According to WHO statistics,423 million people have been infected by Covid-19

K. Honda et al. (Eds.): IUKM 2023, LNAI 14376, pp. 104–116, 2024.
https://doi.org/10.1007/978-3-031-46781-3_10

till February. The nature of the viruses claims that it multiplies in the host's body [1], and their outgrowths depend upon the person's immune system [2]. COVID-19 spreads through air droplets induced whenever a person sneezes or coughs in an open environment. Then these droplets compound with the air in the surroundings and are transferred to the person who makes direct contact with them. Due to this reason, it has impacted the mass expeditiously. Covid-19 has not only affected people physically but has also influenced people's mental health. This alarming situation shook every field of the world. Researchers studied the virus, its cause, prevention techniques, and vaccines. Scientists have proved that wearing masks and hand sanitizers can prevent Covid-19 virus transmission [3]. Regarding health, the government imposed strict restrictions like maintaining social distancing and wearing masks in public places. Face masks became a compulsory accessory to wear in public and cameras were implanted, and guards were appointed to take care of the safety measures. Not wearing masks in public can spread the infection and thus increase the number of Covid patients. In such conditions, Face mask detection became a popular technique to identify such faces without masks. Facemask detection is based on artificial intelligence and machine learning technology. In recent years, machine learning has made predictions, observations, and detection processes more efficient and accurate. Studies claim that Artificial Intelligence was developed in the 1950 s [4] and has made remarkable advancements in artificial intelligence. Machine learning uses algorithms and estimates results based on reference data and past conclusions. Various studies and research have been done on face mask detection. Image detection and object verification is the first step. Research on image detection [6] has helped ease work on image processing and differentiating real and background images. Harzallah et al. proposed techniques for a better understanding image classification with localization [7]. Numerous methods, such as OpenCV DNN [8], Mobilenetv2 [9], and yoloV5 [10] for image processing and object identification [11], using algorithms and concepts. Research by Islam et al. showed that the CNN technique is sensitive to CCTV cameras and detects masks using pixels of the image captured by CCTV [12]. Another step is using the dataset to study and classify faces with and without masks. More et. al proposed the project using KNN [13] in which 2D and 3D masked faces were fed into the database so that the software compares the real-time image with the database and gives an alert if the image mismatches with the database collection. The model proposed by Nagoria et al. [14] works two principles ResNet and decision tree. ResNet is used as a feature extractor, and a decision tree is used for image classification. Proposed model by Jignesh Choudhary et al. using InceptionV3 [8] based on SMFD in which image augmentation method is used for better analysis and accuracy. Research by Reddy et al. [15] contemplated the algorithms with geometry. To properly characterize the mask's orientation, the recovered attributes were computed into several classification models such as Random Forest, Logistic Regression, CNN, Support Vector Machine, AdaBoost, and K-Nearest Neighbors. All these studies reveal that machine learning and deep learning models are widely used in the classification of images. A model like CNN provides excellent results in the classification of the dataset [16]. CNN architecture comprises several layers of convolutions, pooling, fully connected layers, and many other parameters like layer type, connecting nodes, network depth, and the number of neurons. Hence, fundamental CNN architecture requires deep knowledge to tune all the critical parameters to obtain the best results for specific datasets. Automatic design methods for CNN can be highly beneficial rather than infeasible hit-and-trial methods. Hyper-parameter optimization can be considered a model identification problem in machine learning. In other words, hyper-parameter optimization identifies several optimal numbers of neurons, layers, etc. Evolutionary computational techniques can find the optimal set of hyper parameters. In this paper, particle swarm optimization [17] is use to identify the values of hyper-parameters CNN architecture and applied in facemask detection. The main advantages of PSO are that it is easy to implement, has a lower computational cost, and has fewer parameters to adjust. Many researchers have investigated the performance of population-based algorithms in hyper-parameter optimization but still more work needs to be done. The main objective of the paper are as follows:

- Develop a real-time face mask detection system using a convolutional neural network (CNN).
- Apply Particle Swarm Optimization (PSO) to optimize the hyperparameters of the CNN model.
- Improve the accuracy and efficiency of the face mask detection system through hyperparameter optimization.
- Validate the performance of the optimized CNN model on a benchmark dataset and compare it with existing face mask detection methods.

The Local best (L_{best}) and Global best (G_{best}) solutions of PSO are used to tune the five hyper-parameters of CNN, i.e., number of connection layers, Pooling Layer, convolutional layer, and hidden layers. The results provide an improved CNN architecture. The proposed model is applied to classify images wearing masks correctly or incorrectly. Later, the same model is tested on real-time videos for the identification of incorrect wearing of masks.

The rest of the paper is organized in the following manner. In Sect. 2, the Basic background of particle swarm optimization and CNN is described. In Sect. 3, the proposed hyper-parameter-optimized CNN model is discussed. Section 4 explains the experimental setup and dataset used in the analysis. In Sect. 5, proposed model is compared with state of art algorithms in terms of various performance metrics. Section 6 highlights the result and analysis obtained from proposed model. Section 7 concludes the chapter with future scope.

2   Related Terminologies

2.1   Convolutional Neural Network

A Convolutional Neural Network (CNN) is a Machine Supervised learning model created specifically for visual content interaction. It accepts photographs as inputs, recovers and evaluates the image's attributes, and then categorizes them

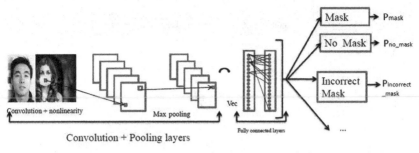

Fig. 1. CNN architecture for face mask detection

using the acquired characteristics. CNN contains several filters [18], each of which captures data from the picture, such as edges and different forms (vertical, horizontal, and round), and then combines all of this information to determine the image. CNN has four layers to process data:

1. The Input layer- As the title suggests, this is the layering in which the input picture is fed, which can be Grayscale or RGB [21]. Pixels with values scaling from 0 to 255 constitute every picture. Before providing them to the model, it has to be normalized, that is, change the scale from 0 to 1.

2. The Convolution layer + Activation Function- It is the layer in which the filter is implemented to the input picture in order to derive or identify its characteristics. A filtration is employed to picture several instances, resulting in a feature map [21] that aids in the classification of the input image and labels. After obtaining the feature map, a nonlinearity-inducing activation function is added to it.

   The activation function in the output layer is called Softmax [22] and calculated as:

   $$S(x_i) = \frac{\exp(x_i)}{\sum_j^n \exp(x_j)} \tag{1}$$

   where $S(x_i)$ is the input value from previous layers, n is number of labels and j is the order of labels.

3. The Pooling Layer- This layer minimises the size of the feature map, which seeks to sustain the pertinent data or aspects of the input picture and also saves estimation time.

4. Fully Connected layer- The input picture is categorised into a label using the Fully connected layer. This layer combines the data from the previous phases to the output layer and, in the end, allocate the input to the appropriate label.

2.2 Particle Swarm Optimization

PSO is based on the intelligent behavior of swarms. PSO is inspired by the food search behavior of swarms introduced by Eberhart et al. [17] in the year 1995.

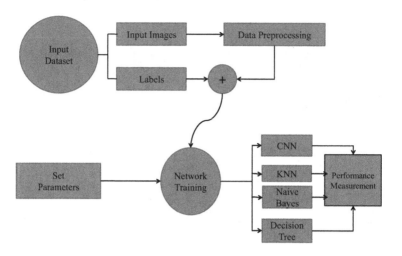

Fig. 2. The basic block diagram of proposed particle swarm optimized CNN

These are the population-based algorithms that find the solutions iteratively. In PSO, individuals are referred to as particles, and the group of particles is known as the Population. Each particle finds the best solution denoted as P_{best} having minimum values. Then the current position vector and velocity vector are used to update the next position as given as

$$V^t = w.V^{t-1} + c_1 r_1 (P_{best}^{t-1} - P^{t-1}) + c_2 r_2 (g_{best}^{t-1} - P^{t-1}) \tag{2}$$

$$P_t = P_t + V_t \tag{3}$$

where w represents inertia weights and c_1 and c_2 are acceleration constant. r_a and r_2 represents the randomly generated values. In PSO, every particle is first initialized which represents the possible solutions. During th process P_{best} and g_{best} interact with every particle and updates the position in search of optimal solution.

3   Proposed PSO Optimized CNN (PCNN)

In this paper, we propose the use of Particle Swarm Optimization (PSO) to optimize the hyperparameters of a CNN for real-time face mask detection. The objective is to enhance the accuracy and efficiency of the face mask detection system by finding the optimal set of hyperparameters through PSO. Particle swarm optimization is used to maximize the accuracy of CNN architecture. The flow diagram of the proposed model is shown in Fig. 2.

3.1   System Model

The Step wise procedure of proposed PCNN is as follows.

Step 1. Data Collection: Collect a dataset of images containing people wearing and not wearing face masks.

Step 2. Preprocessing: Preprocess the images by resizing them and converting them to a suitable format for training the CNN. Convolutional Neural Network (CNN): Design a CNN architecture for face mask detection, consisting of convolutional layers, pooling layers, and fully connected layers.

Step 3. Particle Swarm Optimization (PSO): Utilize PSO algorithm to optimize the hyperparameters of the CNN, such as learning rate, number of filters, and kernel sizes.

Step 4. Hyperparameter Initialization: Initialize a swarm of particles, each representing a potential set of hyperparameters for the CNN.

Step 5. Fitness Evaluation: Evaluate the fitness of each particle by training and testing the CNN using the corresponding hyperparameters on the dataset.

Step 6. Update Particle Velocity and Position: Update the velocity and position of each particle based on its personal best and global best positions, using the PSO equations.

Step 7. Iteration: Repeat the process of fitness evaluation and updating particle positions for a certain number of iterations.

Step 8. Optimal Hyperparameters: Obtain the optimal set of hyperparameters based on the particle with the best fitness value. Evaluation: Evaluate the performance of the optimized CNN model on a separate test set and compare it with existing face mask detection methods.

3.2   Architecture and Working Principal

Initialization of the Population Size. In this phase, population size is defined then particles are randomly generated to reach population size. For each element, the first element is always defined as a convolutional layer. Each element can be filled with a convolutional layer and pooling layer from the second to the maximum fully connected layer. From the maximum fully connected layer to the maximum length, it can be initialized with any four layers until the first fully connected layer is added. The last layer must have the same size as the number of classes. The initialization process is well described in terms of pseudo-code, as shown in Algorithm 1. The particle swarm optimization algorithm is defined as Algorithm 2.

Fitness Calculation. Weight assignment is important step in deep learning framework. A well known Xavier weight initialization strategy is adapted to initialize the weight of of individual particle. after that each individual is decoded in convolutional neural network. After that, K iteration are performed in first phase of training th dataset. Then accuracy of each individual will be calculated and stored as fitness value.

Algorithm 1. Particle initialization algorithm for CNN

1: Input: Population Size N
2: Input: Number of Convolutional Layer N_c
3: Input: Number of Pooling Layer N_p
4: Input: Fully connected Layer N_{fc}
5: Output : Population Initialization P_ϕ
6: **for** x ← P_ϕ **do**
7:     **if** x ≤ N **then**
8:         Convolutional list ← P_ϕ
9:         N_c ← Uniform random number [0 N_c]
10:         **While** Convolutional Unit ≤ N_c
11:         Convolutional Unit ← Convolutional Layer initialize with default settings
12:         Convolutional List ← Convolutional Unit U Convolutional list
13:         **END**
14:         **While** Pool List ≤ N_P **do**
15:         Pool Unit ← Initialize Pool layer randomly
16:         Pool List ← Pool List U Pool Unit
17:         **END**
18:         Fully Connected Layer ← Φ
19:         N_{fc} ← generate uniform integer between [1 N_{fc}]
20:         **While** Fully Connected Layer ≤ N_{fc} **do**
21:         Fully Connected Layer ← Fully Connected Layer **U** Full Unit
22:         **END**
23:         Generate Particle x using Convolutional Unit, Full Unit and Pool Unit
24:     **end if**
25: **end for**

Algorithm 2. Particle Swarm Optimization

1: Input: Intialize Particle x
2: t ← 0
3:   **While** t ≤ Maxiumum Generations **do**
4: Fiteness evaluation for each Particle in x
5: Update Local and Global Position Vectors P_{best} and G_{best}
6: Calculate the Velocity Vector V_i for each particle
7: Update Position Vector
8: t ← t+1
9:   **END**
10: **Return** G_{best} for training

CNN Architecture. A direct encoding technique is used to represent CNN architecture and computational nodes are represented as 2D Grid. Let number of columns and rows are defined as C and R, respectively. Then the number of nodes are given as $R \times C$ and number of input units are defines according to input dataset. The population of particles having varying size comprises of information about type of neuron and connecting nodes. The cth column of neuron must be connected with $(c - l)$ to $(c - 1)$ where l represent level back prameter. Figure 3. represent the different particle with corresponding architecture. In convolutional

CONV	CONV	CONV	Pool	FC	FC	

CONV	CONV	Pool	CONV	Pool	FC	FC

CONV	CONV	CONV	Pool	CONV	FC	FC

Fig. 3. Three different particles comprises of different information.

CONV	CONV	CONV	Pool	FC	FC
Filter 2*2	Filter 2*2	Filter 2*2	Filter 2*2. Stride 2. Type: Max	No of Layer 512	No of Layer 512

Fig. 4. A detailed overview of individual particle.

operation, inputs are padded with zero values to maintain the size of output rows and columns. After convolution, feature map having size M × N is mapped into M' and N'. The Fig. 4 represent the detailed overview of individuals.

4 Experimentation, Result and Analysis

4.1 Dataset Characteristics

The data set was collected from Kaggle.com. There are total 2079 images in the dataset collection. It consists of three types of image categories: Correctly masked face dataset (CMFD), Incorrectly masked face dataset (IMFD) and people not wearing masks and wearing masks in incorrect way. There are 690 images in which people have worn masks, 686 images in which people have not worn masks whereas 703 images in which people have worn masks in an inappropriate manner. This dataset will provide information to the model and hence results can be predicted based on the study of the dataset collection and algorithm. The illustration shows some images of the dataset accumulation.

4.2 Experimental Setup

The proposed model is designed on Python3 software in Intel i7, 16 GB SDD and 32 GB RAM setup. The various algorithmic parameters are selected to find the best results. For the fitness assignment of the candidates in CNN architecture, Gradient descent algorithm trains the CNN with mini batch size of 128. Softmax function is used as loss function. Initial learning rate is 0.1. Adam optimizer is used with parameters value $\beta_1 = 0.9$ and $\beta_2 = 0.99$. The experiment runs for 20 epochs with same int ital learning rate. The dropout = 0.5 is used to avoid over fitting. In PSO, number of particles are 20 with 100 generation size. The inertia coefficient is used as 1. The coefficients C1 and C2 are taken 2.0.

4.3   Performance Evaluation

The primary motive of this paper is to compare the four models. The parameters used for the comparison are Accuracy,Precision, F1-score and Recall.
A confusion matrix is a comprehensive approach of assessing true and false positive and negative outcomes. The confusion matrix offers us estimates of results form of True and false value rather than a decimal reliability.

(TN) True Negative: The model anticipated False, while the real value was False.
(FP) False Positive: The model anticipated True, but the real value was False.
(FN) False Negative: The model anticipated False, but the real value was True.
(TP) True Positive: The model anticipated True and the real value was True.

1. Accuracy : The accuracy informs us how many times the machine learning model accurately anticipates an outcome out of the total number of predictions it has made.

$$Accuracy = \frac{TP + TN}{TP + TN + FP + FN} \tag{4}$$

2. Precision : The model's precision score reflects the potential to effectively forecast the positives out of all the positive predictions it has generated.

$$Precision = \frac{TP}{TP + FP} \tag{5}$$

3. Recall Score The model's capability to reliably forecast positives out of existing positives is measured by the model recall score.

$$Recall Score = \frac{TP}{TP + FN} \tag{6}$$

4. F1-score F1-score is a machine learning model performance statistic that weighs Precision and Recall equally when evaluating accuracy, making it a viable substitute to Accuracy metrics.

$$F1 - score = \frac{2 * Precision * Recall}{Precision + Recall} = \frac{2 * TP}{2 * TP + FP + FN} \tag{7}$$

4.4   Results

It was observed that the camera detects the person's face and tells whether he or she is wearing mask or not. If the person is wearing mask properly then a green colored box will appear saying masked and if the person is not wearing mask or is wearing mask incorrectly, then a red colored box will appear saying no mask. After evaluating all the proposed models, the model will the highest accuracy i.e. CNN can be used to create face mask detector for real-time use. The results are validated with the help of performance measures like Accuracy, Precision, Recall Score and F-1 Score as depicted in Table 1. It is observed from the results that the proposed model has highest accuracy and least error rate. The accuracy plot

Table 1. Classification Report based on KNN

Techniques	Precision	Recall	Recall	F-1 score	Error rate
KNN	92.3	91.5	93.6	95.3	0.23
Decision Tree	94.6	96.2	93.3	96.4	0.15
Naive Bayes	97.5	98.7	95.8	97.2	0.12
PSO-CNN	99.56	98.88	99.2	95.4	0.07

of the proposed PCNN model is as hsown in Fig. 5(a) and 5(b). It is observed from the results that the accuracy is almost near to 100% and maximized with number of epochs and validation loss reaches to zero. The results show that the proposed model correctly identify the masked and non-masked images efficiently.

5 Comparision with State of Art Machine Learning Models

In the study, we have proposed a two stage configuration which will process the result based on the collection of dataset provided and the model implemented. The first stage is the feature extraction in which the software reads and analyses the characteristics of the images provided in the preprocessed dataset. In this, the software examines the images carefully and learns a pattern to detect facemask. Next step is the implementation of the algorithm. The results are compared with three classical machine learning models like K-nearest neighbor [23], Decision Tree [24] and Naive Bayes [25]. This step further bifurcates in two processions: Building and Fitting the algorithm and Performance evaluation. When the data

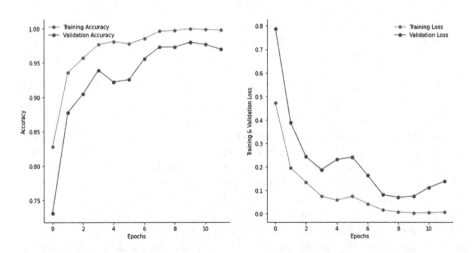

Fig. 5. Graph of Epochs vs. Training and Validation Accuracy/Los

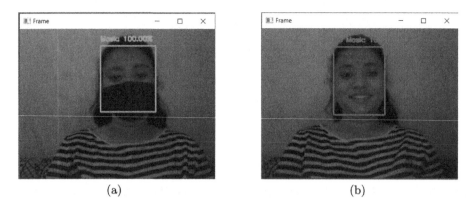

(a)   (b)

Fig. 6. Results obtained from proposed PCNN Model.

is examined, it is rectified using algorithms enforced and then the performance of each calculation is evaluated which helps to give the estimation of the accuracy measurements of loss function is used the proposed algorithms [26].

6 Conclusion and Future Scope

In recent times, face mask detection has gained significant importance due to the global COVID-19 pandemic. The use of Convolutional Neural Networks (CNNs) has shown promising results in various image-based tasks, including face mask detection. However, the performance of CNNs heavily relies on the selection of appropriate hyperparameters. Therefore, optimizing these hyperparameters is crucial to enhance the accuracy and efficiency of face mask detection systems. The objective was to achieve the highest accuracy with the least error in the detection of face masks. To validate the effectiveness of our proposed approach, we will conduct extensive experiments on a benchmark dataset of images containing people wearing and not wearing face masks. The performance of the optimized CNN model will be compared with existing face mask detection methods, demonstrating the superiority of our approach. The utilization of PSO enables the CNN model to efficiently explore the search space and find optimal network configurations, resulting in superior performance. This indicates the effectiveness of the proposed approach in improving the performance of CNN models for this specific task. The use of PSO in optimizing CNN architectures for face mask detection not only enhances the accuracy of the detection process but also reduces the likelihood of false positives or false negatives. This research addresses the critical problem of hyperparameter optimization in CNNs for real-time face mask detection. By leveraging PSO, we aim to contribute to the development of more accurate and efficient face mask detection systems, which have significant implications in ensuring public health and safety during pandemic situations.

References

1. Stern, D.F., Kennedy, S.I.T.: Corona virus multiplication strategy I. Identification and characterization of virus-specified RNA. J. Virol. **34**(3), 665–674 (1980)
2. Phillips, N.: The corona virus is here to stay-here's what that means. Nature **590**(7846), 382–384 (2021)
3. Feng, S., Shen, C., Xia, N., Song, W., Fan, M., Cowling, B.J.: Rational use of face masks in the COVID-19 pandemic. Lancet Respir. Med. **8**(5), 434–436 (2020)
4. Alzubi, J., Nayyar, A., Kumar, A.: Machine learning from theory to algorithms: an overview. J. Phys.: Conf. Series **1142**(1), 012012. IOP Publishing (2018)
5. Oumina, A., El Makhfi, N., Hamdi, M.: Control the Covid-19 pandemic: Face mask detection using transfer learning. In: 2020 IEEE 2nd International Conference on Electronics, Control, Optimization and Computer Science (ICECOCS), pp. 1–5. IEEE (2020)
6. Fathy, M., Siyal, M.Y.: An image detection technique based on morphological edge detection and background differencing for real-time traffic analysis. Pattern Recogn. Lett. **16**(12), 1321–1330 (1995)
7. Harzallah, H., Jurie, F., Schmid, C.: Combining efficient object localization and image classification. In: 2009 IEEE 12th International Conference on Computer Vision ,pp. 237–244. IEEE (2009)
8. Jignesh Chowdary, G., Punn, N.S., Sonbhadra, S.K., Agarwal, S.: Face mask detection using transfer learning of inceptionV3. In: Bellatreche, L., Goyal, V., Fujita, H., Mondal, A., Reddy, P.K. (eds.) Big Data Analytics: 8th International Conference, BDA 2020, Sonepat, India, December 15–18, 2020, Proceedings, pp. 81–90. Springer, Cham (2020). https://doi.org/10.1007/978-3-030-66665-1_6
9. Sanjaya, S.A., Rakhmawan, S.A.: Face mask detection using MobileNetV2 in the era of COVID-19 pandemic. In 2020 International Conference on Data Analytics for Business and Industry: Way Towards a Sustainable Economy (ICDABI), (pp. 1–5). IEEE (2020)
10. Ieamsaard, J., Charoensook, S.N., Yammen, S.: Deep learning-based face mask detection using yoloV5. In: 2021 9th International Electrical Engineering Congress (iEECON), (pp. 428–431). IEEE (2021)
11. Murray, E.A., Bussey, T.J., Hampton, R.R., Saksida, L.M.: The parahippocampal region and object identification. Ann. N. Y. Acad. Sci. **911**(1), 166–174 (2000)
12. Islam, M.S., Moon, E.H., Shaikat, M.A., Alam, M.J.: A novel approach to detect face mask using CNN. In: 2020 3rd International Conference on Intelligent Sustainable Systems (ICISS), (pp. 800–806). IEEE (2020)
13. More, A., Tilloo, H., Kamble, P., Jagtap, B., Bhadane, S.N.: FACE MASK DETECTION USING KNN ALGORITHM
14. Nagoriya, H., Parekh, M.: Live Facemask Detection System (2020)
15. Reddy, S., Goel, S., Nijhawan, R.: Real-time face mask detection using machine learning/deep feature-based classifiers for face mask recognition. In 2021 IEEE Bombay Section Signature Conference (IBSSC), (pp. 1–6). IEEE (2021)
16. Cabani, A., Hammoudi, K., Benhabiles, H., Melkemi, M.: MaskedFace-Net-A dataset of correctly/incorrectly masked face images in the context of COVID-19. Smart Health **19**, 100144 (2021)
17. Kennedy, J., Eberhart, R.: Particle swarm optimization. In: Proceedings of ICNN'95-International Conference on Neural Networks, Vol. 4, pp. 1942–1948. IEEE (1995)

18. Sun, T., Neuvo, Y.: Detail-preserving median based filters in image processing. Pattern Recogn. Lett. **15**(4), 341–347 (1994)

19. Qi, L., Wang, L., Huo, J., Shi, Y., Gao, Y.: Greyreid: a novel two-stream deep framework with RGB-grey information for person re-identification. ACM Trans. Multimed. Comput. Commun. Appl. (TOMM), **17**(1), 1–22 (2021)

20. Milligan, G.W., Cooper, M.C.: A study of standardization of variables in cluster analysis. J. Classif. **5**(2), 181–204 (1988)

21. Gu, J., et al.: Recent advances in convolutional neural networks. Pattern Recogn. **77**, 354–377 (2018)

22. Kouretas, I., Paliouras, V.: Simplified hardware implementation of the softmax activation function. In 2019 8th international conference on modern circuits and systems technologies (MOCAST), (pp. 1–4). IEEE (2019)

23. Peterson, L.E.: K-nearest neighbor. Scholarpedia **4**(2), 1883 (2009)

24. Charbuty, B., Abdulazeez, A.: Classification based on decision tree algorithm for machine learning. J. Appl. Sci. Technol. Trends **2**(01), 20–28 (2021)

25. Berrar, D.: Bayes' theorem and Naive Bayes classifier. Encycl. Bioinform. Comput. Biol.: ABC Bioinform. **403**, 412 (2018)

26. Erickson, B.J., Kitamura, F.: Magician's corner: 9. Performance metrics for machine learning models. Radiology: Artif. Intell. **3**(3), e200126 (2021)

A Computer Vision-Based Framework for Behavior Monitoring and Estrus Detection Through Dynamic Behavioral Analysis

Panisara Kanjanarut[1], Warut Pannakkong[2(✉)], Sun Olapiriyakul[2],
Nuttapong Sanglerdsinlapachai[3], and Shoichi Hasegawa[4]

[1] School of Information, Computer, and Communication Technology,
Sirindhorn International Institute of Technology, Thammasat University,
Khlong Nueng 12121, Pathum Thani, Thailand
[2] School of Manufacturing Systems and Mechanical Engineering, Sirindhorn
International Institute of Technology, Thammasat University,
Khlong Nueng 12121, Pathum Thani, Thailand
warut@siit.tu.ac.th
[3] National Electronics and Computer Technology Center, 112 Thailand Science Park,
Phahon Yothin Road, Klong I, Klong Luang 12120, Pathumthani, Thailand
[4] Tokyo Institute of Technology, 4259 Nagatsuta-cho, Midori-ku, Yokohama, Japan

Abstract. The research field of behavior monitoring and estrus detection in cows has predominantly concentrated on investigating isolated occurrences of mounting behavior, neglecting the analysis of dynamic behaviors exhibited during the estrus period, before and following mounting. To address these limitations, this paper proposes a framework that utilizes computer vision techniques to analyze visual data, classify behavioral features, track the duration of behaviors, and identify potential deviations from normal behavior based on historical data. The part of dynamic behavioral analysis, which encompasses the assessment of behavioral changes and historical behavior data to identify the optimal time window for Artificial Insemination (AI) and potential abnormalities in behaviors, is the main novelty of this framework. Based on our preliminary experiments conducted on a well-known public dataset, the behavior classification model achieves an overall accuracy of 80.7% in accurately classifying various behaviors, including standing, walking, lying down, and feeding. While the model demonstrates proficiency in identifying feeding, lying down, and standing behaviors, there is still room for improvement in accurately recognizing walking behavior. This research contributes to advancing behavior monitoring and estrus detection techniques, providing a way for improved AI practices in the cattle industry.

Keywords: Cow Behavior Monitoring · Estrus Detection · Computer Vision · Machine Learning · Smart Farming

K. Honda et al. (Eds.): IUKM 2023, LNAI 14376, pp. 117–128, 2024.
https://doi.org/10.1007/978-3-031-46781-3_11

1   Introduction

The profitability of cattle farms heavily relies on reproductive performance, as a high pregnancy rate serves as an indicator of success [1–3]. Artificial Insemination (AI) has become the widely adopted approach to address this concern [3]. This technique enables genetic improvement in farm animals by utilizing frozen sperm with high genetic merit, eliminating the logistical challenges associated with natural mating [4]. However, successful insemination must be performed approximately 12–18 h after the cow exhibits standing heat, where a cow or heifer permits other animals to mount her while she remains standing [2,5]. Temperature is also a critical factor, as heat stress can occur. In case of insemination failure, it results in increased costs for farmers, including additional feed costs, animal management labor, extra breeding expenses, and the value of calf loss [1,3]. These complications highlight the necessity for farm workers to have effective decision-making and supporting tools to detect standing heat and determine the proper timing for initiating AI.

Detecting estrus behavior has always been a laborious, subjective, and challenging task in cattle farms. Traditionally, farm workers would frequently observe cows at least twice a day for 30 to 60 min through direct observation or video feeds [2,5]. Besides recognizing standing heat, farmers should also look for additional signs such as increased walking, decreased feeding, chin resting on other cows, and sniffing the vulva region of other cows [2,5–7]. One of the earlier technologies developed to address this task was wearable devices. In 2017, Miura et al. [8] utilized a wireless sensor covered with urethane gel to monitor the surface temperature (ST) at the ventral tail base. By studying the relationship between ST, behavioral estrus expression, ovulation, and hormone profile, detection accuracy ranged from 46% to 71%. Subsequently, Higaki et al. [9] employed a wireless vaginal sensor placed in the cranial vagina to measure vaginal temperature and conductivity for accurate estrous detection. They utilized supervised machine learning and achieved a sensitivity and precision of 94%. Wang et al. [10] introduced a sensor attached to the cow's neck using adjustable nylon straps. This sensor combined acceleration and location data to create an activity index, enhancing estrus recognition through unsupervised learning and self-learning classification models, achieving an accuracy of 90.91%. However, a drawback was that it caused annoyance to the cows and would be costly to adopt on a large scale [1].

Recently, advancements in computer vision and deep learning algorithms have facilitated the development of automated image analysis techniques for continuous monitoring, leading researchers to favor these methods for cows. CowXNet [11] is an estrus detection system that considers mounting behavior and additional signs. It utilizes YOLOv4 for cow detection, a convolutional neural network for body part detection, and a classification algorithm for estrus behavior detection, achieving an accuracy of 83%. Another study [12] presents an improved YOLOv5 model for detecting mounting behavior. This system incorporates complex layers of atrous spatial pyramid pooling, a channel-attention mechanism, and a deep-asymmetric-bottleneck module, resulting in an accuracy of 94.3% with a processing speed of 71 frames per second (fps) in natural breeding scenes. However, both works focus on detecting signs of estrus and fail to consider behavior before and following mounting, thereby leaving the optimal timing for AI uncertain. Behavior classification has also been explored in previous research. Gong et al. [13] propose a system that classifies behavior into standing, walking, and lying down. They utilize the YOLOv4 model for detecting multiple cows and then employ keypoint heatmaps and part affinity fields for classification. The system achieves precision rates of 85% for single cow pose estimation during the day and 78.1% at night. By combining mounting detection and behavior classification, these techniques offer significant advantages by reducing the need for skilled human labor and invasive equipment [1].

Previous research in the field has predominantly focused on examining a point in time when mounting behavior occurs rather than taking into account the overall behavior and considering the events preceding and following the mounting behavior, which is equally important in identifying the estrus period proper insemination time window. Consequently, there is a research need to upgrade the existing computer vision-based framework, specifically the part of behavior monitoring and estrus detection in cows. This research paper addresses this significant research gap by proposing the development and evaluation of a novel computer vision-based framework for cows. The framework harnesses advanced computer vision techniques to analyze video data, accurately classify behavioral features, and effectively query the information for precise estrus detection. Furthermore, it incorporates the capability to suggest the optimal time window for AI. By addressing this research gap, the proposed framework aims to significantly advance the field by providing valuable insights and practical solutions for enhancing reproductive management in cattle farming.

The remainder of the paper is structured as follows: Section 2 presents a detailed description of the materials and methods employed in this research. Section 3 provides the results obtained from the evaluation and offers a discussion. Section 4 outlines the conclusions drawn from this study, highlighting the potential impact of the proposed framework and directions for further research exploring its practical implementation in real-world cattle farming scenarios.

2   Material and Method

This work utilized a publicly available dataset to develop the proposed framework. As there were no real-life experiments involved, direct ethical concerns did not arise. The use of publicly available data adhered to the terms and conditions stipulated by the dataset providers, ensuring compliance with relevant data usage policies and safeguarding intellectual property rights.

2.1   Data

This work used the AnimalPose dataset, which was developed by Chen et al. [14]. This dataset was specially created to address the lack of keypoint-labeled animal datasets and to enable the performance evaluation of animal pose estimation algorithms under weak supervision. It involves five selected mammals: dogs, cats, horses, sheep, and cows. Each animal instance is annotated with 20 keypoints, including eyes, ears, nose, throat, tail base, withers, elbows, knees, and paws, as shown in Fig. 1, The annotation format resembles popular human keypoint formats used in datasets like COCO.

Fig. 1. A cow's keypoint annotation [15].

A custom Python code has excluded all animals other than cows from images and the corresponding annotation files. Subsequently, the retained data was formatted into a CSV (Comma-Separated Values) file format, facilitating the task of annotators in labeling behaviors. Key information, including image file names, bounding box coordinates, and keypoint coordinates, was organized into individual rows within the CSV file. The dataset comprises a total of 504 images and 852 annotations.

To ensure the precision and accuracy of the dataset, the annotators cross-referenced the file names presented in the CSV file with the actual images, verifying the presence of cows and filtering out unrelated images, such as close-ups of cow face and images of the milking process. Consequently, the annotators classified the behavior of the cows using agreed-upon description as listed in Fig. 1. The dataset contains four distinct behaviors. However, a significant proportion of the annotations is primarily the cow's standing behavior. In order to enhance the balance of the dataset, standing annotations are randomly excluded from the training and testing phases. The remaining dataset comprises 219 images and 353 annotations, which have been partitioned into train and test sets using a 75:25 ratio.

2.2   Proposed Framework

The structure of the proposed framework is presented in Fig. 2. The framework begins with the standard practice of detecting the cow from the images and performing keypoint localization. The resulting output is then passed on to the behavior classification step. The following steps, represented by the grey-colored boxes, are the novel components of our framework. The output obtained from

Table 1. Description of cow's behavior.

Behavior	Description
Feeding	A cow is consuming food, typically by grazing on grass, hay, or other feed sources
Lying Down	A cow's posture of resting or reclining on the ground or a surface, with its body horizontally extended
Standing	A cow's legs are actively in a sequential movement of having one or more legs lifted off the ground
Walking	A cow supports its body weight evenly on all four legs

the behavior classification model is the input for our dynamic behavioral analysis part. At the behavior duration calculation step, in cases where a cow is mounted. The system proceeds by appending the behavior and detecting whether the standing behavior persists for longer than 3 s. Other estrus parameters, such as increased walking and decreased feeding, are also evaluated at this stage. Once the cow is identified to be in the estrus period, the system will provide recommendations for the appropriate timing for AI. In cases where the behavior does not involve mounting, the model verifies whether it matches the behavior observed in the previous frame. If it differs, it is appended to the database. This step is implemented to prevent data duplication and manage storage size, as an increased size would complicate the process of querying behavior changes. Any deviation from the normal behavior of the cow can also be evaluated while taking into account its behavior history to identify potential abnormalities in its health The details of all key framework components are described below.

Cow Detection and Keypoint Localization. The methods of cow detection and keypoint localization in previous studies can be categorized into two main types: top-down methods and bottom-up methods. Top-down methods involve detecting objects and getting the bounding box before estimating body parts. Consequently, the accuracy of keypoint localization is high. These works by Lodkaew et al. [11], Wang et al. [12], and Gong et al. [13] all fall into this category. However, it would rely heavily on object detection's success. On the other hand, bottom-up methods primarily focus on detecting body parts as polygonal shapes and grouping them into objects [13,16]. It is generally faster since a regular bounding box is not a natural object representation, and fitting them inside a box will unavoidably include background pixels. However, the main task of this approach in human is to determine which one does the keypoint belong to [16]. This difficulty likely makes the top-down approach more appropriate and will be used since every cow is similar to each other, being quadrupeds, while humans are bipeds [17].

For the cow detection, to demonstrate the testing of the proposed framework, the openMMlab's MMDetecion toolsbox [18] with the Cascade Mask R-CNN model, originally proposed by Radosavovic et al. [19], is selected as it is reported

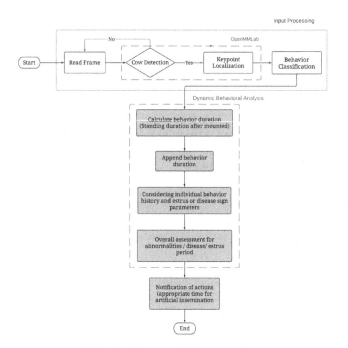

Fig. 2. Proposed Framework.

as one of the best performances from training on ImageNet. Further detailed explanations are discussed in the original paper and information on parameter setting can be found on MMdetection.

OpenMMlab has expanded their work as MMPose, a pose estimation toolsbox that provides a convenient approach to obtain keypoint localization. The model for this task is selected based on the reported performance on the Animal Pose dataset [20], which is a Topdown-HRnet. The Topdown-HRnet incorporates an HRNet backbone architecture which was proposed by Sun et al. [21], and the model predicts heatmaps for keypoint and evaluates by measuring the Mean Squared Error (MSE) between the predicted and ground truth keypoints by calculating the squared Euclidean distance for each keypoint. The loss is obtained by averaging the MSE across all keypoints. Additional explanations are elaborated upon the source materials.

Behavior Classification. For the behavior classification, while the coordinates and visibility of each keypoint provide valuable information about the spatial distribution, those alone may not suffice. To enhance the accuracy of behavior prediction, the angles formed between each keypoint are considered. The calculation process is presented is described in Algorithm 1. It computes the vectors by subtracting coordinates and calculates their magnitudes using the Euclidean norm. After normalizing the vectors, it calculates the dot product to obtain the cosine of the angle. The function checks if the dot product is within the valid range of -1 to 1 and returns 99999 when the keypoint is not visible. The angle is then calculated using the arc cosine function and converted to degrees. Finally, the rounded angle is returned as the output.

$$\text{vector} = \begin{bmatrix} \text{keypoint1}[x] - \text{keypoint2}[x] \\ \text{keypoint1}[y] - \text{keypoint2}[y] \end{bmatrix} \tag{1}$$

$$\text{magnitude} = \sqrt{x^2 + y^2} \tag{2}$$

$$\mathbf{v}_1 \cdot \mathbf{v}_2 = \sum_{i=1}^{n} v_{1i} \cdot v_{2i} \tag{3}$$

Algorithm 1. Calculate Angle

Input: keypoint1, keypoint2, keypoint3
Output: angle_degrees
`calculate_angle` (keypoint1, keypoint2, keypoint3)
Vector 1 ← Calculate vector form between keypoint 1 and keypoint 2;
Vector 2 ← Calculate vector form between keypoint 2 and keypoint 3;
magnitude1 ← Calculate magnitude for Vector 1;
magnitude2 ← Calculate magnitude for Vector 2;
if *magnitude1 = 0 or magnitude2 = 0* **then**
 | **return** *99999*;
end
vector1_norm ← vector1/magnitude1;
vector2_norm ← vector2/magnitude2;
dot_product ← Calculate dot product between Vector 1 and Vector 2;
if *dot_product ≤ -1 or dot_product ≥ 1* **then**
 | **return** *99999*;
end
angle_radians ← arccos(dot_product);
angle_degrees ← angle_radians × $\frac{180}{\pi}$;
return *angle_degrees*;

There are seven additional angle features for this task. The first angle is Left Back Elbow, Tailbase, and Right Back Elbow, and the second angle is Left Front Elbow, Tailbase, and Right Front Elbow. These two angles are used to put emphasize the parallelity of the cow stance. The third angle is Withers, Throat, and Nose, which could help define the cow's neck compared to its back, thereby aiding in determining feeding behavior in which the cow will lower its head. The last four angles are between the Elbow, Knee, and Paws of each leg, allowing for the determination of the bend in each leg. Since AutoML has demonstrated its effectiveness comparable to human performance [22], leveraging Autogluon can significantly assist in validating a framework by saving time and ensuring reproducible results [23]. The possible outputs from this step include chin resting, feeding, lying down, mounting, mounted, standing, and walking. The predicted behavior will be appended to the database if it differs from the previous behavior.

Calculate Behavior Duration. This feature is achieved by setting up variables to track the current behavior and timestamp and then moving to the next frame. If the subsequent behavior differs from the current behavior, the code calculates the time difference between the current and previously recorded timestamps. This time difference is then added to the respective total time variable assigned to each behavior, representing the accumulated time spent in that specific behavior.

Behavior Sign Parameters. In this study, we will primarily focus on parameters related to estrus behavior. Continuous monitoring of cow behavior offers valuable insights into their welfare and health status through the detection of deviations from normal behavior patterns.

For estrus parameters, When the detected behavior is "mounted", the system anticipates whether the next frame depicts "standing" behavior or not. If the following behavior after "mounted" behavior is any other behavior, it indicates that the cow is not in the standing heat state and therefore not ready for AI. The system provides feedback that no immediate "standing" behavior is found after the "mounted" behavior.

If "standing" behavior is found, the system retrieves the frame IDs and indices of the relevant instances. Then, it proceeds to calculate the duration of the "standing" behavior by obtains the frame ID of the first "standing" behavior after "mounted" behavior as the start frame ID of the "standing" behavior. It then considers the frame at which the behavior has changed as the end frame ID.

The time stamps of the starting and ending frames can be retrieved from the database using their IDs. The duration is calculates by subtracting the time stamp of the starting frame from the time stamp of the ending frame. If the duration is equal to or longer than a specific duration (e.g., three seconds), it indicates that the cow is in a standing heat state.

Overall Assessment. To provide the veterinarian with supporting information, the system compares the duration of time the cow spends in each behavior to historical data. The change over time can providing an overall assessment of cows well-being. During the estrus period, cows might show unusual behaviors like walking more or eating less. These changes should be reported to the veterinarian because instead of estrus behavior they could indicate that the cow is sick. The health of cow can affect the success of artificial insemination.

Notification of Actions. If a cow is in the standing heat state, the general guideline for performing AI is 12–18 h after the start of the standing heat state. However, it is important to note that each breed of cow can have a different ovulation period, and therefore the time from standing heat to insemination can vary. It is recommended to consult with a veterinarian to determine the specific timing for insemination based on the breed of the cow.

3   Preliminary Result

In the preliminary stage, the proposed framework is illustrated using publicly available data (AnimalPose dataset [14]). Regarding the outcomes of the Behavior Classification model, it has demonstrated the ability to classify four available classes of behavior: standing, walking, lying down, and feeding, with an overall accuracy of 80.7%. The results of the behavior classification model are summarized in a confusion matrix shown in Table 2.

Table 2. Confusion matrix of behavior classification.

True Behavior	Predicted Behavior				Recall
	Feeding	Lying down	Standing	Walking	
Feeding	21	3	0	1	84.0%
Lying down	1	17	1	0	89.5%
Standing	1	0	23	1	92.0%
Walking	4	0	5	10	52.6%
Precision	77.8%	85.0%	79.3%	83.3%	
Accuracy	80.7%				

The model performs reasonably well in identifying feeding, laying down, and standing behaviors. The recall rates for feeding, laying down, and standing behaviors are 84.0%, 89.5%, and 92.0%, respectively. The precision for these behaviors is also promising, with values of 77.8%, 85.0%, and 79.3%. However, the model shows a lower performance in recognizing walking behavior, with a recall rate of 52.6% and a precision rate of 83.3%. This indicates that the model struggles to distinguish between walking and standing behaviors, leading to misclassifications as also reported by which train with extensive dataset [13]. Furthermore, the detailed result of cow detection, keypoint localization, keypoint localization, and time stamp is shown in Fig. 3.

The anticipated outcome of the framework for practical implementation is presented through an interface, as depicted in Fig. 4. The interface serves as a visual representation of the expected results generated by the framework, demonstrating its potential application in real-world scenarios. Specifically, the interface is designed to showcase the duration of each behavior and facilitate a comparative analysis over time, effectively capturing and displaying data for both standing and feeding behaviors. These behaviors are considered secondary indicators for estrus detection.

Moreover, through continuous monitoring of behaviors, the framework facilitates the identification of potential health issues or patterns that may serve as indicators of estrus [1]. By observing changes in behavior over time, the system can detect abnormalities or irregular behavioral patterns that may signify underlying health problems. This feature enhances the framework's ability to

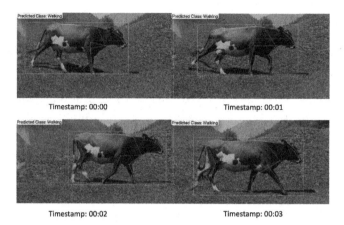

Fig. 3. Predicted Result [24].

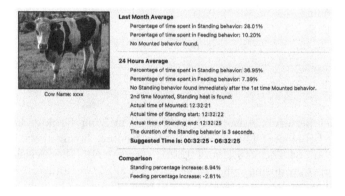

Fig. 4. Result Interface.

provide comprehensive insights into the well-being of the animals and enables timely interventions or veterinary attention when necessary. We recognize that the gray boxes in Fig. 2 portray the novel components of our proposed framework. Their real-world trial hinges on ethical clearance, which we are actively pursuing.

4   Conclusion and Future Work

This paper proposes a computer vision-based framework for behavior monitoring and estrus detection in cows. Instead of focusing on only isolated instances of mounting behavior, this framework considers the dynamic in behavior patterns based on the animal's behavior history. The study highlights the capability of the proposed framework in estrus detection which relies on the dynamic behavioral analysis in suggesting the optimal timing for AI.

From the preliminary result of applying the proposed framework to the public data (AnimalPose dataset [14]), we achieved a commendable overall accuracy of 80.5% in classifying standing, walking, lying down, and feeding behaviors. However, we observed that the model tends to misclassify various poses as walking, which may attribute to the inherent variability in the angles and positions of cow's legs and joints. Additionally, when tested with videos recorded by humans, camera shake and variations in capturing angles can contribute to classification errors.

To address these challenges, an interesting approach could involve combining the methodologies used by McDonagh et al. [17], who employed training with a short video dataset, and Lodkaew et al. [11], who utilized the distance between frames as a feature in their model. However, the lack of datasets containing estrus-related behaviors such as mounting, sniffing, and chin resting poses a limitation, hindering comprehensive testing of our proposed framework, which necessitates further investigation.

In conclusion, our framework is valuable to supporting tools for both cattle farming and veterinarians. This research presents a computer vision-based framework for behavior monitoring and estrus detection in cows, employing the dynamic behavior analysis. By incorporating the evaluation of historical behavior, the framework is expected to enhance the ability of traditional methods in accurately identifying animal abnormalities and suggesting optimal insemination timing.

Acknowledgements. The first author acknowledges the scholarship under the Thailand Advanced Institute of Science and Technology and Tokyo Institute of Technology (TAIST-Tokyo Tech) Program, awarded by the National Research Council of Thailand (NRCT), National Science and Technology Development Agency (NSTDA), and Sirindhorn International Institute of Technology (SIIT), Thammasat University.

References

1. Chen, C., Zhu, W., Norton, T.: Behaviour recognition of pigs and cattle: journey from computer vision to deep learning. Comput. Electron. Agric. **187**, 106255 (2021)
2. Roelofs, J., Lopez-Gatius, F., Hunter, R., Van Eerdenburg, F., Hanzen, C.: When is a cow in estrus? Clinical and practical aspects. Theriogenology **74**(3), 327–344 (2010)
3. Galvão, K., Federico, P., De Vries, A., Schuenemann, G.M.: Economic comparison of reproductive programs for dairy herds using estrus detection, timed artificial insemination, or a combination. J. Dairy Sci. **96**(4), 2681–2693 (2013)
4. Hamid, M., Abduraman, S., Tadesse, B.: Risk factors for the efficiency of artificial insemination in dairy cows and economic impact of failure of first service insemination in and around Haramaya town, Oromia region, eastern Ethiopia. Vet. Med. Int. **2021** (2021)
5. Selk, G.: Artificial insemination for beef cattle. Technical report, Oklahoma Cooperative Extension Service (2004)

6. Kerbrat, S., Disenhaus, C.: A proposition for an updated behavioural characterisation of the Oestrus period in dairy cows. Appl. Anim. Behav. Sci. **87**(3–4), 223–238 (2004)
7. Pahl, C., Hartung, E., Mahlkow-Nerge, K., Haeussermann, A.: Feeding characteristics and rumination time of dairy cows around estrus. J. Dairy Sci. **98**(1), 148–154 (2015)
8. Miura, R., Yoshioka, K., Miyamoto, T., Nogami, H., Okada, H., Itoh, T.: Estrous detection by monitoring ventral tail base surface temperature using a wearable wireless sensor in cattle. Anim. Reprod. Sci. **180**, 50–57 (2017)
9. Higaki, S., et al.: Estrous detection by continuous measurements of vaginal temperature and conductivity with supervised machine learning in cattle. Theriogenology **123**, 90–99 (2019)
10. Wang, J., Zhang, Y., Bell, M., Liu, G.: Potential of an activity index combining acceleration and location for automated estrus detection in dairy cows. Inf. Process. Agric. **9**(2), 288–299 (2022)
11. Lodkaew, T., Pasupa, K., Loo, C.K.: CowXNet: an automated cow estrus detection system. Expert Syst. Appl. **211**, 118550 (2023)
12. Wang, R., et al.: Detection method of cow estrus behavior in natural scenes based on improved YOLOv5. Agriculture **12**(9), 1339 (2022)
13. Gong, C., Zhang, Y., Wei, Y., Du, X., Su, L., Weng, Z.: Multicow pose estimation based on keypoint extraction. PLoS ONE **17**(6), e0269259 (2022)
14. Cao, J., Tang, H., Fang, H.S., Shen, X., Lu, C., Tai, Y.W.: Cross-domain adaptation for animal pose estimation. In: The IEEE International Conference on Computer Vision (ICCV), October 2019
15. Tobi: Brown and white cow. Pexels (2017). https://www.pexels.com/photo/brown-and-white-cow-457447/
16. Zhou, X., Zhuo, J., Krahenbuhl, P.: Bottom-up object detection by grouping extreme and center points. In: Proceedings of the IEEE/CVF Conference on Computer Vision and Pattern Recognition, pp. 850–859 (2019)
17. McDonagh, J., Tzimiropoulos, G., Slinger, K.R., Huggett, Z.J., Down, P.M., Bell, M.J.: Detecting dairy cow behavior using vision technology. Agriculture **11**(7), 675 (2021)
18. Chen, K., et al.: MMDetection: open MMLab detection toolbox and benchmark. arXiv preprint arXiv:1906.07155 (2019)
19. Radosavovic, I., Kosaraju, R.P., Girshick, R., He, K., Dollár, P.: Designing network design spaces (2020)
20. Contributors, M.: OpenMMLab pose estimation toolbox and benchmark (2020). https://github.com/open-mmlab/mmpose
21. Sun, K., Xiao, B., Liu, D., Wang, J.: Deep high-resolution representation learning for human pose estimation. In: Proceedings of the IEEE/CVF Conference on Computer Vision and Pattern Recognition (CVPR), June 2019
22. Hanussek, M., Blohm, M., Kintz, M.: Can AutoML outperform humans? An evaluation on popular OpenML datasets using AutoML benchmark. In: 2020 2nd International Conference on Artificial Intelligence, Robotics and Control, pp. 29–32 (2020)
23. Erickson, N., et al.: AutoGluon-tabular: robust and accurate AutoML for structured data. arXiv preprint arXiv:2003.06505 (2020)
24. Odintsov, R.: A cow walking in a picturesque location. Pexels (2021). https://www.pexels.com/video/a-cow-walking-in-a-picturesque-location-9734068/

Information Fusion for Enhancing Presentation Attack Detection in Iris Recognition

Jittarin Chaivong[1] , Kharidtha Ruangsariyanont[1] ,
Suradej Duangpummet[2]([⊠]) , and Waree Kongprawechnon[1]

[1] Sirindhorn International Institute of Technology, Thammasat University,
99 Moo 18, Paholyothin Highway, Khlong Luang 12120, Pathum Thani, Thailand
waree@siit.tu.ac.th
[2] NECTEC, National Science and Technology Development Agency,
112 Thailand Science Park, Klong Luang 12120, Pathum Thani, Thailand
suradej.dua@nectec.or.th

Abstract. This work investigates data fusion techniques for presentation attack detection (PAD) in iris recognition systems. Although a few PAD methods have been proposed, the uncertainty and inaccuracy of PAD for textured contact lenses remain. Performance variability of classifiers and the quality variation of captured images from different iris scanners complicate the problem. To address these issues, we introduce an iris PAD method that enhances PAD accuracy using information fusion and interval methods. The proposed method fuses scores from two approaches: 2D feature using Gabor wavelet and deep neural networks and 3D feature with decision-making within intervals of the thresholds. Experiments were carried out using a cross-sensor dataset, consisting of two models of iris scanners. Results suggest that the proposed method outperforms the baseline. It was found that our fusion method takes advantage of the unique strengths of each system, enhancing the overall performance and accuracy of PAD methods. These findings highlight the significance of employing deep feature extraction and data fusion techniques to cope with data uncertainty in the ongoing development of biometric security systems to combat new threats.

Keywords: Biometrics · Iris recognition · Presentation attack detection · Information fusion · Transfer learning

1 Introduction

Biometrics refers to the unique physiological or behavioral traits of individuals, such as fingerprint, face, voice, and iris [12,18]. Among these, the iris stands out as the most reliable biometric identifier due to its distinct and changeability patterns across the human population [3,15]. The irises are also almost unchanged and are protected by eyelids. Hence, many advanced security systems have recently utilized iris to identify people or verify an individual for various purposes, such as border controls, secure access controls, national ID, and migrant registrations for vaccination [11,21].

Nevertheless, fraudulent authentication attempts into secure applications have become a vital concern since iris recognition systems are recently vulnerabilities to attacks from fake identities using paper prints, artificial irises, and contact lenses, called presentation attacks or spoofing [1,7,10,20]. Textured-contact lens attacks can be wearing fake-texture contact lenses to conceal his/her identity or printing other iris patterns into contact lenses. There are also a variety of textured contact lenses and many types of iris scanners on the market. Hence, iris-presentation attack detection (PAD) is crucial, and robust and reliable PAD is also challenging. This paper, therefore, focuses on PAD for contact lens attacks in iris recognition systems.

Several PAD methods have been proposed for contact lens attacks [6,9,13]. The iris PAD methods might be classified into active methods and passive methods. An active method, as presented in [13], used a specialized iris scanner to analyze the specular spots of a collimated infrared LED. In another study proposed by Lee *et al.*, the reflectance ratio between the iris and sclera (i.e., the white area of the eye) at 750 nm and 850 nm illumination was measured to differentiate between fake and genuine irises [14]. However, these active methods are limited to specialized systems, while passive methods rely on iris-texture analysis from captured images based on standard iris scanners. Daugman [3] originally invented the most famous iris recognition technique. Later, many methods have been proposed. Ma *et al.* suggested a technique based on frequency analysis for detecting printed irises [15]. He *et al.* proposed a method based on gray level co-occurrence matrix and a support vector machine (SVM) [8].

The state-of-the-art PAD method is based on data fusion technique [5]. This method combines two-dimensional data and three-dimensional properties of the observed iris. It was reported that the 3D data could be obtained from standard iris scanners so that surface analysis was applied to justify wearing contact lenses. For 2D data, the texture of the iris was extracted by using binary statistical image features (BSIF). The support vector machine (SVM) as a classifier was trained to distinguish between irises with and without textured contact lenses. This work suggested the cascade classification. The decision-making was mainly based on 3D data in which the 2D classifier tested the rest trial.

However, from our preliminary study, we found that texture information using BSIF lead to a loss of fine-grained details and potentially affects the accuracy of the spoof detection. To overcome this limitation, we utilized a more robust feature, i.e., Gabor wavelet transform, for extracting 2D information. In addition, we exploit the transfer learning technique for iris texture extraction, the so-called deep feature, to deal with a limited data size but obtain more important information than that of the BSIF feature. Furthermore, for the decision-making, we proposed a score fusion based on the interval of uncertainty and variation of scores from two different sensors.

This paper is organized as follows. Section 2 briefly introduces techniques related to this work. The proposed method is described in Sect. 3. Section 4 presents the experiment and result. Lastly, Sect. 5 discusses and summarizes key points of this study.

2   Backgrounds

This section introduces a conventional process in an iris recognition system with open-source software. Then, the related works used in this study are described, including 3D-PAD based on photometric stereo features and 2D deep feature based on transfer learning.

2.1   Iris Recognition: OSRIS

Figure 1 shows an iris recognition consisting of four steps: segmentation, normalization, template coding or feature extraction, and matching. A few open-source of these processes are available. Masek and Kovesi introduced one of the first open-source re-implementations of Daugman's method [4, 16]. The source code of these processes was later superseded by the so-called open source iris or OSIRIS [17]. The OSRIS steps are as follows:

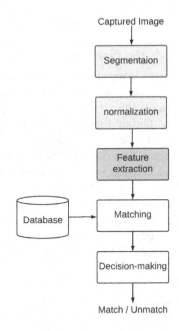

Fig. 1. Iris Recognition.

- **Segmentation**: The first step is to localize part of the image corresponding to the iris. According to the properties of our eyes, two circular boundaries should be detected to locate the iris position, i.e., a large circle separating the outer iris from the sclera and a smaller circle separating the inner iris from the pupil. Daugman's integro-differential operator is used to search these two circles.

- **Normalization**: Iris and pupil circle parameters are required. The polar form is then transformed into a linear form as the image of a rectangle.
- **Feature extraction**: A bank of log-Gabor wavelets is utilized for feature extraction as the following equation:

$$\max_{(r,x_0,y_0)} \left| G_\sigma(r) * \frac{\partial}{\partial r} \int_{r,x_0,y_0} \frac{I(x,y)}{2\pi r}, ds \right|, \tag{1}$$

where "*" represents convolution, $G_\sigma(r)$ represents a smoothing function, $I(x,y)$ represents the image of the eye, (x,y) represents the location of point in image, (x_0,y_0) represents location of center coordinate, and r represents radius.

Gabor filters are a combination of the Gaussian smooth function and the complex-exponential function, and the effective range of the Gabor filter is determined by σ. The Gaussian smooth function is expressed as

$$\frac{1}{2\pi\sigma_y\sigma_x} \exp(-\frac{1}{2}(\frac{x_1^2}{\sigma_y^2} + \frac{y_1^2}{\sigma_y^2})). \tag{2}$$

Given $\frac{1}{2}\left(\frac{x_1^2}{\sigma_x^2} + \frac{y_1^2}{\sigma_y^2}\right) = m^2$, the function's value drops as m increases. The Gabor filter's effective range is an ellipse with the major axis at $2\sqrt{2\sigma_x}$ and the minor axis at $2\sqrt{2\sigma_y}$. The relationship between filter scale and $\sigma_x, 0_y$ would be established because the attenuation of the Gaussian function limits the effective range of the Gabor filter [17]. OSIRIS offers six hand-crafted Gabor kernels. These kernels consist of three pairs at different scales: 9×15, 9×27 and 9×51. The feature (iris code) is the convolution operation between the normalized image and kernels, which has a length of 1536.

- **Matching**: Fractional Hamming distance is used for calculating matching scores from two templates. To compute a matching score between the two iris codes for all conceivable shifts because normalization cannot eliminate shift effects in the polar coordinates. This will lengthen the matching time while increasing the recognition rate. The matching score is determined as follows [19].

$$Score = \max\{\frac{1}{|S_t|}\sum_{y \in S_t}[c_t(y)c_q(y-x) \cap m_t(y)m_q(y-x)]\}, \tag{3}$$

where t and q are template and query, $c_t(y)$ and $c_q(y)$ are the iris code and $m_t(y)$ and $m_q(y)$ are the occlusion template. $|S_t|$ is the effective code length (excluding the occlusion section).

2.2 3D-PAD Method

The 3D-PAD method based on stereo features was proposed by [2]. This technique utilizes two near-infrared (NIR) illuminators placed at different positions relative to the lens. Since the surface of the iris is not perfectly flat, we can exploit the other shadows from the left and right images.

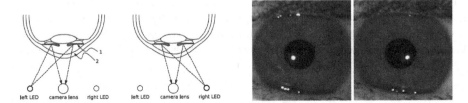

Fig. 2. A 3D-PAD technique based on two near-infrared (NIR) illuminators placed at different positions relative to the lens (Left). Example images from the two light positions on the same eye [5] (Right).

The photometric stereo approach concludes with the estimation of normal vectors, which should not differ considerably from the object's average normal vector. The images show shadows produced by the partially opaque texture printed on the lens, which can be seen in various locations depending on the illuminator used. Aside from significant shadows detected in places, we also note changes in how the printed texture forms visual features under different illumination angles. As a result of the irregular and noisy surface being estimated, the photometric stereo will produce various normal vectors.

The process involves the calculation of occlusion masks, denoted as m_{left} and m_{right}, corresponding to the left and right iris images, I_{left} and I_{right}, respectively. For iris pixels that are not occluded, m is 1, otherwise 0 for background. Consider $\bar{n}$ to be the average normal vector within the non-occluded iris area. When comparing an approximately flat iris to an irregular object comprised of an iris and a textured contact lens, the Euclidean distances between the normals and their average tend to be smaller for the flat iris.

Therefore, we can use the variance(var) of the Euclidean distance between the normal and their average, calculated within the non-occluded iris area. The PAD score is defined as:

$$q = var\|n_{x,y} - \bar{n}\|, \tag{4}$$

where $n_{x,y}$ is the normal vector at each location, where $\bar{n}$ is the average normal vector within the non-occluded area, where $\|\cdot\|$ is the l^2 (Euclidean) norm of x.

$$v = \frac{1}{N} \sum_{x,y} n_{x,y}, \tag{5}$$

where N is the number of non-occluded iris points, and $(x, y) : m_{(}x, y) = 1$. From Eq. (5) and Figs. 5 - 6, it was found that the normal vector, v, of the textured contact lenses images have a higher variance than those of natural eye and transparent contact lenses [5].

3   Proposed Method

This work aims to improve the accuracy of PAD in iris recognition systems by fusing 2D and 3D data. The objective is to achieve better results compared to previous work in the field. This section presents a comprehensive explanation of fusing the OSPAD-2D and 3D using the open-source iris processing as mentioned above. The proposed method is shown in Fig. 3.

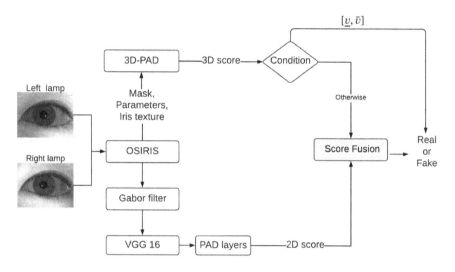

Fig. 3. Diagram of the proposed method.

3.1   For 2D

Since we first explored the BSIF feature with the CNN, we found that BSIF has limited performance, as shown in Fig. 4 below shows the comparison of the natural eye and eye with contact lenses that make the classification error.

To classify the iris image as an authentic iris or a textured contact lens, we employ the Gabor wavelet as the feature extractor and use a CNN with deep feature extraction based on the VGG16 model for a classifier.

3.2   For 3D

We utilized the NDiris3D dataset, which provided a diverse collection of 3D iris images. Our purpose was to analyze this dataset and determine the authenticity of the iris samples. Thus, we employed Gabor filters for feature extraction

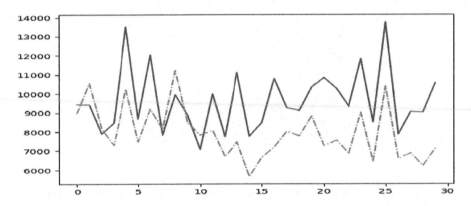

Fig. 4. BSIF feature from the test dataset: Blue color represents the BSIF from the real eye while orange represents the eye with contact lens.

using the 3D-OSPAD method. This analytical approach involved evaluating the extracted features and calculating a score. This score served as an indicator of the probability that the data was real or spoofed. We considered two different iris sensors: the AD100 and LG4000. Our algorithm determines the threshold conditions from the different sensors based on the file extension of the iris images. The AD100 sensor captures eye images into a bitmap file (*.bmp), while the LG4000 sensor uses TIFF format (*.tiff). Hence, it allows us to effectively match the iris images to their respective sensors and leverage the specific characteristics of each sensor during our work.

3.3   Fusion

For the AD100 sensor, we make the condition by using this equation. From Fig. 6 and Fig. 5 data, we observed that the histogram indicates errors predominantly occurring within 0.1 to 0.3.

$$\text{Final score} = \begin{cases} \text{3D score} & v < 0.1 \, or \, v > 0.3 \\ \text{Score fusion} & \text{Otherwise} \end{cases} \tag{6}$$

For the LG4000 sensor, the histogram shown in Fig. 7 indicates that the errors predominantly occur within 0.2 to 0.6. Thus, the condition is defined as in Eq. (7).

$$\text{Final score} = \begin{cases} \text{3D score} & v < 0.2 \, or \, v > 0.6 \\ \text{Score fusion} & \text{Otherwise} \end{cases} \tag{7}$$

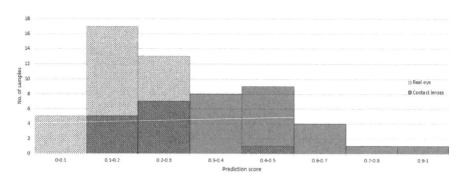

Fig. 5. The histogram of 3D score from iris images captured by AD100 sensor and their labels.

4   Experiment and Result

4.1   Dataset

Since we aim to discern between authentic and manipulated eye images acquired by scanners from different senseors, we leverage the invaluable NDiris3D dataset[1], comprising a rich collection of images from the AD100 and LG4000 sensors. The NDiris3D consists of 88 pairs of eye images, including images with and without contact lenses from three brands: Johnson & Johnson, Ciba Vision, and Bausch & Lomb. The dataset encompasses contact lenses with regular and irregular patterns. The dataset was divided by 70% for training and 30% testing the performance of the proposed method.

4.2   Evaluation Matrices

This section presents a thorough analysis of the performance of the textured-contact lens PAD. Thus, we evaluated and compared the performance of our work with the baseline using the following matrices.

- False Positive Rate (FPR) shows the ratio of detected fake samples as real with total fake samples. FPR is defined as:

$$FPR = \frac{FP}{FP + TN},\tag{8}$$

  where FP is false positive or detecting fake as real, and TN is true negative or detecting real as real.
- False Negative Rate (FNR) shows the ratio of detected real samples as fake with total real samples. FNR is defined as:

$$FNR = \frac{FN}{FN + TP},\tag{9}$$

---

[1] https://cvrl.nd.edu/projects/data/.

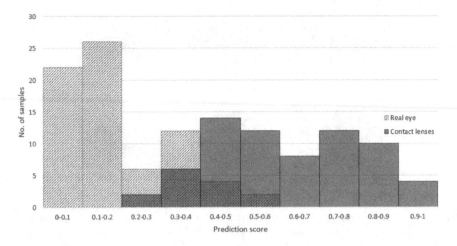

Fig. 6. The histogram of 3D scores from iris images captured by LG4000 sensor and their labels.

where FN is False negative or several detecting real as fake, and TP is True positive or the number of detecting fake as fake.

- Accuracy (Acc) shows the ratio of detected samples wrong with the total samples. Accuracy is defined as:

$$Accuracy\,(Acc.) = \frac{TP + TN}{TP + TN + FP + FN}. \tag{10}$$

Table 1. Comparison results between the proposed method and the baseline [5].

Method	Sensor					
	LG 4000			AD 100		
	Acc. (%)	FPR (%)	FNR (%)	Acc. (%)	FPR (%)	FNR (%)
OSPAD-fusion [5]	94.76	6.36	4.12	91.14	11.84	5.92
Proposed method	**98.57**	2.70	**0.00**	85.92	**6.67**	19.51

Table 1 presents a comprehensive comparison of the performance achieved by the fusion method. Here, we employed score fusion, which effectively combines the scores obtained from 3D and 2D modalities. This fusion approach results in significantly higher accuracy in iris recognition. Notably, our experimental results using the LG4000 sensor surpassed the performance reported in [5], demonstrating the effectiveness of our approach.

It was found that the accuracy achieved with the AD100 sensor is still slightly lower compared to previous studies. However, the overall increase in accuracy across both sensors indicates the success of our proposed method. We have

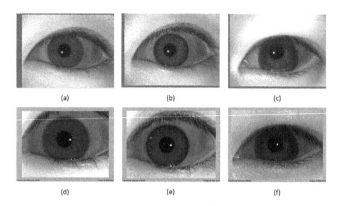

Fig. 7. Example images from the NDiris3D dataset [2]. (Top): image captured by LG 4000 sensor. (Bottom): image captured by AD100 sensor. (a) and (d) are real eye images. (b) and (e) is an eye with a regular contact lens image. Images (c) and (f) are an eye with an irregular contact lens image.

improved iris recognition accuracy by leveraging the benefits of score fusion and optimizing the integration of 3D and 2D scores. These findings further validate the effectiveness and viability of our approach, showcasing its potential for enhancing the performance of iris recognition systems.

5   Discussion and Conclusion

This research investigated the accuracy improvement of presentation attack detection in iris recognition systems by employing score fusion from 2D mm and 3D and advanced feature extraction techniques, including OSIRIS and VGG16. Our work aimed to develop a robust presentation attack detection (PAD) algorithm capable of detecting fake irises, particularly those created using texture contact lenses. By integrating 2D and 3D data, we combined high-resolution iris images with depth information to identify inconsistencies and anomalies associated with fake irises. This approach yielded promising results in differentiating between genuine and manipulated iris patterns, enhancing the security and reliability of biometric systems. The proposed method significantly improves PAD accuracy for iris recognition systems by training and testing different feature extraction techniques on various datasets. These findings emphasize the significance of utilizing advanced feature extraction techniques and other data types in developing biometric security systems to address new threats effectively. By incorporating these sophisticated methodologies, we can enhance the strength and efficacy of our security systems, ensuring resilient protection against evolving risks.

References

1. Agarwal, R., Jalal, A., Arya, K.: A review on presentation attack detection system for fake fingerprint. Mod. Phys. Lett. B **34**(05), 2030001 (2020)
2. Czajka, A., Fang, Z., Bowyer, K.: Iris presentation attack detection based on photometric stereo features. In: 2019 IEEE Winter Conference on Applications of Computer Vision (WACV), pp. 877–885. IEEE (2019)
3. Daugman, J.: How iris recognition works. In: The Essential Guide to Image Processing, pp. 715–739. Elsevier (2009)
4. Daugman, J.G.: High confidence visual recognition of persons by a test of statistical independence. IEEE Trans. Pattern Anal. Mach. Intell. **15**(11), 1148–1161 (1993)
5. Fang, Z., Czajka, A., Bowyer, K.W.: Robust iris presentation attack detection fusing 2D and 3D information. IEEE Trans. Inf. Forensics Secur. **16**, 510–520 (2020)
6. Galbally, J., Toth, A.B.: Anti-spoofing: iris databases. In: Li, S., Jain, A. (eds.) Encyclopedia of Biometrics, pp. 1–7. Springer, Boston (2009). https://doi.org/10.1007/978-3-642-27733-7_9050-2
7. Gragnaniello, D., Poggi, G., Sansone, C., Verdoliva, L.: Contact lens detection and classification in iris images through scale invariant descriptor. In: 2014 Tenth International Conference on Signal-Image Technology and Internet-Based Systems, pp. 560–565. IEEE (2014)
8. He, X., An, S., Shi, P.: Statistical texture analysis-based approach for fake iris detection using support vector machines. In: Lee, S.-W., Li, S.Z. (eds.) ICB 2007. LNCS, vol. 4642, pp. 540–546. Springer, Heidelberg (2007). https://doi.org/10.1007/978-3-540-74549-5_57
9. He, X., Lu, Y., Shi, P.: A new fake iris detection method. In: Tistarelli, M., Nixon, M.S. (eds.) ICB 2009. LNCS, vol. 5558, pp. 1132–1139. Springer, Heidelberg (2009). https://doi.org/10.1007/978-3-642-01793-3_114
10. Hsieh, S.H., Li, Y.H., Wang, W., Tien, C.H.: A novel anti-spoofing solution for iris recognition toward cosmetic contact lens attack using spectral ICA analysis. Sensors **18**(3), 795 (2018)
11. Kaur, N., Juneja, M.: A review on iris recognition, pp. 1–5 (2014). https://doi.org/10.1109/RAECS.2014.6799603
12. Kohli, N., Yadav, D., Vatsa, M., Singh, R., Noore, A.: Detecting medley of iris spoofing attacks using desist. In: 2016 IEEE 8th International Conference on Biometrics Theory, Applications and Systems (BTAS), pp. 1–6. IEEE (2016)
13. Lee, E.C., Park, K.R., Kim, J.: Fake iris detection by using Purkinje image. In: Zhang, D., Jain, A.K. (eds.) ICB 2006. LNCS, vol. 3832, pp. 397–403. Springer, Heidelberg (2005). https://doi.org/10.1007/11608288_53
14. Lee, S.J., Park, K.R., Kim, J.: Robust fake iris detection based on variation of the reflectance ratio between the iris and the sclera. In: 2006 Biometrics Symposium: Special Session on Research at the Biometric Consortium Conference, pp. 1–6. IEEE (2006)
15. Ma, L., Tan, T., Wang, Y., Zhang, D.: Personal identification based on iris texture analysis. IEEE Trans. Pattern Anal. Mach. Intell. **25**(12), 1519–1533 (2003)
16. Masek, L.: Matlab source code for a biometric identification system based on iris patterns. The School of Computer Science and Software Engineering (2003)
17. Othman, N., Dorizzi, B., Garcia-Salicetti, S.: Osiris: an open source iris recognition software. Pattern Recogn. Lett. **82**, 124–131 (2016)

18. Pala, F., Bhanu, B.: Iris liveness detection by relative distance comparisons. In: Proceedings of the IEEE Conference on Computer Vision and Pattern Recognition Workshops, pp. 162–169 (2017)
19. Rathgeb, C., Uhl, A., Wild, P., Hofbauer, H.: Design decisions for an iris recognition SDK. In: Handbook of Iris Recognition, pp. 359–396 (2016)
20. Rigas, I., Komogortsev, O.V.: Eye movement-driven defense against iris print-attacks. Pattern Recogn. Lett. **68**, 316–326 (2015)
21. The Thai Red Cross Society: The thai red cross society and network partners provide the proactive covid-19 vaccination rollout for displaced persons. https://english.redcross.or.th/news/7036/. Accessed 15 June 2023

Score-Level Fusion Based on Classification Tree for Improving Iris Verification Systems

Oranus Kotsuwan[1]![ID], Puntika Leepagorn[1]![ID], Suradej Duangpummet[2]![ID],
and Jessada Karnjana[2]([✉])![ID]

[1] Sirindhorn International Institute of Technology, Thammasat University,
99 Moo 18, Paholyothin Highway, Khlong Luang 12120, Pathum Thani, Thailand
[2] NECTEC, National Science and Technology Development Agency,
112 Thailand Science Park, Klong Luang 12120, Pathum Thani, Thailand
{suradej.dua,jessada.kar}@nectec.or.th

Abstract. Iris verification has long been considered the most reliable biometric verification technology. The iris verification system compares a feature extracted from the iris image with one previously stored in the database during the enrollment procedure to accept or reject the claimed identity from the image. Conventionally, the decision is made based on a comparison score obtained from a distance measurement and a pre-defined threshold, which considerably affects the system's performance. This study aims to enhance the accuracy of iris verification by optimizing the false rejection rate (FRR) and false acceptance rate (FAR) near the threshold. Our proposed method utilizes four distance measurements: Hamming distance (HD), Jaccard distance (JD), Tanimoto dissimilarity index (TDI), and weighted Euclidean distance (WED), and combines them with a decision tree. We evaluated our approach using the CASIA-IrisV2 dataset and observed improvement compared to conventional methods. The experimental results show that the proposed method's accuracy, precision, and F1 score are improved by at least 2.20%, 4.50%, and 5.66%, respectively. These findings highlight the potential of our work for real-world applications.

Keywords: Iris verification · Score-level fusion · Classification tree

1 Introduction

Iris verification is widely recognized as a highly efficient and reliable method for biometric verification. Compared to other biometrics, human's iris tend to be the most consistent one which suits for the purpose of verification [4]. In addition, its exceptional accuracy and resistance to forgery have made it a preferred choice in various applications, including access control systems and national security databases.

In recent years, significant efforts have been made in the field of iris verification and the development of matching algorithms for accurate verification. Many researchers attempted to reach a higher performance algorithm using fusion

K. Honda et al. (Eds.): IUKM 2023, LNAI 14376, pp. 141–152, 2024.
https://doi.org/10.1007/978-3-031-46781-3_13

[7,9,10,18–20]. For example, the fusion of image multichannel using color iris images characterized by three spectral channels - red, green, and blue. Quality scores are employed to select two channels of a color iris image which are fused at the image level using a redundant discrete wavelet transform (RDWT). The fused image is then used in a score-level fusion framework along with the remaining channel [19].

Furthermore, there is work about a fusion approach to unconstrained iris verification which combines many methods such as comparison maps, domain analysis, and etc. [18]. Additionally, there is a study that focuses on the score-level fusion of the widely used distance measurement, Hamming distance, with a bit that has its value changed across iris code, fragile bit [10]. Moreover, the fusion of iris templates generated from multiple iris images of the same eye has also been explored [7]. These fusion strategies have demonstrated improved accuracy compared to using a single modality. However, the accuracy of these algorithms might be hindered by the difficulty of obtaining an optimal threshold. It is crucial to strike a balance between accepting genuine matches and rejecting false matches based on the calculated distance.

The iris verification system follows a four-step process: preprocessing, feature extraction, matching, and decision. During preprocessing, the iris image undergoes pupil region detection, iris localization, normalization, and removal of eyelash noise. Feature extraction is then performed using a Log-Gabor filter, which extracts distinctive features and encodes them into an iris template. In the matching, the template is compared to the template of the claimed person stored in the database, using matching algorithms. Finally, in the decision step, the comparison score obtained from the matching algorithm is used to determine if the input iris image belongs to the claimed identity.

In this research, we aim to overcome the limitation of traditional threshold-based approaches, which is the uncertainty in confidence level selecting a threshold, by exploring interval score-level fusion. A three-way decision-making approach can be used to deal with an uncertain decision between acceptance and rejection [21]. Aizi *et al.* has proposed his work on score-level fusion based on zones of interest and successfully improved the performance of biometric verification [1]. Nevertheless, the aspect of iris left alone for verification is still has not yet been explored. Having only an iris dataset, we aim to reduce accepting imposters or rejecting authorities.

In other words, it is to reduce false rejection rate (FRR) and false acceptance rate (FAR). Therefore, instead of relying solely on a single distance measurement algorithm, we combine the outputs of four different distance measurement algorithms within an interval score fusion framework. This approach allows us to consider a range of thresholds, thereby accounting for the inherent uncertainty in selecting a single threshold.

The rest of this paper is structured as follows. Section 2 presents the background of iris verification. Section 3 introduces the distance measurements employed in this study. Section 4 presents details of the proposed method. The experiment and result are made in Sect. 5. Discussion and conclusion can be found in Sect. 6 and Sect. 7, respectively.

2   Iris Verification

The iris verification system takes an iris image and a claimed identity and decides whether the iris image is of the claimed person [16]. In general, it consists of four steps, as shown in Fig. 1: preprocessing, feature extraction, matching algorithm, and decision.

The preprocessing step comprises four subprocesses [6], as illustrated in the top-right of Fig. 1. First, the Hough transform is employed to detect the pupil region in the input iris image. Second, an active contour method is applied to localize and detect the position of the iris. Third, the iris region is normalized by mapping its pixel values to a standardized representation. Last, noise caused by eyelashes in the normalized iris image is separated.

Feature extraction is performed on the preprocessed image using a Log-Gabor filter, a filter used in image processing to analyze local orientation characteristics of an image [8]. The image is convolved with the filter to extract distinctive features, which are then encoded into a binary dataset, called an iris template.

In the matching algorithm, the template obtained from the feature extraction step is compared to the template of claimed person, which is stored in the database during the enrollment, to compute similarity or dissimilarity between them, i.e., the matching algorithm returns a comparison score. There are many matching algorithms, e.g., Hamming distance [15], direct-grey scale surface [5], and phase-based image matching [14]. Each has a unique performance characteristic suitable for different datasets and purposes. Some well-known matching algorithms are introduced in Sect. 3.

Lastly, in the decision step, the comparison score is used to decide if the input iris image is of the claimed identity, i.e., if both templates belong to the same person.

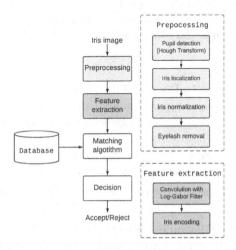

Fig. 1. Iris verification. Subprocesses of preprocessing and feature extraction are shown in the dashed-line boxes on the right.

3   Distance Measurement

In our iris verification method, we employ various distance measures or matching algorithms to quantify the similarity or dissimilarity between iris templates. Distance measurement plays a vital role in the decision-making process to accept or reject an iris sample. We utilize the following distance measures: Jaccard distance, Hamming distance, Tanimoto dissimilarity index, and weighted Euclidean distance. This section concisely explains them for the purpose of thoroughness.

Jaccard Distance (JD) quantifies the similarity between two binary templates by calculating the ratio of the intersection to the union of their respective feature sets [11]. A higher score indicates a higher dissimilarity between templates. The Jaccard distance (JD) of two boolean arrays of length n, i.e., the target template T and the query template Q, is defined by the following equation.

$$JD = \frac{c_{01} + c_{10}}{c_{01} + c_{10} + c_{11}}, \tag{1}$$

where c_{ij}, for i and j are in $\{0,1\}$, is the number of occurrences of $T[k] = i$ and $Q[k] = j$ for $k = 1$ to n.

Hamming Distance (HD) represents the dissimilarity between two binary templates by counting the number of differing bits. A lower HD signifies a higher degree of similarity between templates [3].

$$HD = \frac{\sum_{k=0}^{n}(T[k] \oplus Q[k]) \cap (T^*[k] \cap Q^*[k])}{n - \sum_{k=0}^{n}(T^*[k] \cup Q^*[k])}, \tag{2}$$

where $T^*[k]$ and $Q^*[k]$ are binary sequences for $k=1$ to n, in which 1 represents noise, and 0 means no noise at the index k in the target template and the query template, respectively. n is the total number of template bits, and $\oplus$ denotes an exclusive-or operation.

Tanimoto Dissimilarity Index (TDI) assesses the dissimilarity between two binary templates by comparing the number of common features to the total number of features [13]. A higher TDI indicates a higher dissimilarity between templates.

$$TDI = \frac{2(c_{10} + c_{01})}{c_{11} + c_{00} + 2(c_{10} + c_{01})}, \tag{3}$$

where c_{11}, c_{10}, c_{01}, and c_{00} are the same as those defined in JD.

Weighted Euclidean Distance (WED) calculates the dissimilarity between two templates by considering the weighted Euclidean distance between their feature vectors [12]. WED is given by the following equation. Note that, to compute WED, the query template Q of size n is firstly reshaped to an R-by-L

matrix M, where $R \cdot L = n$ and $M[r, l]$ (an entry at row r and column l) is $Q[k]$, such that $k = L \cdot (r-1) + l$, where k is the template index running from 1 to n, r is a row index of the matrix, running from 1 to R, and l is a column index of the matrix, running from 1 to L. Then, we compute each row's standard deviation, denoted by ς_r, representing the weight for WED. Let σ_k be the weight at k. Hence, $\sigma_{L \cdot (r-1) + l} = \varsigma_r$ for $r = 1$ to R and $l = 1$ to L. In this work, R and L are set to 64 and 800, which are the first and second dimensions of the template, respectively.

$$\text{WED} = \sqrt{\frac{\Sigma_{k \in I}(T[k] - Q[k])^2 \cdot \sigma_k^2}{|I|}}, \tag{4}$$

where $I = \{p | T^*[p] = 0 \text{ or } Q^*[p] = 0\}$.

4 Proposed Method

Given an iris biometric template extracted from the input iris image, the proposed method compares it against the stored templates in the database for deciding an individual's identity. Traditionally, in template comparison, various matching algorithms are used to measure the similarity or dissimilarity between the templates. The well-known and widely-used algorithms include the Hamming distance, Euclidean distance, Jaccard distance, etc., as described in Sect. 3. Each method measures the similarity between two vectors or two sets or sequences in different dimensions. In other words, each has different pros and cons. Since the proposed method aims to maximize the effectiveness of template comparison, it combines the advantages of various matching algorithms by fusing comparison results obtained from them.

The foundation of the proposed method is straightforward and can be illustrated as follows. Once a matching algorithm takes a pair of templates, it gives a score as the comparison result. Based on the score and a predefined threshold value (τ), the iris verification system decides whether to accept or reject the individual. Without losing generality, let us assume that the score covers a range from 0 to 1, where 0 represents a 100% match between two templates, and 1 is a 100% mismatch between them, as shown in Fig. 2. In general situation, making the decision around the threshold value is more complicated than doing it when the comparison score is far away from the threshold. In other words, the uncertainty near the threshold is higher than that far from it. Put differently, false-positive results decrease if the threshold value is shifted toward the 0-end, and false-negative results drop if the threshold value is moved toward the 1-end.

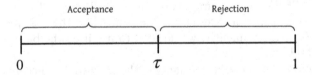

Fig. 2. Score scale with a threshold value (τ).

Therefore, instead of using only one threshold value to divide the score scale into two parts, i.e., match (acceptance) and mismatch (rejection), the performance of template comparison can be improved by reducing false positives and false negatives through dividing the score scale into three parts, i.e., match, mismatch, and uncertainty, as shown in Fig. 3. Let τ_1 and τ_2 be two threshold values that define the boundaries between divided parts on the score scale, where τ_1 and τ_2 are real numbers in the interval $[0, 1]$ and $\tau_1 \leq \tau_2$. Note that when $\tau_1 = \tau_2$, they are set to the same threshold value (τ) in a typical template comparison method. According to this proposed scheme, as illustrated in Fig. 3, when the comparison score from a metric, called a base metric in this work, is lower than τ_1, the iris verification system decides to accept; when the score is higher than τ_2, the individual is rejected. When the score is in the interval $[\tau_1, \tau_2]$, a proposed score fusion algorithm is to take action in making a decision. It means that deciding on a score between τ_1 and τ_2 pushes the template comparison method based on a single matching algorithm into a high uncertainty situation. Hence, combining many matching algorithms may reduce such tension. Details of the proposed method are elaborated in the following subsections.

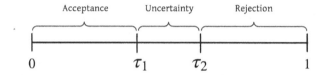

Fig. 3. Score scale with threshold values τ_1 and τ_2.

4.1 Preliminary Investigation on Matching Algorithms

Four matching algorithms are chosen for preliminary investigation because of their popularity and celebrity, which include Hamming distance, Tanimoto dissimilarity index, weighted Euclidean distance, and Jaccard distance. The dataset used for this initial experiment is taken from CASIA-IrisV2 [2]. It consists of 2380 pairs of templates to be matched, in which 1200 pairs are truly matched and 1180 are truly mismatched. Those pairs are divided into a training set and a test set with a ratio of 4:1.

Given a few different threshold values around a point at which the false acceptance rate (FAR) considerably equals the false rejection rate (FRR) for each algorithm, the numbers of true positives, true negatives, false positives, and false negatives, together with FAR, FRR, true acceptance rate (TAR), true rejection rate (TRR), and accuracy, are shown in Table 1. It can be seen that the Jaccard distance with a threshold of 0.651 provides the best performance in terms of accuracy. Consequently, we decided to use it as the base metric of our proposed method.

Then, we investigated the performance of the Jaccard distance further by moving the threshold toward 0 and 1 until the numbers of false positives and false negatives were considerably low, respectively. The result is shown in Table 2. Table 2

Table 1. Comparison of TAR, TRR, FAR, FRR, and accuracy among matching algorithms with different threshold values (τ) around a point at which FAR considerably equals FRR.

τ	Method	TAR	TRR	FAR	FRR	Accuracy
0.451	HD	68.20	100	0	31.80	84.10
0.631	JD	75.31	99.56	0.43	24.68	87.44
0.215	WED	49.79	100	0	50.21	74.89
0.622	TDI	68.20	100	0	31.79	84.10
0.461	HD	76.15	99.56	0.43	23.85	87.86
0.641	JD	84.10	97.39	2.61	15.89	90.74
0.225	WED	68.20	100	0	31.79	81.10
0.632	TDI	76.56	99.56	0.44	23.43	88.06
0.471	HD	87.03	90.28	9.72	12.97	88.69
0.651	JD	92.47	91.09	8.91	7.53	91.77
0.235	WED	86.61	91.09	8.91	13.39	88.89
0.642	TDI	89.12	89.07	10.03	10.88	89.09
0.481	HD	93.72	63.04	36.96	6.28	78.38
0.661	JD	97.91	79.13	20.87	2.09	88.52
0.245	WED	97.90	23.04	76.96	2.09	60.48
0.652	TDI	99.58	14.60	85.40	0.42	57.09
0.491	HD	97.91	20.00	80.00	2.09	58.95
0.671	JD	99.58	67.39	32.60	0.42	83.49
0.255	WED	100	0	100	0	50.00
0.662	TDI	100	0	100	0	50.00

shows the numbers of true positives (TP), true negatives (TN), false positives (FP), and false negatives (FN), together with accuracy, in three regions (i.e., acceptance, uncertainty, and rejection) partitioned by the interval $[\tau_1, \tau_2]$, as indicated in the top row. Therefore, we can define the interval $[\tau_1, \tau_2]$ and choose the base metric from this investigation. Specifically, $[\tau_1, \tau_2]$ is $[0.631, 0.671]$. That is, when the Jaccard distance is lower than 0.631, the iris verification system accepts the individual; when the coefficient is greater than 0.671, the individual is rejected; otherwise, the score fusion algorithm described below handles the case.

4.2 Score Fusion Based on Classification Tree

Our proposed score fusion method uses a decision tree that takes a score vector as the input. The score vector consists of five entries and is denoted by $[\text{HD JD WED TDI } v]^{\text{T}}$, where HD is a Hamming distance, JD is a Jaccard

148   O. Kotsuwan et al.

distance, TDI is Tanimoto dissimilarity index, WED is a weighted Euclidean distance, and $v \in \{0,1\}$, where 0 is rejection and 1 is acceptance, is a result from a modified majority vote. The modified majority vote works as follows. It agrees with the majority of the four metrics. However, to break a tie in voting, we always choose the positive (or acceptance). Because as shown in Table 1, three out of four metrics are better in positive than negative prediction. Therefore, the positive is statistically preferable when two votes for the positive against the other two for the negative.

The proposed framework is sketched in Fig. 4. Accordingly, the flowchart of the proposed framework can be summarized in Fig. 5.

Table 2. Comparison of TP, TN, FP, FN, and accuracy (%) when the score scale is partitioned by different intervals.

	[0.611,0.691]			[0.621,0.681]			[0.631,0.671]			[0,1]		
	Acceptance	Uncertainty	Rejection	Acceptance	Uncertainty	Rejection	Acceptance	Uncertainty	Rejection	Acceptance	Uncertainty	Rejection
TP	137	79	0	159	58	0	180	40	0	0	218	0
TN	0	176	58	0	111	124	0	75	161	0	236	0
FP	4	9	0	5	7	0	5	6	0	0	11	0
FN	0	23	0	0	23	0	0	18	1	0	21	0
Accuracy	97.16	88.85	100	96.95	85.35	100	97.29	82.73	99.38	-	93.41	-
	92.59			93.00			93.82			93.41		

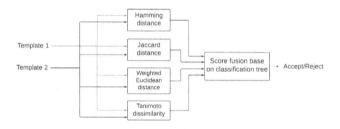

Fig. 4. Proposed framework.

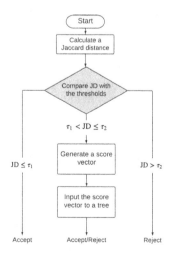

Fig. 5. Flowchart of the proposed method.

5   Experiment and Result

To evaluate the performance of the proposed method, we conducted experiments on the same dataset described in Sect. 4.1. Also, JD is chosen as the base metric, based on the preliminary investigation of the training dataset mentioned above, and the decision interval $[\tau_1, \tau_2]$ is $[0.631, 0.671]$. Therefore, the template pairs in the training set, with JD falling in the interval $[0.631, 0.671]$, were used to compute HD, WED, TDI, and the voting results and form score vectors for training a classification tree. The decision tree classifier from scikit-learn was used in this experiment [17]. We compared the best performance obtained from every traditional method with the proposed method. The evaluation results of the training set and the test set are shown in Table 3. It can be seen that the performance of the proposed method is improved from those of the conventional way in almost all aspects, except that the FRR of the proposed method is negligibly slightly higher than that of JD, and its TAR is negligibly marginally lower than that of JD. Compared with JD, which is the best for a single method, the proposed method improves its accuracy, precision, and F1 by 2.20%, 4.50%, and 5.66%, respectively.

Table 3. Performance comparison among the traditional methods and our proposed method.

	Training set					Test set				
	Proposed	HD	JD	WED	TDI	Proposed	HD	JD	WED	TDI
TAR	97.38	83.99	90.38	83.26	85.59	92.05	87.03	92.47	86.61	89.12
TRR	98.82	93.65	93.98	94.62	92.04	95.55	90.28	91.09	91.09	89.07
FAR	1.17	6.34	6.01	5.37	7.95	4.45	9.72	8.91	8.91	10.03
FRR	2.61	16.00	9.61	16.75	14.40	7.95	12.97	7.53	13.39	10.88
Accuracy	98.09	88.79	92.15	88.92	88.78	93.83	88.69	91.77	88.89	89.09
Precision	98.83	93.07	93.91	93.98	91.67	95.24	89.65	90.95	90.39	88.75
F1-score	97.38	88.30	92.11	88.29	88.52	93.62	91.70	88.32	88.46	88.93

6   Discussion

There are two points to discuss regarding the performance and improvement of the proposed method in this section. First, since the basic idea of the proposed method is to reduce the uncertainty when making a decision around the threshold value of any base metric by considering comparison scores of the other distance measurement, it is logical to say that its performance depends upon the choice of the base metric. This work selects the base metric from four well-known distance measures. Therefore, we do not claim that JD is the best one. It is worth expanding the investigation to more metrics, which will be done in the future.

Second, from the proposed framework and Table 2, the performance is also affected by choice of the decision interval $[\tau_1, \tau_2]$. The numbers of false positives and false negatives in the acceptance and rejection regions decrease as the interval $[\tau_1, \tau_2]$ increases, i.e., the uncertainty region increases. When the interval $[\tau_1, \tau_2]$ is extended to the full scale $[0, 1]$, it means the classification tree is applied to all data without using the base metric. Interestingly, based on our simulation, using only the classification tree does not maximize the accuracy or minimize the number of false positives and false negatives. Therefore, finding the optimal decision interval is another issue to study further. The interval $[\tau_1, \tau_2]$ used in our work was from the trend we saw in the preliminary investigation.

7   Conclusion

In this study, we proposed a method combining comparison scores from various matching algorithms using a classification tree. It aims to improve the performance of the iris verification system by minimizing FRR and FAR around the threshold. Instead of following the conventional way that uses a single threshold value to divide the comparison score space into two regions: acceptance and rejection, we proposed two thresholds partitioning the score space into three areas: acceptance, rejection, and uncertainty, where the uncertainty area is an area around the traditional threshold. Then, the classification tree is deployed to decide if a claimed identity should be accepted or rejected. The matching algorithms utilized in this study include Hamming distance, Jaccard distance, Tamimoto dissimilarity index, and weighted Euclidean distance. We evaluated the proposed method using the CASIA-IrisV2 dataset. The experimental results show that the accuracy, precision, and F1 score of the proposed method are increased by 2.20%, 4.50%, and 5.66%, respectively, compared with the Jaccard distance, which is the best among the four measures studied. However, it should be noted that the performance of the proposed method depends upon the base metric and two thresholds that define the partitioning boundaries. We have yet to optimize those boundaries and the base metric in this study. Therefore, the performance is likely to improve further once optimized.

Acknowledgements. We would like to extend our appreciation to Mr. Krit Anegsiripong for his valuable advising about the iris recognition framework. His assistance and guidance have been instrumental in the success of this project, and we are truly grateful for his support.

References

1. Aizi, K., Ouslim, M.: Score level fusion in multi-biometric identification based on zones of interest. J. King Saud Univ. Comput. Inf. Sci. **34**(1), 1498–1509 (2022). https://doi.org/10.1016/j.jksuci.2019.09.003. https://www.sciencedirect.com/science/article/pii/S1319157819303696
2. National Laboratory of Pattern Recognition Institute of Automation: CASIA v2 database (2002). https://biometrics.idealtest.org

3. Bookstein, A., Kulyukin, V.A., Raita, T.: Generalized hamming distance. Inf. Retrieval **5**, 353–375 (2002)
4. Castellà, L., Carlos, J.: Human iris biometry (2016)
5. Chen, X., Wu, C., Xiong, L., Yang, F.: The optimal matching algorithm for multi-scale iris recognition. Energy Procedia **16**, 876–882 (2012). https://doi.org/10.1016/j.egypro.2012.01.140. https://www.sciencedirect.com/science/article/pii/S1876610212001506. 2012 International Conference on Future Energy, Environment, and Materials
6. Daugman, J.: New methods in iris recognition. IEEE Trans. Syst. Man Cybern. Part B (Cybern.) **37**(5), 1167–1175 (2007). https://doi.org/10.1109/TSMCB.2007.903540
7. Desoky, A.I., Ali, H.A., Abdel-Hamid, N.B.: Enhancing iris recognition system performance using templates fusion. Ain Shams Eng. J. **3**(2), 133–140 (2012). https://doi.org/10.1016/j.asej.2011.06.003. https://www.sciencedirect.com/science/article/pii/S2090447911000177
8. Du, Y.: Using 2D Log-Gabor spatial filters for iris recognition. In: Flynn, P.J., Pankanti, S. (eds.) Biometric Technology for Human Identification III, vol. 6202, p. 62020F. International Society for Optics and Photonics, SPIE (2006). https://doi.org/10.1117/12.663834
9. Gad, R., Abd El-Latif, A.A., Elseuofi, S., Ibrahim, H.M., Elmezain, M., Said, W.: IoT security based on iris verification using multi-algorithm feature level fusion scheme. In: 2019 2nd International Conference on Computer Applications & Information Security (ICCAIS), pp. 1–6 (2019). https://doi.org/10.1109/CAIS.2019.8769483
10. Hollingsworth, K.P., Bowyer, K.W., Flynn, P.J.: Improved iris recognition through fusion of hamming distance and fragile bit distance. IEEE Trans. Pattern Anal. Mach. Intell. **33**(12), 2465–2476 (2011). https://doi.org/10.1109/TPAMI.2011.89
11. Kosub, S.: A note on the triangle inequality for the jaccard distance. Pattern Recogn. Lett. **120**, 36–38 (2019). https://doi.org/10.1016/j.patrec.2018.12.007. https://www.sciencedirect.com/science/article/pii/S0167865518309188
12. Lee, Y., Filliben, J.J., Micheals, R.J., Phillips, P.J.: Sensitivity analysis for biometric systems: a methodology based on orthogonal experiment designs. Comput. Vis. Image Underst. **117**(5), 532–550 (2013)
13. Lipkus, A.H.: A proof of the triangle inequality for the tanimoto distance. J. Math. Chem. **26**(1–3), 263–265 (1999)
14. Miyazawa, K., Ito, K., Aoki, T., Kobayashi, K., Nakajima, H.: An effective approach for iris recognition using phase-based image matching. IEEE Trans. Pattern Anal. Mach. Intell. **30**(10), 1741–1756 (2008). https://doi.org/10.1109/TPAMI.2007.70833
15. Murugan, A., Savithiri, G.: Feature extraction on half iris for personal identification. In: Proceedings of the 2010 International Conference on Signal and Image Processing, ICSIP 2010 (2010). https://doi.org/10.1109/ICSIP.2010.5697468
16. Omidiora, E., Adegoke, B., Falohun, S., Ojo, D.: Iris recognition systems: technical overview. IMPACT Int. J. Res. Eng. Technol. **3**, 63–72 (2015)
17. Pedregosa, F., et al.: Scikit-learn: machine learning in Python. J. Mach. Learn. Res. **12**, 2825–2830 (2011)
18. Santos, G., Hoyle, E.: A fusion approach to unconstrained iris recognition. Pattern Recognit. Lett. **33**(8), 984–990 (2012). https://doi.org/10.1016/j.patrec.2011.08.017. https://www.sciencedirect.com/science/article/pii/S0167865511002686. Noisy Iris Challenge Evaluation II - Recognition of Visible Wavelength Iris Images Captured At-a-distance and On-the-move

19. Vatsa, M., Singh, R., Ross, A., Noore, A.: Quality-based fusion for multichannel iris recognition, pp. 1314–1317 (2010). https://doi.org/10.1109/ICPR.2010.327
20. Wild, P., Hofbauer, H., Ferryman, J., Uhl, A.: Segmentation-level fusion for iris recognition. In: 2015 International Conference of the Biometrics Special Interest Group (BIOSIG), pp. 1–6 (2015). https://doi.org/10.1109/BIOSIG.2015.7314620
21. Yao, Y.: An outline of a theory of three-way decisions, vol. 7413 (2012). https://doi.org/10.1007/978-3-642-32115-3_1

Controlling LIME Kernel Width to Achieve Comprehensible Explanations on Tabular Data

Hai Duong[1,2], Lam Hoang[1,2], and Bac Le[1,2(✉)]

[1] Faculty of Information Technology, University of Science, Ho Chi Minh City, Vietnam
lhbac@fit.hcmus.edu.vn
[2] Vietnam National University, Ho Chi Minh City, Vietnam

Abstract. LIME [9] is an Explainable AI (XAI) method that can offer local explanation for any Machine Learning model prediction. However, the design of LIME often leads to controversial problems in the explanation, which are mostly due to the randomness of LIME's neighborhood generating process. In this paper, we contribute a method that can help LIME deliver comprehensible explanation by optimizing feature attribution and kernel width of the generating process. Our method ensures high level of features attribution while keeping kernel width lower than the default setting to remain high locality in the explanation. The study will focus mainly on LIME for tabular data.

Keywords: LIME · XAI · tabular data · locality

1 Introduction

A very challenging aspects of deploying machine learning models in real world settings is understanding how these models make their predictions. Despite many state-of-the-art models on different tasks, they often lack interpretability, making it difficult for practitioners to know why certain decisions were made or which features contributed to a particular outcome. To address this challenge, recent work has focused on developing techniques that allow us to explain and visualize the decision processes of machine learning models.

Among most popular explanation technique, Local Interpretable Model-Agnostic Explanations (LIME [9]) can produce local explanation which highlights the local feature attribution. Feature attribution is a mean to quantify how important the set of features used is to the prediction of the black-box model. By means of feature importance, users can easily understand what aspects the model is mostly focusing on, then decide whether to trust the prediction or not.

LIME achieve locality by defining a new neighborhood of the instance we want to explain, then weights each new data point based on the kernel width it was provided with and labels it with the black-box model. Lastly, LIME trains

K. Honda et al. (Eds.): IUKM 2023, LNAI 14376, pp. 153–164, 2024.
https://doi.org/10.1007/978-3-031-46781-3_14

a new model with much lower complexity on the generated neighborhood and define feature importance by some feature selection methods and the coefficient of each feature.

Such a design ensures that our explanation are faithful enough to the original model. However, many problems lie in the fact that the generating process are random and the definition of neighborhood is hard to achieved in most cases. Specifically, LIME is reported to "deliver degenerated explanation when kernel width is set too small" [3] due to the lack of control in kernel width in the generating process. The authors in the [3] proposed an approach to treat every sample point equally according to the kernel width so that LIME can adapt better to the new neighborhood. This approach requires modifying LIME's structure and therefore is hard to be applied on general usage.

In this paper, we will first introduce the concept of LIME and some extensions of it as related works in Sect. 2. In Sect. 3, we will study on how choosing different kernel width can lead to either comprehensible or incomprehensible explanations, then propose a method to select an appropriate kernel width so that the explanation is meaningful and concordant to the original model. Finally, we will test the method on different datasets and discuss the results in Sect. 4, then summarize the research in Sect. 5. The study also focus on the use of LIME with tabular data.

2   Related Works

2.1   LIME Framework

LIME was first introduced in 2016, providing the ability to explain a prediction of a black-box model in term of its local neighborhood. To actualize its purpose, LIME is first fed in the training data to learn the statistical distribution of each used features, then uses it to generate an artificial neighborhood of the instance we want to explain. The new neighborhood will be labelled by the black-box model we are examining and consequently, we end up with a new training set in the black-box ML surface. LIME will perform some feature selection methods before training a simple white-box model on that new training set with the selected features. The explanation will finally be presented in the form of feature attribution, which is the coefficient of each feature in the whitebox model.

2.2   LIME Extensions

The fact that each part in the LIME's design are separate encourages many extensions to improve the quality of LIME's explanation.

To begin with, some studies focus on how LIME's explanation might be unstable. In fact, two explanations with the same input might indicate completely different feature attribution. To tackle the problem, DLIME [15] avoids the sampling step and instead choose the neighborhood from the training dataset by using Hierarchical Clustering [8] and K-Nearest Neighbour [2] to find data points that are closest to that we want to explain. The authors of ALIME [10] compare the standard deviations of the Ridge [5] coefficients and use a denoising auto-encoder [11] to define a more accurate neighborhood. OptiLIME [12] focused the inverse impact of kernel width on LIME's stability and use Bayes Optimization to find the optimal kernel width that can achieve stability.

Other groups of researchers find different ways to measure the quality of LIME's predictions. Firstly, CSI and VSI [13] are proposed to measure the stability of feature attribution by. VSI (Variables Stability Index) checks if the selected features in different LIME calls are the same, and CSI (Coefficients Stability Index) examines the similarity of the coefficient of each features. In addition, LEAF [1] framework offers us a mean to evaluate stability, local concordance, fidelity and prescriptivity of local explanation from LIME and SHAP [7] on binary classification model. The authors also study how those metrics are different when we choose different number of features to explain.

Finally, s-LIME [3]'s authors indicate that low kernel width can result in null in feature attributions, and solve it by considering kernel width in the sampling step.

2.3 Our Insights

As carefully reviewed, we comprehend that previously proposed methods had either enriched the components inside LIME (different sampling method, hyper-parameters lending between components...) or tested the impact of one setting on the performance of the algorithm. However, methods in the first group requires modifying LIME's structure and sometimes hard to reproduce the results. On the other side, we consider the second group to be more faithful to how LIME was originally designed, therefore can be applied on general usage. We want to tackle the problem that s-LIME in the first group had indicated, with the method of the second group: find the optimal kernel width that can solve the incomprehensible explanation.

3 Proposed Method

3.1 The Impact of Choosing Different Kernel Width on LIME's Explanation

LIME Optimization Problem. With $x \in \mathbb{R}^d$, the explanation of LIME is obtained by:

$$\xi(x) = \operatorname*{argmin}_{g \in G} \mathcal{L}(f, g, \pi_x) + \Omega(g) \tag{1}$$

Our black-box model is defined as a function $f : \mathbb{R}^d \to \mathbb{R}$, and $g \in G$ where G is the class of potentially interpretable models. $\Omega(g)$ measures the complexity of model g. We perturb x to achieve samples $z' \in \{0,1\}^d$. Then, we recover z' in the space of f to obtain $z \in \mathbb{R}^d$ and $f(z)$ as labels for the new data. The weighting

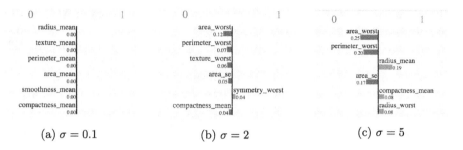

(a) $\sigma = 0.1$ (b) $\sigma = 2$ (c) $\sigma = 5$

Fig. 1. LIME explanations with different kernel widths on the same instance of Breast Cancer [14] dataset ($k = 6$)

kernel $\pi_x(z) = \exp(-D(x, z)^2/\sigma^2)$ with some distance function D and kernel width σ will then be applied on the perturbed sample. The weighted square loss will then be defined as:

$$\mathcal{L}(f, g, \pi_x) = \sum \pi_x(z)(f(z) - g(z'))^2 \tag{2}$$

Kernel width σ controls the locality of the explanation so that smaller values give more weight to the instances that lie close to x, **hence increase locality**.

Incomprehensible Explanation with Small Kernel Width σ. We tested the impact of different σ on LIME's explanation of a Logistic Regression model trained on Breast Cancer [14] dataset. The result was similar to that of s-LIME where no feature attribution could be found on very low σ. Increasing σ, however, might raise the coefficient of each feature significantly. Figure 1 shows different results on $\sigma = 0.1$, 2 and 5 with increasing feature attribution on Breast Cancer [14] dataset.

s-LIME [3] also addressed that the original design set σ to

$$0.75 \cdot \sqrt{\textbf{number of features}}$$

with no further explanation. Random as it may seem, we think LIME's authors tested the method and resorted to the most stable default setting that LIME can output meaningful explanation regardless dimensionality of the dataset. To examine our anticipation, we quantify feature attribution of LIME's explanation by summing all coefficient of each feature used by $g(z')$. With $g(z') = \beta + \sum_{i \in d} w_i \cdot z_i$, feature attribution of one explanation can be calculated as

$$\mathbb{F}(\xi(x)|\sigma) = \sum_{i \in d} |w_i| \tag{3}$$

We tested the variation of $\mathbb{F}$ when σ changes on 3 datasets: Heart Disease [6], Breast Cancer [14] and Arrhythmia [4] with 13, 30 and 297 features, respectively. The result in Fig. 2 shows that LIME's default setting had picked a kernel width that can achieve high feature attribution in all 3 cases. However, this does not

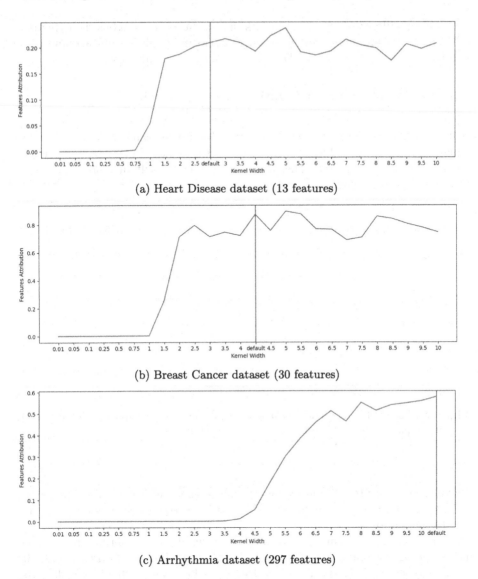

(a) Heart Disease dataset (13 features)

(b) Breast Cancer dataset (30 features)

(c) Arrhythmia dataset (297 features)

Fig. 2. Variation of $\mathbb{F}$ on different σ value for different datasets, under same LIME setting

ensure that positive result can occur in all circumstances, and sometimes the picked kernel width can be smaller but still can achieve the same result.

3.2   What Is an Optimal Kernel Width

Default LIME's σ might offer good feature attribution, but it does not guarantee that σ is low enough to remain its locality. As can be observed in Fig. 2, feature attribution will increase from null to its significant level and stops rising further. Therefore, the minimum kernel width that ensures high feature attribution $\mathbb{F}$ can be considered the optimal kernel width σ value in our problem.

3.3   Methodology to Find Optimal Kernel Width

Although our the target value is $\mathbb{F}$, we propose $\lambda = \mathbb{F}/\sqrt{\sigma}$ to be our optimizing metric. Firstly, λ can be higher for lower σ despite the same $\mathbb{F}$ value. Secondly, square rooting σ can reduce the importance of σ in case $\mathbb{F}$ increases very slowly. Figure 3 shows that picking the highest $\mathbb{F}/\sqrt{\sigma}$ will result in more desirable $\mathbb{F}$ than picking the highest $\mathbb{F}/\sigma$.

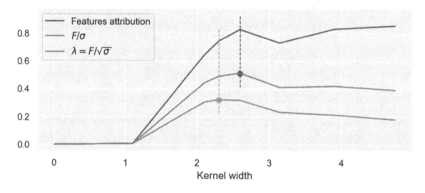

Fig. 3. Variation of $\mathbb{F}$, $\mathbb{F}/\sigma$ and $\mathbb{F}/\sqrt{\sigma}$ when σ increases, with impact on feature attribution compared through highlighted peaks of each metric

Kernel Width Step. Finding the optimal kernel width σ requires increasing σ by some amount after each step. The relative kernel width range might vary greatly among different dataset, so setting a fixed σ step will not be ideal and sometimes is time-consuming if the step is too small and the optimal point is too large. We do not only want the step to be dynamic, but also correlate with the varying speed of $\mathbb{F}$. To quickly reach the desired σ point without exceeding too far away from it, the design is to have big step when $\mathbb{F}$ changes insignificantly and small step when $\mathbb{F}$ varies greatly from the previous one. Using slope between current and previous step is an ideal way to measure how much $\mathbb{F}$ is varying from σ.

$$slope = \frac{\mathbb{F}_t - \mathbb{F}_{t-1}}{\sigma_t - \sigma_{t-1}} \tag{4}$$

We want the step size to have inverse relationship with the size of the slope. Since the slope can be negative, using its exponent will not only make it non-negative, but also amplify the significance of the slope. The increase amount can now be $step \cdot \dfrac{1}{slope^2}$. Note that slope can be 0, so we can choose to add the amount of initial step size to the denominator to avoid division by 0. Using $step \cdot \dfrac{1}{slope^2 + step}$ will limit our step to the maximum of 1 when the slope diverges to 0, making it scalable without exceeding too far from the optimal range. Finally, we do not want the step to make no difference at all when our slope becomes too large, we decide to set the minimum amount that σ must increase to $step/2$. The new σ can be calculated as

$$\sigma_t = \sigma_{t-1} + \max\left(\frac{step}{slope^2 + step}, step/2\right) \tag{5}$$

Figure 4 shows that we achieve dynamic σ step with initial step size $= 0.1$.

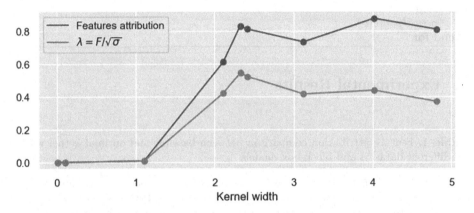

Fig. 4. Dynamic kernel width survey steps based on how fast feature attribution are varying compared to previous step, highlighted by dots

Early Stopping. We attempt to stop surging kernel width at the point where feature attribution no longer increases significantly. As shown in Fig. 2, $\mathbb{F}$ will rise sharply to a certain level then only fluctuate afterwards. Therefore, the design of the method will add in a patience n hyper-parameter so that we can add a break condition for that recent n steps. According to Fig. 4, λ refuse to rise further despite gaining a bit of improvement in $\mathbb{F}$. After this observation, we decide to break the test based on how many times λ decreases since that last n patience steps. Specifically, the number of decreases in λ should occupy no less than 50% of last n patience steps. All the steps can be demonstrated as Algorithm 1.

Algorithm 1. Early stopping λ

Require: Instance x to be explained, patience n, number of iterations num_iter
  $some\_bucket \leftarrow array$
  $\sigma \leftarrow 0.1$
  for $i \in$ range from 1 to num_iter **do**
    $\lambda \leftarrow \mathbb{F}(\xi(x)|\sigma)/\sqrt{\sigma}$
    $some\_bucket.$append(λ)
    $decrease\_cnt \leftarrow 0$
    for $j \in$ last n elements of $some\_bucket$ **do**
      if $some\_bucket[j]$ is decrease **then**
        $decrease\_cnt \leftarrow decrease\_cnt + 1$
      end if
    end for
    if $decrease\_cnt \geq n/2$ **then**
      Return σ with highest λ in $some\_bucket$
    end if
    $\sigma \leftarrow Update(\sigma)$
  end for

---

4   Experimental Results

Table 1. Feature attribution comparison between baseline and optimal kernel width on different datasets and black-box models

Dataset	Feat	Model	Kernel width σ		feature attribution $\mathbb{F}$		Runtime[1]
			Chosen	Default	Chosen	Default	
BreastCancer	30	LogReg	**2.64**	4.11	**0.81 ± 0.05**	0.79 ± 0.07	3.51
		RanFor	**2.11**	4.11	**0.24 ± 0.02**	0.23 ± 0.02	3.00
Arrhythmia	297	LogReg	**6.29**	12.94	0.48 ± 0.03	**0.51 ± 0.04**	29.1
		RanFor	**5.98**	12.94	0.09 ± 0.01	**0.11 ± 0.01**	17.50
HousePrice	61	LinReg	**2.02**	5.86	170k ± 18k	**276k ± 16k**	3.12
		KNN	**3.67**	5.86	18k ± 1.5k	**29k ± 1.9k**	4.1

[1]in seconds

We carried out the test on Logistic Regression with Breast Cancer and Arrhythmia datasets with the ratio of 7 for training set and 3 for test set (random state = 42). We fit training set on LIME using $num\_samples = 100$, $num\_features = 6$ to explain instance 25th of each test set. The initial step size of our algorithm is set to 0.1 and patience $n = 5$.

On Breast Cancer dataset, we managed to pick the optimal kernel width through the algorithm, as shown in Fig. 5a. The algorithm captured a good level of feature attribution $\mathbb{F}$ at a very low kernel width σ, whereas the default setting could also find an ideal $\mathbb{F}$ value but with twice as much σ, resulting in very low locality. The result held true to Arrhythmia dataset as well (Fig. 5b). It can be observed that the optimal kernel width σ will change among different datasets,

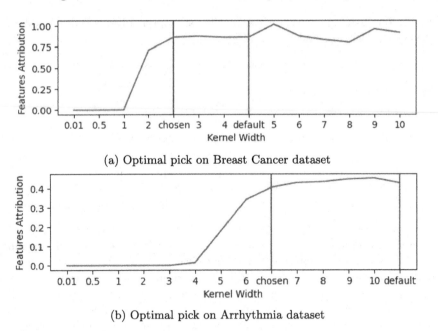

(a) Optimal pick on Breast Cancer dataset

(b) Optimal pick on Arrhythmia dataset

Fig. 5. Result of the kernel width picked by our algorithm vs the default setting of LIME

as well as the level of feature attribution. Our algorithm successfully captured ideal kernel width despite difference in dimensionality, proven to be able to apply on different tabular datasets.

Runtime is another aspect to be considered. On an Apple M1 8 cores CPU laptop. The process takes 3.51 s for Breast Cancer dataset and 29.1 s for Arrhythmia dataset. We acknowledge that the computation time can be different based on the complexity of the black-box model and the dimensionality of the dataset.

To further prove the capability of the algorithm, we proceeded to experiment on different black-box methods, both classification and regression. The datasets used were Breast Cancer [14], Arrhythmia [4] and House Price[1]. For classification, we used Logistic Regression (LogReg) and Random Forest (RanFor); for regression, Linear Regression (LinReg) and K-Nearest Neighbors (KNN) were taken into account. All settings were the same to the previous experiment. The result from Table 1 indicates that for classification tasks, we managed to remain locality by reducing σ compares to baseline, without significant difference feature attribution $\mathbb{F}$. Our successful metric λ won on both tasks, as followed in Table 2. Regression task, on the other hand, showed a large gap in the feature attribution achieved by our method comparing to the default setting.

---

[1] House Price is A Kaggle competition dataset. Despite giving good example for regression task, it may not be reliable for further scientific research.

Table 2. Successful metric λ between baseline and optimal kernel width on different datasets and black-box models

Dataset	Feat	Model	$\lambda = \mathbb{F}/\sqrt{\sigma}$	
			Chosen	Default
BreastCancer	30	LogReg	**0.5**	0.39
		RanFor	**0.17**	0.11
Arrhythmia	297	LogReg	**0.19**	0.14
		RanFor	**0.04**	0.03
HousePrice	61	LinReg	**120k**	114k
		KNN	9k	**12k**

We can observe the level of feature attribution in regression task are much higher (over hundred thousand) than classification task. The reason is that for classification, the output of a model is a continuous probability ranging from 0 to 1, whereas the output on regression task is not limited to this constraint. The method had difficulty in finding a more desirable set of $\sigma, \mathbb{F}$ and λ because Eq. 5 did not take the distribution of $\mathbb{F}$ into account, consequently poorly define the increasing speed when $\mathbb{F}$ range is too large. However, doing that requires more steps and computational time, while feature attribution achieved is already high enough for the explanation to be comprehensible, which is our main goal of the problem.

Overall, the method succeeded in achieving an optimal kernel width σ that offers ideal feature attribution $\mathbb{F}$ in classification tasks, having successful metric $\lambda = \mathbb{F}/\sqrt{\sigma}$ winning the baseline on LIME's default setting for σ. Despite the successful metric could not be proven to beat baseline of regression task, $\mathbb{F}$ was high enough to make the explanation comprehensible.

5 Conclusion

In this paper we have introduced a method to control the kernel width of LIME's neighborhood generating process to achieve good level of feature attribution, which is the sum of coefficients of every feature used by LIME's white-box model. The method is mainly based on defining dynamic step size for kernel width and early stopping for the relationship between feature attribution and kernel width. By doing this, we managed to solve the problem mentioned in s-LIME [3] that LIME can generate poor explanation that has null feature attribution. Our proposed method did not modify the original LIME's implementation, making it applicable to general usage. The experiments on tabular datasets with different number of features indicate that the method can guarantee high feature attribution while keeping kernel width low enough to remain locality, despite different dimensionality of the datasets. In the end, our main purpose to achieve comprehensible explanation was fulfilled.

The study is limited to the relationship between locality (represent by kernel width) and feature attributions of LIME. In future work, we would like to investigate more on the impact of kernel width on different metrics like stability and fidelity. Moreover, the method should be studied further to adapt to different feature attribution distribution of different dataset. In the future, we want to develop this method so that it can be applied on image and text datasets.

Acknowledgement. This research was funded by Vingroup Innovation Foundation (VINIF) under project code VINIF.2021.JM01.

References

1. Amparore, E., Perotti, A., Bajardi, P.: To trust or not to trust an explanation: using LEAF to evaluate local linear XAI methods. PeerJ Comput. Sci. **7**(e479), e479 (2021)
2. Cover, T., Hart, P.: Nearest neighbor pattern classification. IEEE Trans. Inf. Theor. **13**(1), 21–27 (2006). https://doi.org/10.1109/TIT.1967.1053964
3. Gaudel, R., Galárraga, L., Delaunay, J., Rozé, L., Bhargava, V.: s-LIME: reconciling locality and fidelity in linear explanations. In: Bouadi, T., Fromont, E., Hüllermeier, E. (eds.) IDA 2022. LNCS, vol. 13205, pp. 102–114. Springer, Cham (2022). https://doi.org/10.1007/978-3-031-01333-1_9
4. Guvenir, H., Acar, B., Muderrisoglu, H., Quinlan, R.: Arrhythmia. UCI Machine Learning Repository (1998). https://doi.org/10.24432/C5BS32
5. Hoerl, A.E., Kennard, R.W.: Ridge regression: biased estimation for nonorthogonal problems. Technometrics **12**(1), 55–67 (1970). https://doi.org/10.1080/00401706.1970.10488634
6. Andras, J., William, S., Pfisterer, M., Detrano, R.: Heart Disease. UCI Machine Learning Repository (1988). https://doi.org/10.24432/C52P4X
7. Lundberg, S.M., Lee, S.I.: A unified approach to interpreting model predictions. In: Proceedings of the 31st International Conference on Neural Information Processing Systems, NIPS 2017, pp. 4768–4777. Curran Associates Inc., Red Hook (2017)
8. Murtagh, F.: Hierarchical clustering. In: Lovric, M. (ed.) International Encyclopedia of Statistical Science, pp. 633–635. Springer, Heidelberg (2011). https://doi.org/10.1007/978-3-642-04898-2_288
9. Ribeiro, M.T., Singh, S., Guestrin, C.: "Why should i trust you?": explaining the predictions of any classifier. In: Proceedings of the 22nd ACM SIGKDD International Conference on Knowledge Discovery and Data Mining, KDD 2016, pp. 1135–1144. Association for Computing Machinery, New York (2016). https://doi.org/10.1145/2939672.2939778
10. Shankaranarayana, S.M., Runje, D.: ALIME: autoencoder based approach for local interpretability. In: Yin, H., Camacho, D., Tino, P., Tallón-Ballesteros, A.J., Menezes, R., Allmendinger, R. (eds.) IDEAL 2019. LNCS, vol. 11871, pp. 454–463. Springer, Cham (2019). https://doi.org/10.1007/978-3-030-33607-3_49
11. Vincent, P., Larochelle, H., Bengio, Y., Manzagol, P.A.: Extracting and composing robust features with denoising autoencoders. In: Proceedings of the 25th International Conference on Machine Learning, ICML 2008, pp. 1096–1103. Association for Computing Machinery, New York (2008). https://doi.org/10.1145/1390156.1390294

12. Visani, G., Bagli, E., Chesani, F.: Optilime: optimized LIME explanations for diagnostic computer algorithms. In: Conrad, S., Tiddi, I. (eds.) Proceedings of the CIKM 2020 Workshops co-located with 29th ACM International Conference on Information and Knowledge Management (CIKM 2020), Galway, Ireland, 19–23 October 2020. CEUR Workshop Proceedings, vol. 2699. CEUR-WS.org (2020). https://ceur-ws.org/Vol-2699/paper03.pdf

13. Visani, G., Bagli, E., Chesani, F., Poluzzi, A., Capuzzo, D.: Statistical stability indices for lime: obtaining reliable explanations for machine learning models. J. Oper. Res. Soc. **73**(1), 91–101 (2022). https://doi.org/10.1080/01605682.2020.1865846

14. William, W., Mangasarian, O., Street, N., Street, W.: Breast Cancer Wisconsin (Diagnostic). UCI Machine Learning Repository (1995). https://doi.org/10.24432/C5DW2B

15. Zafar, M.R., Khan, N.: Deterministic local interpretable model-agnostic explanations for stable explainability. Mach. Learn. Knowl. Extr. **3**(3), 525–541 (2021). https://doi.org/10.3390/make3030027. https://www.mdpi.com/2504-4990/3/3/27

Pattern Classification and Data Analysis

Rockfall Isolation Technique Based on DC-DBSCAN with k-Means Clustering and k-Nearest Neighbors Algorithm

Thanakon Augsondit[1]($\boxtimes$) , Thanaphat Khemniwat[1] ,
Pannathorn Sathirasattayanon[1] , Patthranit Kaewcharuay[1] ,
Kasorn Galajit[2] , Jessada Karnjana[2] , and Sasiporn Usanavasin[1]

[1] Sirindhorn International Institute of Technology, Thammasat University,
Pathum Thani, Thailand
{m6222040336,6422770345,6422782316,6422782241}@g.siit.tu.ac.th,
sasiporn.us@siit.tu.ac.th
[2] NECTEC, National Science and Technology Development Agency,
Pathum Thani, Thailand
{kasorn.galajit,jessada.karnjana}@nectec.or.th
https://www.siit.tu.ac.th/

Abstract. Recently, spatial-clustered point clouds have been applied
to various applications, such as glacier movement and rockfall detection, which are crucial for ensuring human safety. The density-based
spatial clustering of applications with noise (DBSCAN) is a well-known
spatial clustering algorithm. It is effective but requires two predefined
parameters needed to be appropriately set. The suitable values of these
parameters depend on the distribution of the input point cloud. Thus, to
address this issue, we previously proposed a non-parametric DBSCAN
based on a recursive approach and called it divide-and-conquer-based
DBSCAN or DČ-DBSCAN. Even though it outperformed the traditional
DBSCAN, the performance of the previous DC-DBSCAN or DBSCAN
is limited when two groups or two clusters are too close. Therefore, this
study proposes an improved version of DC-DBSCAN that utilizes the k-
means clustering algorithm to further cluster some groups resulting from
DC-DBSCAN. To determine which groups are to be clustered further, a
k-nearest neighbors algorithm is used. The experimental results demonstrate that the proposed method enhances the impurity and normalized
mutual information (NMI) scores compared with DBSCAN and DC-
DBSCAN. The purity score of the proposed method is 97.91%, and the
NMI score is 96.48%. Compared to DC-DBSCAN, our proposed method
achieves a 12.37% improvement in purity and a 3.61% improvement in
NMI. Also, it can spatially cluster some groups that DBSCAN and DC-
DBSCAN cannot do.

Keywords: DBSCAN · k-means clustering · divide-and-conquer-based
DBSCAN · k-nearest neighbors · rockfall detection

© The Author(s), under exclusive license to Springer Nature Switzerland AG 2024
K. Honda et al. (Eds.): IUKM 2023, LNAI 14376, pp. 167–178, 2024.
https://doi.org/10.1007/978-3-031-46781-3_15

1   Introduction

A rockfall is the downward movement of a piece of bedrock detached from a cliff or steep slope. It can be described as the swifter form of a landslide and has the potential to cause harm to both properties and human lives. Its impact disrupts transportation and commerce, such as rockfalls that block highways and waterways. Also, falling rocks cause direct casualties.

Different approaches can be used to understand rockfall incidents, for example, looking at historical records, assessing vulnerability, estimating how often it happens, evaluating the danger level, and gauging the associated risks. Recently, researchers have employed a terrestrial laser scanner (TLS) to investigate geological phenomena due to its benefit in gaining high-resolution data. TLS has also been applied in studying rockfalls. A common framework for identifying rockfalls from two sets of point-cloud data consists of three parts which are preprocessing, clutter removal, and rockfall isolation. Recently, the preprocessing and clutter removal processes have been improved continuously and significantly. However, there remains scope for enhancing the efficiency of the rockfall isolation process. In general, density-based spatial clustering of applications with noise (DBSCAN) is a popular clustering algorithm for labeling a group of data based on density [3,9]. DBSCAN algorithm requires two parameters which are ε and minPts. These parameters play an important role in clustering performance. The ε parameter is the largest Euclidean distance that allows two points to be in the same neighborhood, and the parameter minPts is the minimum number of points in an ε-neighborhood circle centered at a point to be considered as a core point.

In the literature, there is a wide range of proposed methods for estimating parameters. For instance, McInnes *et al.* presented an innovative variant of DBSCAN, known as hierarchical DBSCAN or HDBSCAN [8]. Karami and Johansson employed the differential evolution (DE) algorithm to optimize the parameters of DBSCAN [4]. An optimizer based on multiverse optimization (MVO) was applied by Lai *et al.* for DBSCAN parameter estimation [6]. However, both DE and MVO consumed processing time. The non-parametric DBSCAN algorithm called DC-DBSCAN was proposed by Pitisit Dillon *et al.*, which applies the recursive method to find the most suitable parameter ε for each cluster [2]. However, the non-parametric DBSCAN algorithm poses a problem when clusters are located closely, which is assumed to be the limitation of DC-DBSCAN. To overcome this limitation, this study focuses on improving the algorithm by integrating multiple clustering techniques, including k-means clustering and k-nearest neighbors algorithm.

2   Background

This section provides the background knowledge required for understanding our proposed method briefly. It includes point cloud preprocessing, density-based spatial clustering of applications with noise (DBSCAN), DBSCAN with a divide-and-conquer approach (DC-DBSCAN), k-means clustering, and k-nearest neighbors (k-NN) algorithm.

2.1   Point Cloud Preprocessing

The rockfall detection framework that analyzes data obtained from terrestrial laser scanners (TLS) consists of three steps: preprocessing, clutter removal, and clustering or rockfall isolation. The first two steps prepare the point cloud for the isolation process. The first step, preprocessing, consists of two sub-processes: registration and subtraction. The registration process geometrically aligns two point clouds, and the subtraction process differentiates them to obtain a difference. In the second step, the clutter removal process is applied to remove noise in the difference. The noise-removed difference is then spatially clustered for rockfall event isolation.

2.2   DBSCAN

DBSCAN is a clustering algorithm that groups data points based on density. It clusters the data points by grouping those close to each other [3,9]. In other words, DBSCAN identifies regions with a high density of points and determines them as clusters. Let p and q indicate points in a point cloud $\mathbf{P}$, and the Euclidean distance between them is $d(p,q)$. Let $N_\varepsilon(p)$ denote a set of points with distances to p less than or equal to ε, i.e., $N_\varepsilon(p) = \{q \in \mathbf{P} \mid d(p,q) \leq \varepsilon\}$, and be called the ε-*neighborhood* of the point p. If $|N_\varepsilon(p)|$ is greater than a predefined integer minPts, the point p is a *core point*. The DBSCAN algorithm clusters the data points by first identifying core points and collecting all *density-reachable* points from each core point. Points p_1 and p_n are *density-reachable* if there exists a sequence of points $(p_1, p_2, ..., p_n)$ that any two successive points p_i and p_{i+1} for $i = 1$ to n in the sequence satisfies the condition $p_i \in N_\varepsilon(p_{i+1})$. Points that do not belong to any cluster are labeled as noise. It is worth emphasizing that DBSCAN has two parameters strongly affecting its performance, i.e., ε and minPts.

2.3   DC-DBSCAN

DC-DBSCAN [2] is a non-parametric variant of DBSCAN. Technically, it searches for one optimal DBSCAN parameter while making another constant. The general concept of DC-DBSCAN can be summarized as follows. Initially, a preprocessed point cloud undergoes grid analysis to obtain an appropriate initial ε. The grid analysis divides the point cloud into a 3-by-3 grid, i.e., nine equally sized boxes, and counts points in each grid box. If a stopping criterion concerning the median and variance of point counts is met, the maximum between the box's width and height is the parameter ε. Otherwise, the whole process is repeated iteratively to the grid box with the maximum count.

Then, the DBSCAN algorithm is performed on the input point cloud with ε obtained from the grid analysis. As a result, DBSCAN returns at least one cluster with a noise group. On this basis, DBSCAN is executed on each cluster separately with a smaller ε. This process is repeated iteratively to smaller and smaller clusters until a stopping criterion, i.e., broadly speaking, the number of points labeled as noise outnumbers the other non-noise groups, is satisfied. It is worth noting that even though DC-DBSCAN is non-parametric, it fixes minPts while searching ε, and there can be more than one value of ε used in clustering.

2.4 k-Means Clustering and k-Nearest Neighbors Algorithm

The k-means clustering is an unsupervised machine learning algorithm that aims to partition observations or data into k clusters, where k is the main parameter for the algorithm. The k-means clustering algorithm minimizes the within-cluster variance or inertia, defined as the sum of squared distances between each data point and its corresponding cluster centroid.

The k-nearest neighbors (k-NN) algorithm is a non-parametric supervised learning method for classification or regression. In this work, it is used for classification. The algorithm predicts based on the similarity between the input and neighboring data points.

3   Proposed Method

The proposed method takes a preprocessed point cloud, as described in Sect. 2.1, as an input and returns labeled clusters as the output. It operates as follows. First, the input point cloud is fed into grid analysis similar to that used in DC-DBSCAN, as described in Sect. 2.3, to get an initial ε. Then, the conventional DBSCAN with the initial ε is used to cluster the input point cloud. If it cannot split the point cloud into more than one cluster, the value of ε is reduced by 0.01, and the DBSCAN is performed again on the point cloud with the updated ε. This process of reducing ε and executing DBSCAN is repeated until multiple clusters are obtained.

Once multiple clusters are achieved, the clustering result is checked to determine whether it is acceptable. The criterion is that the clustering is accepted if the number of noise points from the DBSCAN algorithm is less than the number of points in the smallest split cluster. When accepted, each split cluster is fed into a variant of DC-DBSCAN with k-means clustering and k-NN algorithm, referred to as KK-DC-DBSCAN, utilizing the most-updated ε.

If the clustering result does not meet the acceptability criterion, the DBSCAN is deemed unsuccessful in effectively clustering the input point cloud. In this scenario, an alternative method based on k-means clustering is employed. The input point cloud, previously used as input for DBSCAN, is utilized to extract features related to statistical quantities (details will be provided in the subsequent subsection). These features are then fed into a k-NN binary classifier, determining whether further clustering by k-means is necessary for the point cloud of interest.

Suppose the classifier determines that applying the k-means algorithm is unnecessary. In that case, the cluster is saved and combined with other labeled clusters that have completed clustering. On the other hand, if the classifier suggests further clustering using the k-means algorithm, an iterative k-means process is used to identify the most suitable value of k. Lastly, the point cloud undergoes clustering via the k-means algorithm, and the resulting clusters are saved and combined with other labeled clusters. The proposed method can be depicted in Fig. 1.

3.1   Training k-NN Binary Classifier

The k-NN binary classifier decides whether or not an unaccepted point cloud should be clustered further by k-means clustering. It takes a feature derived from statistical quantities of the point cloud as the input. The feature extraction pipeline is shown in Fig. 2 and summarized as follows.

First, the point cloud is normalized using the min-max scaler algorithm. Then, it is used to construct two feature vectors, F_1 and F_2, as shown by the left and right branches, respectively, in Fig. 2. On the left, the grid analysis is applied to the point cloud to get the populations of all nine grid boxes. Statistical quantities from these populations are used to form F_1, including mean, mode, median, minimum, maximum, difference between maximum and minimum, population and sample standard deviations, and population and sample variances. On the right, the distance to the nearest point of every point is calculated first. Then, the same statistical quantities, excluding the mode, of the distances are used to construct the feature vector F_2.

The concatenation of F_1 and F_2 is then a candidate for the feature to be input into the k-NN classifier. Note that the candidate is a vector in 19-dimensional space. We reduce the dimension by using the following strategy. First, we calculated Pearson correlation coefficients between each entry and the target. The entry is removed if the correlation between itself and the target is less than 0.2. Then, we computed correlation coefficients between any two entries. One entry of a pair with a correlation coefficient higher than 0.95 is removed, and we remove the one with a smaller correlation coefficient between that entry and the target. Under these conditions, the dimension is reduced to four: population standard deviation from F_1, population variance from F_1, the difference between maximum and minimum from F_2, and maximum from F_2. These four entries are four components of the feature to be input into the k-NN classifier.

The k-NN classifier classifies the point cloud's feature into two distinct groups: one that requires further clustering and another that does not. In the context of this study, we train the classifier on features with targets from 251 point clouds (some contain a single cluster, whereas the others have more than one). To make the target variables in training, we individually apply those 251 point clouds to DC-DBSCAN and compare the clustering results with ground-truth clusters. If DC-DBSCAN works correctly for any point clouds, those point clouds are labeled with 'no further classification.' Otherwise, further classification is required.

Note also that for those 251 point clouds, 85 point clouds require further clustering, and 166 do not. Therefore, to handle this unbalance, the random oversampling technique is utilized to upsampling those 85 point clouds to 166.

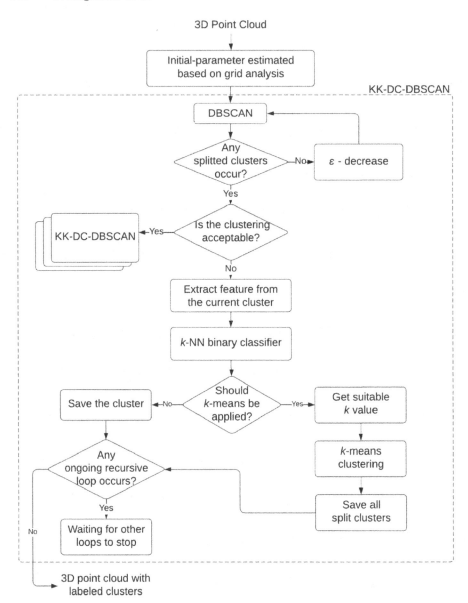

Fig. 1. Proposed method.

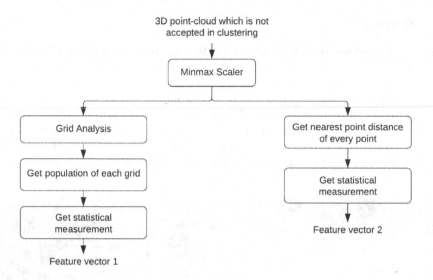

Fig. 2. Features extraction.

3.2   Iterative k-Means Clustering

The iterative k-means clustering method is employed to determine the most suitable k. The central concept of this algorithm involves applying k-means clustering and reducing the value of k iteratively until obtaining a suitable one, which is the first k satisfying the following condition:

$$\frac{Q_{k-1}}{Q_k} > 2, \tag{1}$$

and

$$Q_k = \sum_{i=1}^{n}(x_i - x_{k,i}^*)^2 + \sum_{i=1}^{n}(z_i - z_{k,i}^*)^2, \tag{2}$$

where x_i are z_i are coordinates of point i on the x-axis and z-axis, respectively, and $x_{k,i}^*$ and $z_{k,i}^*$ are an x-axis coordinate and z-axis coordinate of the center of the cluster nearest to the point i when the k-means clustering is applied to the point cloud of n data points.

4   Experiment and Results

This section provides details about the simulation and evaluation of the proposed method. Evaluation metrics and the dataset used in the experiment are also introduced.

4.1   Dataset

An open dataset generated from a terrestrial laser scanner (Optech's Intelligent Laser Ranging and Imaging System or ILRIS3D), provided by Abellan *et al.* [1,10] was chosen to conduct the experiments to evaluate the proposed method. The dataset contains 3D data points representing the surface of the Puiggcercos cliff, Catalonia, Spain, as shown in Fig. 3. In total, four point clouds were used for the experiment: two point clouds are generated by using CloudCompare, and the other two point clouds are provided by Pitisit Dillon *et al.* [2]. Each point cloud consists of eight rockfall events, as shown in Fig. 4.

Fig. 3. 3D point cloud (left) of the surface of Puiggcercos cliff (right), Catalonia, Spain.

4.2   Evaluation Metrics

The effectiveness of the rockfall isolation method can be evaluated using two metrics: purity score and normalized mutual information (NMI).

The purity measure how well the clusters match. However, it is not the best evaluation for clustering 3D point clouds, because this measure does not consider the number of exceed clusters as a penalty, and having more clusters can increase purity level. The purity λ is defined as

$$\lambda = \frac{1}{n} \sum_{h=1}^{k} \max_{1 \leq j \leq l} \{n_h^1, n_h^2, n_h^3, ..., n_h^j\}, \tag{3}$$

where n represents the overall number of points, l denotes the total number of true clusters, k is the total number of clusters identified by the algorithm, and n_h^j represents the count of points labeled by the algorithm as cluster h that also belong to the ground-truth cluster j.

The normalized mutual information (NMI) score is defined as follows [5,7]. Let X be a discrete random variable representing a ground-truth label (or ground-truth cluster ID) of points in the point cloud. Let the ground truth contain $u + 1$ clusters, the possible outcomes of X are $0, 1, ..., u$ where each outcome is a ground-truth cluster-ID. Let Y be a discrete random variable representing a cluster ID labeled by the clustering algorithm. Considering that the algorithm

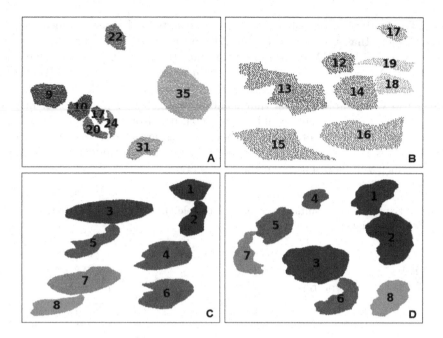

Fig. 4. Rockfall events in each scenario.

organizes the point cloud into $v + 1$ clusters, where the outcomes for Y range from 0 to v and each outcome represents a cluster ID. The NMI score $I^*(X;Y)$ is defined as follows:

$$I^*(X;Y) = \frac{2 \cdot I(X;Y)}{H(X) + H(Y)}, \qquad (4)$$

where $I^*(X;Y)$ is the mutual information between X and Y, $H(X)$ is the entropy of X, and $H(Y)$ is the entropy of Y.

4.3 Experimental Results

This work compared the proposed method to the DC-DBSCAN and ground truth. The experimental evaluations of the proposed method on four point clouds are shown in Table 1. The first column displays event numbers. The second and third columns show points clustered by the proposed method and DC-DBSCAN, respectively. The last column shows the ground-truth point count in each event. Two rows under each table show both methods' purity and NMI scores. It can be seen from the result that the proposed method outperformed the DC-DBSCAN method.

The proposed method's average purity and NMI score were 97.91% and 96.48%, respectively, while the DC-DBSCAN's average purity and NMI score were 85.60% and 92.87%, respectively. Moreover, in all four scenarios, the proposed method's purity and NMI score were better than the DC-DBSCAN method's. At the bottom left of Table 1, the results of the point clouds in scenario C show that the proposed method's purity and NMI score were much better than DC-DBSCAN. After investigating, the result turned out that the DC-DBSCAN algorithm is not doing well in separating the clusters that the little parts of their border are contiguous as the proposed method. The results of scenario C in Table 1 shows that some events got zero point which means that some contiguous events could not be separated from each other. However, in some situations, the proposed method's scores are slightly lower than the DC-DBSCAN method's, and some point clouds were separated into too many events, which will be discussed in the next section.

Table 1. Experimental results for scenarios A (top-left), B (top-right), C (bottom-left), and D (bottom-right).

Event number	The number of points			Event number	The number of points		
	Proposed method	DC-DBSCAN	Ground truth		Proposed method	DC-DBSCAN	Ground truth
1	290	0	304	1	248	247	247
2	176	176	176	2	1025	1025	1025
3	589	589	589	3	480	758	513
4	2283	2283	2283	4	855	855	855
5	419	419	419	5	931	931	931
6	465	755	451	6	159	159	159
7	749	749	749	7	278	0	245
8	159	159	159	8	201	201	202
Purity score	99.73%	94.07%		Purity score	99.19%	94.11%	
NMI score	99.31%	96.99%		NMI score	98.49%	96.76%	

Event number	The number of points			Event number	The number of points		
	Proposed method	DC-DBSCAN	Ground truth		Proposed method	DC-DBSCAN	Ground truth
1	1623	0	1516	1	3151	0	2868
2	1437	3060	1544	2	4859	8010	5142
3	4186	5834	3922	3	5283	5283	5283
4	3006	3006	3006	4	981	981	981
5	1648	0	1912	5	2821	4666	2711
6	3008	3008	3008	6	1279+1130	2409	2409
7	2545	4982	3255	7	1845	0	1955
8	2437	0	1727	8	2910	2910	2910
Purity score	94.53%	74.08%		Purity score	98.18%	80.12 %	
NMI score	92.82%	87.34%		NMI score	95.28%	90.39%	

5   Discussion

Even though the proposed method can improve the efficiency of the clustering algorithm, especially in the case that two events are located closely, this proposed method still has an error due to the classifier model. The result of scenario D is shown in Fig. 5, showing an addition event, which is over-clustering. Moreover, in the point clouds with no event border contiguous or nearby, the DC-DBSCAN method outperformed the proposed method because the classifier model sometimes determines incorrectly and results in over-clustering. This problem can be solved by increasing the number of the dataset, using a more diverse dataset, and balancing the dataset.

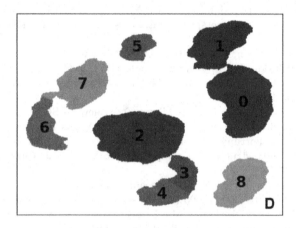

Fig. 5. Rockfall events of scenario D.

6   Conclusion

The research article introduced a modified version of DC-DBSCAN which improves the efficiency of clustering rockfall events that are in close proximity. The proposed KK-DC-DBSCAN method incorporates four key processes: estimating the initial ε parameter through grid analysis, employing DC-DBSCAN, utilizing a k-NN binary classifier to determine whether a cluster should be clustered with k-means clustering, and applying k-means clustering to the clusters identified by the k-NN classifier. In some cases, the clustering process in the DC-DBSCAN method terminates prematurely, resulting in a large ε. As a consequence, it becomes challenging to differentiate and separate closely located events. The results of the experiment demonstrated that the proposed method achieved higher purity and NMI scores compared to the DC-DBSCAN method when clustering two or more rockfall events in close proximity.

Acknowledgements. This research is financially supported by the Thailand Advanced Institute of Science and Technology (TAIST), the National Science and

Technology Development Agency (NSTDA), the Tokyo Institute of Technology, Sirind-horn International Institute of Technology, Thammasat University, and the Natioanl Research Council of Thailand (NRCT) under the TAIST-Tokyo Tech program.

References

1. Abellán, A., Calvet, J., Vilaplana, J.M., Blanchard, J.: Detection and spatial pre-diction of rockfalls by means of terrestrial laser scanner monitoring. Geomorphol-ogy **119**(3–4), 162–171 (2010)
2. Dillon, P., Aimmanee, P., Wakai, A., Sato, G., Hung, H.V., Karnjana, J.: A novel recursive non-parametric DBSCAN algorithm for 3D data analysis with an appli-cation in rockfall detection. J. Disaster Res. **16**(4), 579–587 (2021)
3. Ester, M., Kriegel, H.P., Sander, J., Xu, X., et al.: A density-based algorithm for discovering clusters in large spatial databases with noise. In: KDD, vol. 96, pp. 226–231 (1996)
4. Karami, A., Johansson, R.: Choosing DBSCAN parameters automatically using differential evolution. Int. J. Comput. Appl. **91**(7), 1–11 (2014)
5. Kvålseth, T.O.: On normalized mutual information: measure derivations and prop-erties. Entropy **19**(11), 631 (2017)
6. Lai, W., Zhou, M., Hu, F., Bian, K., Song, Q.: A new DBSCAN parameters deter-mination method based on improved MVO. IEEE Access **7**, 104085–104095 (2019)
7. Lancichinetti, A., Fortunato, S., Kertész, J.: Detecting the overlapping and hier-archical community structure in complex networks. New J. Phys. **11**(3), 033015 (2009)
8. McInnes, L., Healy, J., Astels, S.: hdbscan: hierarchical density based clustering. J. Open Source Softw. **2**(11), 205 (2017)
9. Schubert, E., Sander, J., Ester, M., Kriegel, H.P., Xu, X.: DBSCAN revisited, revisited: why and how you should (still) use DBSCAN. ACM Trans. Database Syst. (TODS) **42**(3), 1–21 (2017)
10. Tonini, M., Abellan, A.: Rockfall detection from terrestrial lidar point clouds: a clustering approach using R. J. Spatial Inf. Sci. **8**, 95–110 (2014)

A Novel Noise Clustering Based on Local Outlier Factor

Yukihiro Hamasuna[1,2(✉)] and Yoshitomo Mori[3]

[1] Faculty of Informatics, Kindai University, 3-4-1 Kowakae, Higashiosaka, Osaka 577-8502, Japan
[2] Cyber Informatics Research Institute, Kindai University, Higashiosaka, Japan
yhama@info.kindai.ac.jp
[3] Graduate School of Science and Engineering, Kindai University, 3-4-1 Kowakae, Higashiosaka, Osaka 577-8502, Japan
2333340435v@kindai.ac.jp

Abstract. Reducing the impact of outliers is an essential issue in machine learning, including clustering. There are two main approaches to reducing the impact of outliers: one is to build robust models, and the other is to remove outliers through preprocessing. In this paper, we propose a new noise clustering method that combines noise clustering, which builds a model robust to outliers, and local outlier factor, which removes outliers as a preprocessing step. The proposed method is an optimization problem of noise clustering with a weighting of dissimilarities by LOF. Numerical experiments were conducted using four artificial datasets to verify the effectiveness of the proposed method. In the experiments, the proposed method was compared with k-medoids clustering, DBSCAN, and noise clustering. The results show that the proposed method yields good results regarding both clustering performances and detecting outliers. The guideline for determining k and ε among the three parameters D, k, and ε required by the proposed method was also suggested.

Keywords: Noise clustering · Local outlier factor · Outlier · Robustness

1 Introduction

Machine learning is a core technology for leveraging large amounts of stored data and has been applied to many real-world problems [1]. Machine learning can be broadly classified into supervised and unsupervised learning. Clustering is one of the leading unsupervised learning methods, dividing a set of objects into groups called clusters [2,3]. Clustering aims to classify similar objects into the same cluster and dissimilar objects into different clusters. In machine learning, including clustering, where large amounts of data are handled, datasets may contain outliers. Outliers are data with characteristics that differ from other objects and can negatively impact learning processes and predictions. Minimizing the impact of outliers in machine learning requires the removal of outliers

K. Honda et al. (Eds.): IUKM 2023, LNAI 14376, pp. 179–191, 2024.
https://doi.org/10.1007/978-3-031-46781-3_16

through data preprocessing and constructing models that are less susceptible to outliers [4].

As with many machine learning methods, the impact of outliers is significant in clustering. The c-means or fuzzy c-means method is a method of cluster partitioning by dissimilarity from the representative points of the clusters [2,3]. When using the c-means or fuzzy c-means method, cluster representatives attracted to outliers are calculated. As a result, sparse clusters containing outliers may be generated. Noise clustering [5] and possibilistic clustering [6] have been proposed as robust clustering methods. Noise clustering is a technique that adds noise cluster to the objective function, which is intended to consist only of outliers [5]. Possibilistic clustering differs from c-means and fuzzy c-means in that it does not use constraints on the membership degree [6]. In addition to these methods, DBSCAN is also known to be less susceptible to outliers [7]. DBSCAN is a method that sequentially constructs clusters based on the density of objects. Reducing the influence of outliers and obtaining better cluster partitions have been recognized as an essential issue in the past.

As discussed above, possibilistic clustering, noise clustering, and DBSCAN are approaches to building models that are less susceptible to outliers. On the other hand, the Local outlier factor (LOF) has been proposed as a method to evaluate the degree of an outlier of an object based on the distribution of data [8]. LOF is a density-based anomaly detection method that assigns a local density of each object based on the distance from neighboring objects, and computes the degree of outliers of the objects by comparing them with the local densities of neighboring objects. The outlier degree calculated here takes a value around one when a certain number of data objects exist in the neighborhood. On the other hand, a value much higher than one is calculated for data objects that are outliers in the data group. Outlier detection can be performed by using the value as a preprocessing of the data. Since it can also be applied to anomaly detection, LOF is used as a typical method in outlier detection. LOF is unique in that it calculates the outlier degree for each object in the dataset.

As has been discussed, two approaches exist for reducing the impact of outliers. One approach is to build a model that is less sensitive to outliers. The other approach is to preprocess the data using outlier detection techniques such as LOF. Noise clustering is a method of constructing a model that is less susceptible to outliers by adding noise clusters [5]. However, it does not take into account the characteristics of individual objects. Also, LOF considers the local distribution of objects but does not consider the broad structure in the dataset [8]. It is not that difficult a task to treat data objects that are clearly out of the data distribution as outliers. In this paper, we propose a new clustering method, noise clustering based on LOF (NCLOF), which combines noise clusters and LOF. The proposed method penalizes the dissimilarity of data objects with a high outlier degree using LOF. This idea would make it possible to cluster objects at the edges of a distribution that is difficult to handle into appropriate clusters. The effectiveness of the proposed method is demonstrated by evaluating the clustering performance on artificial and benchmark data. In addition, the impact of the parameters used in the proposed method on cluster partitioning is also shown.

2 Preliminaries

A set of objects to be clustered is given, and it is denoted by $X = \{x_l \in \Re^p \mid l = 1 \sim n\}$. Each object x_l is a vector in the p-dimensional Euclidean space $\Re^p$, that is, an object $x_l \in \Re^p$. A cluster is denoted by G_i, and a collection of clusters is denoted by $\mathcal{G} = \{G_1, \ldots, G_c\}$. A cluster center of G_i is denoted by $v_i \in \Re^p$, and a set of v_i is given by $V = \{v_1, \ldots, v_c\}$. The membership degree of x_k belonging to G_i and the partition matrix are denoted as u_{ki}, and $U = (u_{ki})_{1 \leq k \leq n,\ 1 \leq i \leq c}$, respectively.

2.1 Noise Clustering

The objective function of noise clustering is denoted as follows:

$$J_{nc}(U, V) = \sum_{i=1}^{c} \sum_{l=1}^{n} u_{li}\|x_l - v_i\|^2 + \sum_{l=1}^{n} u_{l0} D.$$

$D > 0$ is called a noise parameter. Here, the zeroth cluster is the noise cluster. Constraints for membership degree $\mathcal{U}_{nc}$ is as follows:

$$\mathcal{U}_{nc} = \left\{ (u_{li}) : u_{li} \in \{0, 1\},\ \sum_{i=0}^{c} u_{li} = 1,\ \forall l \right\}. \tag{1}$$

Optimal solutions for v_i and u_{li} are as follows:

$$v_i = \frac{\sum_{l=1}^{n} u_{li} x_l}{\sum_{l=1}^{n} u_{li}}, \tag{2}$$

$$u_{li} = \begin{cases} 1 & (\min_{0 \leq i \leq c} d_{li}) \\ 0 & (\text{otherwise}) \end{cases}, \tag{3}$$

where, the dissimilarity d_{li} is described as follows:

$$d_{li} = \begin{cases} D & (i = 0) \\ \|x_l - v_i\|^2 & (\text{otherwise}) \end{cases}.$$

2.2 Local Outlier Factor

LOF is an anomaly detection method based on density calculated using neighborhood data [8]. LOF uses an idea called local density to calculate the degree of outlier by comparing the local density of data objects with that of their neighbors. LOF is both the name of the method and the name of the value representing the outlier degree to be calculated. LOF is calculated using k-distance, reachability distance, k-distanced neighborhood, and local reachability density. These details are described below.

k is any positive integer and represents the number of neighborhood objects in the LOF. k-distance(x) of data $x \in X$ is the distance $d(x, o)$ to data $o \in X$ satisfying the following conditions [8]:

– for at least k objects $o' \in X \setminus \{x\}$ it holds that $d(x, o') \leq d(x, o)$
– for at most $k - 1$ objects $o' \in X \setminus \{x\}$ it holds that $d(x, o') < d(x, o)$

The value of k-distance is expressed as the distance from x to the k-th nearest data; it is defined this way to account for the case where there are multiple data for which k-distance has the same value.

A k-distance neighborhood $N_k(x)$ is a set of objects that are within k-distance of specific data. The k-distance neighborhood of object x is defined by (4).

$$N_k(x) = \{q \in X \setminus \{x\} \mid d(x, q) \leq k\text{-}distance(x)\} \tag{4}$$

The reachability distance is the actual distance between the objects if they are far apart, but if they are close enough, it is replaced by the k-distance. The reachability distance of object x is defined by (5).

$$reach\text{-}dist_k(x, o) = \max\{k\text{-}distance(o), d(x, o)\} \tag{5}$$

The local reachability density is defined using reachability distance. The local reachability density means that the larger the value, the closer the neighbor objects are. It is calculated as the inverse of the average of the reachability distances of the objects in the k-distance neighborhood of object x. The local reachability density of data x is defined as follows (6):

$$lrd_k(x) = \frac{|N_k(x)|}{\displaystyle\sum_{o \in N_k(x)} reach\text{-}dist_k(x, o)} \tag{6}$$

The LOF value is calculated as the average of the ratios of the local reachability densities of the k-distance neighbors. The smaller the object's local reachability density and the larger the k-distance neighbor's local reachability density, the larger the value. The value of LOF for data x is defined as follows:

$$LOF_k(x) = \frac{\sum_{o \in N_k(x)} \frac{lrd_k(o)}{lrd_k(x)}}{|N_k(x)|} \tag{7}$$

Using the values $LOF_k(x)$ calculated above, the LOF detects outliers. LOF is used for data preprocessing because it can be used to detect outliers from a given dataset.

3   Proposed Method

The proposed method, NCLOF, is an extension of noise clustering using LOF. As noted in the introduction, two approaches exist for reducing the impact of outliers. Noise clustering is an approach to building models that reduce the impact of outliers. LOF is an approach that focuses on the outlier degree of objects and detects outliers through preprocessing. Noise clustering can be thought of as an approach that focuses on the model of clusters, while LOF is an approach that focuses on objects. The proposed method, NCLOF, combines both approaches to provide more robust clustering for datasets with outliers. The optimization problem of NCLOF is expressed as follows:

$$\min \quad J(U, V) = \sum_{i=1}^{c}\sum_{l=1}^{n} u_{li}\gamma_{l}\|x_l - v_i\|^2 + \sum_{l=1}^{n} u_{l0}D, \tag{8}$$

$$\text{s. t.} \quad \mathcal{U}_{nc} = \left\{ (u_{li}) : u_{li} \in \{0,1\}, \ \sum_{i=0}^{c} u_{li} = 1, \ ^\forall l \right\}.$$

where the γ_l in (8) is the value to which the LOF of x_l is transformed by the following equation:

$$\gamma_l = \begin{cases} 1 & (LOF_k(x_l) \le 1 + \varepsilon) \\ LOF_k(x_l) & (\text{otherwise}) \end{cases}. \tag{9}$$

$\varepsilon > 0$ is a parameter indicating the upper limit of LOF, which is converted to 1 when LOF exceeds $1 + \varepsilon$. The value of LOF expressed in (7) takes a value well above 1 for objects considered outliers and around 1 for objects not considered outliers. (9) represents that the γ_l takes the value of LOF for objects considered outliers, while the γ_l is 1 for objects considered not to be outliers. In NCLOF, the value of LOF is used for assignment to noise clusters. Therefore, for individuals with larger LOF values, no upper limit is set for γ_l in order to make it easier for them to be assigned to noise clusters due to the influence of larger weights. This transformation is performed for objects where the value of LOF (7) is greater than $1 + \varepsilon$ in order to obtain a larger dissimilarity. By using γ_l, the dissimilarity of between x_l and cluster centers is $\gamma_l\|x_l - v_i\|^2$. By introducing gamma to noise clustering, dissimilarity to noise clusters is represented as $\gamma_l\|x_l - v_i\|^2$.

Optimal solutions for v_i and u_{li} of NCLOF are as follows:

$$v_i = \frac{\sum_{l=1}^{n} u_{li}\gamma_l x_l}{\sum_{l=1}^{n} u_{li}\gamma_l}, \tag{10}$$

$$u_{li} = \begin{cases} 1 & (\min_{0 \le i \le c} d_{li}) \\ 0 & (\text{otherwise}) \end{cases}, \tag{11}$$

where, the dissimilarity d_{li} is described as follows:

$$d_{li} = \begin{cases} D & (i = 0) \\ \gamma_l\|x_l - v_i\|^2 & (\text{otherwise}) \end{cases}. \tag{12}$$

The optimal solution for $\boldsymbol{v}_i$ is in the form of (2) with the addition of γ_l. The optimal solution for u_{li} has the same form as (3), but the d_{li} is changed to (12).

The NCLOF algorithm is summarized in **Algorithm 1**:

Algorithm 1. NCLOF

NCLOF1 Set cluster number c, parameters k, D, ε, and initial cluster centers $\boldsymbol{v}_i \in \boldsymbol{V}$ by choosing objects at random.

NCLOF2 Calculate γ_l using (9).

NCLOF3 Calculate $u_{li} \in \boldsymbol{U}$ using (11).

NCLOF4 Calculate $v_i \in \boldsymbol{V}$ using (10).

NCLOF5 If the convergence criterion is satisfied, stop. Otherwise, return to **NCLOF3**.

The number of repetitions, convergence of each variable, or convergence of an objective function is used as the convergence criterion in NCLOF.

4   Numerical Experiments

We conducted numerical experiments with four artificial datasets to verify the effectiveness of NCLOF. First, we describe the calculation conditions of the numerical experiments. Second, we compare the results among NCLOF and three conventional methods: k-means clustering [3], DBSCAN [7], and noise clustering [5]. Third, we summarize the results and the features of the proposed method.

4.1   Experimental Setup

The above four methods are compared using four artificial datasets regarding the evaluated value of the adjusted rand index (ARI) [9]. ARI is a measure of similarity between two cluster partitions. The value of ARI is 1 when the two cluster partitions match entirely. For all methods, the maximum number of iterations of the algorithm was set to 100. The artificial datasets are visualized in Figs. 1, 2, 3 and 4. Each object is color-coded in these figures by the cluster it belongs to. Outliers are also shown in black. The number of objects in each cluster and outliers, are shown in the legends of Figs. 1, 2, 3 and 4.

4.2   Experimental Results

The results of clustering the four artificial datasets using each method and evaluating them with ARI are summarized in Tables 1 through 4. The ARI values in Tables 1 through 4 are calculated from the cluster partition obtained using each method and given cluster labels. The bold values in the max column are the highest ARI values for each dataset. The first column of each table shows the clustering method. The values of the parameters used in each method are

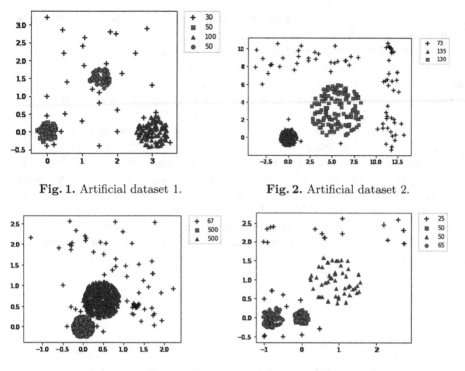

Fig. 1. Artificial dataset 1.

Fig. 2. Artificial dataset 2.

Fig. 3. Artificial dataset 3.

Fig. 4. Artificial dataset 4.

listed in parentheses after the name of each method. The second column shows
the maximum ARI, the third column shows the minimum ARI, and the fourth
column shows the average and standard deviation of the ARI. Tables 1, 2, 3, and
4 show the results for artificial dataset 1, artificial dataset 2, artificial dataset 3,
and artificial dataset 4, respectively. The results for each dataset are examined.

For artificial dataset 1, relatively good results were obtained except for k-
means. Noise clustering clustered two outliers in the lower middle green cluster
and two outliers in the upper left of the lower right blue cluster. DBSCAN
clustered one outlier in the cluster at the bottom right of the bottom right blue
cluster. NCLOF ($D = 0.22, k = 10, \varepsilon = 0.3$) assigns outliers that were included
in clusters by noise clustering and DBSCAN to noise clusters as outliers. NCLOF
performs better than other methods for artificial dataset 1.

For artificial dataset 2, relatively good results were obtained except for k-
means. Noise clustering clustered two outliers in the cluster around the small
cluster on the left. DBSCAN clustered outliers in the cluster to the lower right of
the small cluster on the left. The ARI was slightly lower because the objects in
the large cluster on the right were classified as outliers. NCLOF ($D = 10.3, k =
30, \varepsilon = 0.3 \, or \, \varepsilon = 1.5$) only clustered outliers in the lower right of the small
left cluster in the cluster. Therefore, it showed a higher ARI when compared to

noise clustering and DBSCAN. NCLOF performed slightly better than the other methods for the artificial dataset 2.

For artificial dataset 3, DBSCAN showed promising results. Noise clustering included outliers around small clusters and multiple objects on the border of large clusters in clusters. DBSCAN clustered objects on the edges of the large clusters as outliers. NCLOF ($D = 10.3, k = 50, \varepsilon = 0.3$) clustered multiple objects on the border of the large cluster in the cluster. NCLOF clustered objects around the small clusters assigned to clusters by noise clustering as outliers. NCLOF resulted in a result that was greater than noise clustering but less than DBSCAN.

For artificial dataset 4, NCLOF showed promising results. Noise clustering clustered outliers around the bottom left two clusters in the cluster. Like noise clustering, DBSCAN also clustered outliers around the bottom left two clusters in its clusters, as did NC. The results are better than noise clustering because fewer outliers were included in the clusters than noise clustering. NCLOF ($D = 0.54, k = 9, \varepsilon = 0.09 or 0.3$) assigns outliers that were included in clusters by noise clustering and DBSCAN to noise clusters. NCLOF performs better than the other methods for artificial dataset 4.

Table 1. ARI results for artificial dataset 1.

method	max	min	ave. $\pm$ sd
k-means	0.821	0.351	0.751 ± 0.161
DBSCAN (Eps $= 0.17$, MinPts $= 10$)	0.969	0.969	0.969 ± 0.000
Noise clustering ($D = 0.22$)	0.941	0.260	0.658 ± 0.212
NCLOF ($D = 0.22, k = 10, \varepsilon = 0.3$)	**1.000**	0.259	0.693 ± 0.236
NCLOF ($D = 0.22, k = 10, \varepsilon = 0.09$)	0.976	0.260	0.691 ± 0.212
NCLOF ($D = 0.22, k = 10, \varepsilon = 1.5$)	0.941	0.259	0.670 ± 0.224
NCLOF ($D = 0.22, k = 100, \varepsilon = 0.3$)	0.954	0.260	0.672 ± 0.208

Table 2. ARI results for artificial dataset 2.

method	max	min	ave. $\pm$ sd
k-means	0.523	0.523	0.523 ± 0.000
DBSCAN (Eps $= 1$, MinPts $= 10$)	0.984	0.984	0.984 ± 0.000
NC ($D = 10.3$)	0.984	0.063	0.730 ± 0.237
NCLOF ($D = 10.3, k = 5, \varepsilon = 0.3$)	0.984	0.064	0.733 ± 0.237
NCLOF ($D = 10.3, k = 30, \varepsilon = 0.09$)	0.976	0.229	0.725 ± 0.229
NCLOF ($D = 10.3, k = 30, \varepsilon = 0.3$)	**0.992**	0.071	0.759 ± 0.214
NCLOF ($D = 10.3, k = 30, \varepsilon = 1.5$)	**0.992**	0.258	0.769 ± 0.258
NCLOF ($D = 10.3, k = 100, \varepsilon = 0.3$)	0.896	0.056	0.589 ± 0.221

Table 3. ARI results for artificial dataset 3.

method	max	min	ave. ± sd
k-means	0.765	0.765	0.765 ± 0.000
DBSCAN (Eps = 0.08, MinPts = 13)	**0.987**	0.987	0.987 ± 0.000
NC ($D = 0.22$)	0.935	0.641	0.881 ± 0.090
NCLOF ($D = 0.22, k = 50, \varepsilon = 0.09$)	0.916	0.625	0.860 ± 0.096
NCLOF ($D = 0.22, k = 50, \varepsilon = 0.3$)	0.943	0.063	0.889 ± 0.121
NCLOF ($D = 0.22, k = 50, \varepsilon = 1.5$)	0.937	0.625	0.890 ± 0.082
NCLOF ($D = 0.22, k = 10, \varepsilon = 0.3$)	0.939	0.637	0.903 ± 0.064
NCLOF ($D = 0.22, k = 500, \varepsilon = 0.3$)	0.935	0.632	0.889 ± 0.080

Table 4. ARI results for artificial dataset 4.

method	max	min	ave. ± sd
k-means	0.741	0.372	0.626 ± 0.142
DBSCAN (Eps = 0.28, MinPts = 5)	0.953	0.953	0.953 ± 0.000
NC ($D = 0.54$)	0.904	0.336	0.740 ± 0.181
NCLOF ($D = 0.54, k = 9, \varepsilon = 0.09$)	**1.000**	0.273	0.728 ± 0.205
NCLOF ($D = 0.54, k = 9, \varepsilon = 0.3$)	**1.000**	0.309	0.779 ± 0.189
NCLOF ($D = 0.54, k = 9, \varepsilon = 1.5$)	0.977	0.268	0.758 ± 0.206
NCLOF ($D = 0.54, k = 50, \varepsilon = 0.3$)	0.870	0.309	0.696 ± 0.157

4.3 Parameters Dependency

In this section, we examine the impact of parameters in NCLOF on cluster partitioning. NCLOF has D, which is the dissimilarity of noise clusters, k (a parameter in LOF) used in the process of calculating γ_l, and γ_l (used in (9)), which gives weight to the dissimilarity. Organizing the impact of each parameter on cluster partitioning is an important guideline. Artificial datasets 2 and 4 are used to show the change in ARI when each parameter is varied. In Figs. 5, 6, 7, 8, 9 and 10, The vertical axis shows the value of ARI, and the horizontal axis shows the value of ε. The green "▲" represents the average value of ARI.

First, let us discuss the parameter ε. Figure 5 shows a box plot for artificial dataset 2 with NCLOF set to $k = 5, D = 0.2$ and varying ε. Similarly, Fig. 6 shows a box plot of artificial dataset 4 with NCLOF set to $k = 5, D = 0.2$ and ε varied. From Figs. 5 and 6, it can be suggested that there is no significant difference in the results where the value of ε is greater than 0.3. Too small values of ε tend to show lower values at the maximum ARI. It is thought to be because there is more opportunity to assign the $LOF_k(x_l)$ as it is to the variable that penalizes dissimilarity with the cluster center. If the value is too large, there are more opportunities to assign 1 to the variable that penalizes dissimilarity with the cluster center, resulting in a cluster partition that is no different from conventional noise clustering. Therefore, it is considered necessary to set a value of ε that is neither too small nor too large. In this experiment, setting a value between 0.3 and 0.6 tended to give good results.

Next, let us discuss the parameter k. Figure 7 shows a box plot of NCLOF for artificial dataset 2 with $\varepsilon = 0.3, D = 0.2$, and k varied. Similarly, Fig. 8 shows a box plot for artificial dataset 4 when NCLOF is set to $\varepsilon = 0.3, D = 0.2$, and k is varied. Increasing the value of k does not change the result much. The suitable value depends on the dataset. However, if the value is too large, increasing the value will not change the result much. In addition, the amount of calculation may increase. From the above, it is considered better not to make the value too large. When k was set to a small value, such as 2 or 3, the maximum value of ARI tended to decrease. Therefore, it is considered that values of k, such as 2 and 3 should be avoided. In this experiment, it can be confirmed that good results are obtained when k is set to 6 or higher.

Finally, let us discuss the parameter D. Figure 9 shows a plot for artificial dataset 2 with NCLOF set to $k = 5, \varepsilon = 0.3$, and D varied. Similarly, Fig. 10 shows a box plot for artificial dataset 4 when NCLOF is set to $k = 10, \varepsilon = 0.6$, and D is varied. The value of D must be determined experimentally since even a change of 0.1 tends to change the results significantly. This is considered to be the same as the feature of conventional noise clustering.

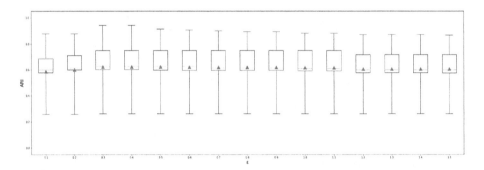

Fig. 5. Box plot showing the change in ARI when NCLOF$(k = 5, D = 0.2)$ is executed for artificial dataset 2 by varying ε.

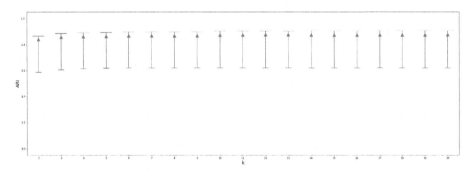

Fig. 6. Box plot showing the change in ARI when NCLOF$(k = 5, D = 0.2)$ is executed for artificial dataset 4 by varying ε.

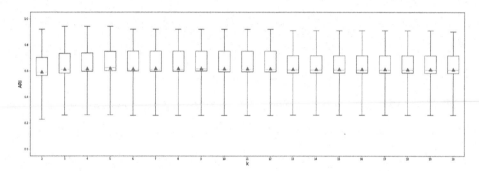

Fig. 7. Box plot showing the change in ARI when NCLOF($\varepsilon = 0.3, D = 0.2$) is executed for artificial dataset 2 by varying k.

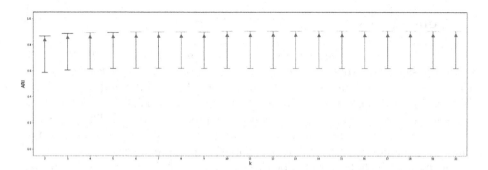

Fig. 8. Box plot showing the change in ARI when NCLOF($\varepsilon = 0.3, D = 0.2$) is executed for artificial dataset 4 by varying k.

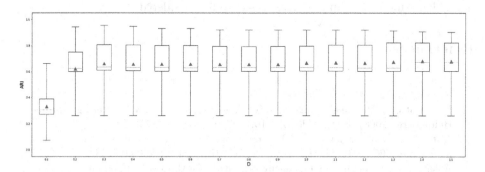

Fig. 9. Box plot showing the change in ARI when NCLOF($k = 5, \varepsilon = 0.3$) is executed for artificial dataset 2 by varying D.

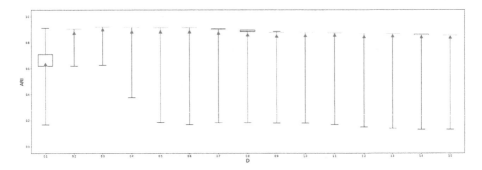

Fig. 10. Box plot showing the change in ARI when NCLOF($k = 10, \varepsilon = 0.6$) is executed for artificial dataset 4 by varying D.

5   Conclusions

In this paper, a new clustering method, noise clustering based on LOF(NCLOF), was proposed. The proposed method is an extension of noise clustering and is a fusion of an approach that builds an outlier-robust model and an approach that handles each object as an outlier. Numerical experiments on four artificial datasets were suggested the effectiveness of the proposed method. The properties of the three parameters included in the proposed method are also verified. For ε and k, guidelines for parameter determination was provided.

Future work includes verification through numerical experiments using outlier detection datasets [10]. In addition, comparative experiments with clustering methods that are robust against outliers are also needed. Application to the automatic estimation of the number of clusters by sequential extraction using noise clustering is also an important issue to be considered.

References

1. Han, J., Kamber, M., Pei, J.: Data Mining: Concepts and Techniques, Morgan Kaufmann, (2012)
2. Miyamoto, S., Ichihashi, H., Honda, K.: Algorithms for Fuzzy Clustering. Springer, Heidelberg (2008). https://doi.org/10.1007/978-3-540-78737-2
3. Jain, A.K.: Data clustering: 50 years beyond K-means. Pattern Recogn. Lett. **31**(8), 651–666 (2010). https://doi.org/10.1016/j.patrec.2009.09.011
4. Askari, S.: Fuzzy c-means clustering algorithm for data with unequal cluster sizes and contaminated with noise and outliers: review and development. Expert Syst. Appl. **165**(1), 113856 (2021). https://doi.org/10.1016/j.eswa.2020.113856
5. Davé, R.N.: Characterization and detection of noise in clustering. Pattern Recogn. Lett. **12**, 657–664 (1991). https://doi.org/10.1016/0167-8655(91)90002-4
6. Krishnapuram, R., Keller, J.M.: A possibilistic approach to clustering. IEEE Trans. Fuzzy Syst. **1**(2), 98–110 (1993). https://doi.org/10.1109/91.227387
7. Ester, M., Kriegel, H.-P., Sander, J., Xu, X.: A density-based algorithm for discovering clusters in large spatial databases with noise. In: Proceedings of the Second

International Conference on Knowledge Discovery and Data Mining (KDD 1996), pp. 226–231 (1996). https://doi.org/10.1145/335191.335388

8. Breunig, M.M., Kriegel, H.-P., Ng, R.T., Sander, J.: LOF: identifying density-based local outliers. ACM SIGMOD Rec. **29**, 93–104 (2000). https://doi.org/10.1145/335191.335388

9. Hubert, L., Arabie, P.: Comparing partitions. J. Classif. **2**(1), 193–218 (1985)

10. Rayana, S.: ODDS Library, Stony Brook, NY: Stony Brook University, Department of Computer Science (2016). http://odds.cs.stonybrook.edu

On Some Fuzzy Clustering Algorithms Based on Series Models

Tomoki Nomura and Yuchi Kanzawa$^{(\boxtimes)}$ (iD)

Shibaura Institute of Technology, 3-7-5 Toyosu, Koto, Tokyo 135-8548, Japan
{ma23151,kanzawa}@shibaura-it.ac.jp

Abstract. Fuzzy c-means is a basic and general fuzzy clustering algorithm for vectorial data and several variants of this algorithm have been proposed in the literature. However, research on fuzzy clustering for series data is not as advanced as that for vectorial data. To the best of our knowledge, no fuzzy clustering algorithms for series data using series models have been proposed. In this paper, we proposed several fuzzy clustering algorithms based on a combination of three types of series models, namely the autoregressive and moving average model, hidden Markov model, and linear Gaussian state space model, as well as three types of fuzzification techniques, namely Kullback-Leibler divergence regularization, Bezdek-type fuzzification, and q-divergence basis. Numerical experiments demonstrate the superiority of the proposed methods compared to the conventional methods in terms of clustering accuracy.

Keywords: Fuzzy clustering · Series data · Series models

1 Introduction

Fuzzy c-means (FCM) [1] is a basic and general fuzzy clustering algorithm for vectorial data. Several variants of this algorithm have been proposed based on different fuzzification techniques, including the Entropy-regularized FCM (EFCM) [2] and Tsallis entropy-regularized FCM (TFCM) [3]. Differentiating it from other variants the traditional FCM is referred to as the Bezdek-type FCM (BFCM) in this paper. None of these algorithms explicitly consider cluster size. To address this issue, several FCM variants with variables for controlling cluster size have been proposed, including Kullback-Leibler divergence-regularized FCM (KLFCM) [4], modified BFCM (mBFCM) [4], and q-divergence-based FCM (QFCM) [5]. The corresponding fuzzification methods are referred to as KL-type, B-type, and Q-type fuzzification in this paper.

Series data analysis is important for science, business, finance, and economics. One method for series data analysis is a series model. Examples of such models include the autoregressive and moving average (ARMA) model, hidden Markov model (HMM), and linear Gaussian state space model (LGSSM). The ARMA model combines two types of series data processes, such as autoregressive and moving average processes. The state space model is one in which state variables

© The Author(s), under exclusive license to Springer Nature Switzerland AG 2024
K. Honda et al. (Eds.): IUKM 2023, LNAI 14376, pp. 192–204, 2024.
https://doi.org/10.1007/978-3-031-46781-3_17

exist behind the observed values and the observed values are determined by state variables. The HMM is a model in which the state variables are discrete and LGSSM is a model in which the noise terms follow Gaussian distributions in the state space model.

Various fuzzy clustering algorithms for series data have also been proposed, including the auto-correlation-based fuzzy c-means [6] and quantile cross-spectral density based fuzzy c-means clustering model [7]. These algorithms are feature-based approach. Additional clustering algorithms have also been proposed using model-based approach. Xiong et al. proposed the ARMA mixtures [8], and this algorithm is based on ARMA model. Alon et al. proposed the mixture of HMMs [9], and this algorithm is based on HMM. Umatani et al. proposed the mixtures of LGSSMs [10], and this algorithm is based on LGSSM. These algorithms are mixture models for the ARMA model, HMM, and LGSSM. However, to the best of our knowledge, no fuzzy clustering algorithms based on these series models have been proposed.

In this paper, we focus on fuzzy clustering algorithms based on series models. We proposed nine fuzzy clustering algorithms based on three series models: ARMA, HMM, and LGSSM. First, we propose FCM with KL-type fuzzification based on each series model. Each algorithm is described as ARMA model-based KLFCM (ARMA-KLFCM), HMM-based KLFCM (HMM-KLFCM), and LGSSM-based KLFCM (LGSSM-KLFCM). Second, we proposed FCM with B-type fuzzification based on each series model. Each algorithm is described as ARMA model-based BFCM (ARMA-BFCM), HMM-based BFCM (HMM-BFCM), and LGSSM-based BFCM (LGSSM-BFCM). Third, we proposed FCM with Q-type fuzzification based on each series model. Each algorithm is described as ARMA model-based QFCM (ARMA-QFCM), HMM-based QFCM (HMM-QFCM), and LGSSM-based QFCM (LGSSM-QFCM).

The remainder of this paper is organized as follows: In Sect. 2, we introduce the conventional clustering algorithms. In Sect. 3, we describe the nine proposed clustering algorithms based on series models. In Sect. 4, we present the results of numerical experiments. In Sect. 5, we describe the conclusions of this paper.

2   Preliminaries

2.1   Clustering for Vectorial Data

We consider partitioning a dataset of D-dimensional objects denoted as $X = \{x_k \mid x_k \in \mathbb{R}^D, k \in \{1, \ldots, N\}\}$ into C clusters. Let $v = \{v_i \in \mathbb{R}^D \mid i \in \{1, \ldots, C\}\}$ be the set of centroids for the clusters. Let $u_{i,k}$ ($i \in \{1, \ldots, C\}, k \in \{1, \ldots, N\}$) be the membership degree of x_k in cluster #i. Let π_i ($i \in \{1, \ldots, C\}$) be the variables controlling the i-th cluster size. u and π satisfy the following constraints:

$$\sum_{i=1}^{C} u_{i,k} = 1, u_{i,k} \in [0,1], \quad \sum_{i=1}^{C} \pi_i = 1, \pi_i \in (0,1). \tag{1}$$

Clustering methods for the KLFCM, mBFCM, and QFCM were constructed
based on the following optimization problems:

$$\underset{u,v,\pi}{\text{minimize}} \sum_{i=1}^{C}\sum_{k=1}^{N} u_{i,k} d_{i,k}^{(\text{FCM})} + \lambda^{-1}\sum_{i=1}^{C}\sum_{k=1}^{N} u_{i,k}\ln\left(\frac{u_{i,k}}{\pi_i}\right), \tag{2}$$

$$\underset{u,v,\pi}{\text{minimize}} \sum_{i=1}^{C}\sum_{k=1}^{N} (\pi_i)^{1-m}(u_{i,k})^m d_{i,k}^{(\text{FCM})}, \tag{3}$$

$$\underset{u,v,\pi}{\text{minimize}} \sum_{i=1}^{C}\sum_{k=1}^{N} (\pi_i)^{1-m}(u_{i,k})^m d_{i,k}^{(\text{FCM})} + \frac{\lambda^{-1}}{m-1}\sum_{i=1}^{C}\sum_{k=1}^{N} (\pi_i)^{1-m}(u_{i,k})^m, \tag{4}$$

subject to Eq. (1), where $d_{i,k}^{(\text{FCM})} = \|x_k - v_i\|_2^2$ denotes the dissimilarity between
object x_k and cluster $\#i$, and $\lambda > 0, m > 1$ are fuzzification parameters.

2.2   Clustering for Series Data

Let $X = \{x_{k,t} \mid x_{k,t} \in \mathbb{R}^D, k \in \{1,\ldots,N\}, t \in \{1,\ldots,T\}\}$ be a dataset of N
series, where each series is a set of D-dimensional vectors of length T, where
$D = 1$ for the ARMA model.

First, we consider a clustering method based on the ARMA model. The
coefficients of ARMA are denoted as $\phi = \{\phi_i|\phi_i \in \mathbb{R}^{p+1}, i \in \{1,\ldots,C\}\}$ and
$\psi = \{\psi_i|\psi_i \in \mathbb{R}^q, i \in \{1,\ldots,C\}\}$, where p and q are the autoregressive and
moving average orders, respectively. The sequence of independent and identi-
cally distributed Gaussian white noise is denoted as $e = \{e_{i,k,t}|e_{i,k,t} \in \mathbb{R}, i \in \{1,\ldots,C\}, k \in \{1,\ldots,N\}, t \in \{1,\ldots,T\}\}$ and e is estimated using the follow-
ing form:

$$e_{i,k,t} = x_{k,t} - \phi_{i,0} - \sum_{j=1}^{p}\phi_{i,j}x_{k,t-j} - \sum_{j=1}^{q}\psi_{i,j}e_{i,k,t-j}. \tag{5}$$

The set of variances of Gaussian white noise for clusters $\{1,\ldots,C\}$ is denoted
as $\sigma^2 = \{\sigma_i^2|\sigma_i^2 \in \mathbb{R}, i \in \{1,\ldots,C\}\}$. The ARMA mixtures is constructed based
on the following optimization problem:

$$\underset{u,\phi,\psi,\sigma^2,\pi}{\text{minimize}} \sum_{i=1}^{C}\sum_{k=1}^{N} u_{i,k} d_{i,k}^{(\text{ARMA})} + \sum_{i=1}^{C}\sum_{k=1}^{N} u_{i,k}\ln\left(\frac{u_{i,k}}{\pi_i}\right), \tag{6}$$

subject to Eq. (1), where,

$$d_{i,k}^{(\text{ARMA})} = \frac{T}{2}\ln(2\pi) + \frac{T}{2}\ln(\sigma_i^2) + \frac{1}{2\sigma_i^2}\sum_{t=1}^{T} e_{i,k,t}^2. \tag{7}$$

Second, we consider a clustering method based on the HMM. Let L be the
number of states in the HMM. Let $\omega = \{\omega_i|\omega_i \in \mathbb{R}^L, i \in \{1,\ldots,C\}\}$ and

$A = \{A_i | A_i \in \mathbb{R}^{L \times L}, i \in \{1, \ldots, C\}\}$ be the set of initial state probability distributions for the clusters and the set of state transition probability distributions for the clusters, respectively. ω and A have the constraints

$$\sum_{\ell=1}^{L} \omega_{i,\ell} = 1, \omega_{i,\ell} \in [0,1], \quad \sum_{\ell=1}^{L} A_{i,\ell,\ell'} = 1, A_{i,\ell,\ell'} \in [0,1]. \tag{8}$$

The set of means of Gaussian observation probability density for the clusters $\{1, \ldots, C\}$ and states $\{1, \ldots, L\}$ is denoted as $\nu = \{\nu_{i,\ell} | \nu_{i,\ell} \in \mathbb{R}^D, i \in \{1, \ldots, C\}, \ell \in \{1, \ldots, L\}\}$. The set of covariance matrices of Gaussian probability observation density for the clusters $\{1, \ldots, C\}$ and states $\{1, \ldots, L\}$ is denoted as $\Psi = \{\Psi_{i,\ell} | \Psi_{i,\ell} \in \mathbb{R}^{D \times D}, i \in \{1, \ldots, C\}, \ell \in \{1, \ldots, L\}\}$. The probability of being in state $\#\ell$ of the t-th order for the i-th cluster given $x_{k,1}, x_{k,2}, \ldots, x_{k,t}$ is denoted as $\alpha_{i,k,t,\ell}$ and the set of $\alpha_{i,k,t,\ell}$ ($i \in \{1, \ldots, C\}, k \in \{1, \ldots, N\}, t \in \{1, \ldots, T\}, \ell \in \{1, \ldots, L\}$) is denoted as α. The probability of being in state $\#\ell$ of the t-th order for the i-th cluster given $x_{k,t+1}, x_{k,t+2}, \ldots, x_{k,T}$ is denoted as $\beta_{i,k,t,\ell}$ and the set of $\beta_{i,k,t,\ell}$ ($i \in \{1, \ldots, C\}, k \in \{1, \ldots, N\}, t \in \{1, \ldots, T\}, \ell \in \{1, \ldots, L\}$) is denoted as β. The probability of being in state $\#\ell$ of the t-th order for the i-th cluster given x_k is denoted as $\gamma_{i,k,t,\ell}$ and the set of $\gamma_{i,k,t,\ell}$ ($i \in \{1, \ldots, C\}, k \in \{1, \ldots, N\}, t \in \{1, \ldots, T\}, \ell \in \{1, \ldots, L\}$) is denoted as γ. The probability of being in state $\#\ell$ of the $t-1$-th order and state $\#\ell'$ of the t-th order for the i-th cluster given x_k is denoted as $\xi_{i,k,t,\ell,\ell'}$ and the set of $\xi_{i,k,t,\ell,\ell'}$ ($i \in \{1, \ldots, C\}, k \in \{1, \ldots, N\}, t \in \{1, \ldots, T\}, \ell \in \{1, \ldots, L\}, \ell' \in \{1, \ldots, L\}$) is denoted as ξ. The values of α, β, γ, and ξ can be obtained using the forward-backward algorithm with $\{\omega_i, A_i, \nu_i, \Psi_i\}$ ($i \in \{1, \ldots, C\}$) and x_k ($k \in \{1, \ldots, N\}$). The mixture of HMMs is constructed based on the following optimization problem:

$$\underset{u,\omega,A,\nu,\Psi,\pi}{\text{minimize}} \sum_{i=1}^{C} \sum_{k=1}^{N} u_{i,k} d_{i,k}^{(\text{HMM})} + \sum_{i=1}^{C} \sum_{k=1}^{N} u_{i,k} \ln\left(\frac{u_{i,k}}{\pi_i}\right), \tag{9}$$

subject to Eqs. (1) and (8), where,

$$d_{i,k}^{(\text{HMM})} = -\ln\left(\sum_{\ell=1}^{L} \alpha_{i,k,t,\ell} \beta_{i,k,t,\ell}\right) \quad (\forall t \in \{1, \ldots, T\}). \tag{10}$$

Third, we consider a clustering method based on the LGSSM. Let M be the dimensionality of the state variables. The set of state variables behind each series $x_{k,t}$ for the clusters $\{1, \ldots, C\}$ is denoted as $Z = \{z_{i,k,t} | z_{i,k,t} \in \mathbb{R}^M, i \in \{1, \ldots, C\}, k \in \{1, \ldots, N\}, t \in \{1, \ldots, T\}\}$. The set of means of initial state distributions for the clusters $\{1, \ldots, C\}$ is denoted as $\mu = \{\mu_i | \mu_i \in \mathbb{R}^M, i \in \{1, \ldots, C\}\}$. The set of covariance matrices of initial state distributions for the clusters $\{1, \ldots, C\}$ is denoted as $P = \{P_i | P_i \in \mathbb{R}^{M \times M}, i \in \{1, \ldots, C\}\}$. The set of transition matrices for the clusters $\{1, \ldots, C\}$ is denoted as $G = \{G_i | G_i \in \mathbb{R}^{M \times M}, i \in \{1, \ldots, C\}\}$. The set of transition covariance matrices for the clusters $\{1, \ldots, C\}$ is denoted as $\Gamma = \{\Gamma_i | \Gamma_i \in \mathbb{R}^{M \times M}, i \in \{1, \ldots, C\}\}$. The set of observation matrices for the clusters $\{1, \ldots, C\}$ is denoted as $F = \{F_i | F_i \in \mathbb{R}^{D \times M}, i \in \{1, \ldots, C\}\}$. The set of observation covariance matrices for the clusters $\{1, \ldots, C\}$ is denoted as $\Sigma = \{\Sigma_i | \Sigma_i \in \mathbb{R}^{D \times D}, i \in \{1, \ldots, C\}\}$. The expected value of $z_{i,k,t}$ with respect to $z_{i,k,t}$ given $x_{k,1}, x_{k,2}, \ldots, x_{k,j}$ and $\{\mu_i, P_i, G_i, \Gamma_i, F_i, \Sigma_i\}$ is denoted as $\hat{\mu}_{i,k,t|j}$ and the set of $\hat{\mu}_{i,k,t|j}$ ($i \in \{1, \ldots, C\}, k \in \{1, \ldots, N\}, t \in \{1, \ldots, T\}, j \in \{0, \ldots, T\}$) is denoted as $\hat{\mu}$. The expected value of $(z_{i,k,t} - \hat{\mu}_{i,k,t|j})(z_{i,k,t} - \hat{\mu}_{i,k,t|j})^\top$, with respect to $z_{i,k,t}$ given $x_{k,1}, x_{k,2}, \ldots, x_{k,j}$ and $\{\mu_i, P_i, G_i, \Gamma_i, F_i, \Sigma_i\}$ is denoted as $\hat{V}_{i,k,t|j}$ and the set of $\hat{V}_{i,k,t|j}$ ($i \in \{1, \ldots, C\}, k \in \{1, \ldots, N\}, t \in \{1, \ldots, T\}, j \in \{0, \ldots, T\}$) is denoted as $\hat{V}$. The value of $\hat{\mu}$ and $\hat{V}$ can be obtained using the Kalman filter and smoother [12]. The mixtures of LGSSMs is constructed based on the following optimization problem:

$$\underset{u, \mu, P, G, \Gamma, F, \Sigma, \pi}{\text{minimize}} \sum_{i=1}^{C} \sum_{k=1}^{N} u_{i,k} d_{i,k}^{(\text{LGSSM})} + \sum_{i=1}^{C} \sum_{k=1}^{N} u_{i,k} \ln \left(\frac{u_{i,k}}{\pi_i} \right), \qquad (11)$$

subject to Eq. (1), where,

$$d_{i,k}^{(\text{LGSSM})} = \frac{DT}{2} \ln(2\pi) + \frac{1}{2} \sum_{t=1}^{T} \ln \left(\det \left(F_i \hat{V}_{i,k,t|t-1} F_i^\top + \Sigma_i \right) \right)$$

$$+ \frac{1}{2} \sum_{t=1}^{T} \left(x_{k,t} - F_i \hat{\mu}_{i,k,t|t-1} \right)^\top \left(F_i \hat{V}_{i,k,t|t-1} F_i^\top + \Sigma_i \right)^{-1} \left(x_{k,t} - F_i \hat{\mu}_{i,k,t|t-1} \right).$$

$$(12)$$

The details of the algorithms associated with conventional methods have been omitted for the sake of brevity.

3   Proposed Methods

3.1   Concept

In this paper, we introduce nine fuzzy clustering methods based on series models such as ARMA, HMM, and LGSSM.

First, we propose the ARMA-KLFCM, HMM-KLFCM, and LGSSM-KLFCM. These objective functions were formulated by incorporating a fuzzification parameter into the KL-divergence term within the lower bound of the log-likelihood function of each mixture model. The ARMA-KLFCM, HMM-KLFCM, and LGSSM-KLFCM were constructed based on the following optimization problems:

$$\underset{u,\phi,\psi,\sigma^2,\pi}{\text{minimize}} \sum_{i=1}^{C} \sum_{k=1}^{N} u_{i,k} d_{i,k}^{(\text{ARMA})} + \lambda^{-1} \sum_{i=1}^{C} \sum_{k=1}^{N} u_{i,k} \ln \left(\frac{u_{i,k}}{\pi_i} \right), \tag{13}$$

$$\underset{u,\omega,A,\nu,\Psi,\pi}{\text{minimize}} \sum_{i=1}^{C} \sum_{k=1}^{N} u_{i,k} d_{i,k}^{(\text{HMM})} + \lambda^{-1} \sum_{i=1}^{C} \sum_{k=1}^{N} u_{i,k} \ln \left(\frac{u_{i,k}}{\pi_i} \right), \tag{14}$$

$$\underset{u,\mu,P,G,\Gamma,F,\Sigma,\pi}{\text{minimize}} \sum_{i=1}^{C} \sum_{k=1}^{N} u_{i,k} d_{i,k}^{(\text{LGSSM})} + \lambda^{-1} \sum_{i=1}^{C} \sum_{k=1}^{N} u_{i,k} \ln \left(\frac{u_{i,k}}{\pi_i} \right). \tag{15}$$

subject to Eq. (1) for the ARMA-KLFCM and LGSSM-KLFCM, and Eqs. (1) and (8) for the HMM-KLFCM. These optimization problems are regarded as replacing $d_{i,k}^{(\text{FCM})}$ with $d_{i,k}^{(\text{ARMA})}$, $d_{i,k}^{(\text{HMM})}$, and $d_{i,k}^{(\text{LGSSM})}$ in the KLFCM optimization problem.

Second, we propose the ARMA-BFCM, HMM-BFCM, and LGSSM-BFCM. These optimization problems were constructed by replacing $d_{i,k}^{(\text{FCM})}$ with $d_{i,k}^{(\text{ARMA})}$, $d_{i,k}^{(\text{HMM})}$, and $d_{i,k}^{(\text{LGSSM})}$ in the mBFCM optimization problem. The ARMA-BFCM, HMM-BFCM, and LGSSM-BFCM were constructed based on the following optimization problems:

$$\underset{u,\phi,\psi,\sigma^2,\pi}{\text{minimize}} \sum_{i=1}^{C} \sum_{k=1}^{N} (\pi_i)^{1-m} (u_{i,k})^m d_{i,k}^{(\text{ARMA})}, \tag{16}$$

$$\underset{u,\omega,A,\nu,\Psi,\pi}{\text{minimize}} \sum_{i=1}^{C} \sum_{k=1}^{N} (\pi_i)^{1-m} (u_{i,k})^m d_{i,k}^{(\text{HMM})}, \tag{17}$$

$$\underset{u,\mu,P,G,\Gamma,F,\Sigma,\pi}{\text{minimize}} \sum_{i=1}^{C} \sum_{k=1}^{N} (\pi_i)^{1-m} (u_{i,k})^m d_{i,k}^{(\text{LGSSM})}. \tag{18}$$

subject to Eq. (1) for the ARMA-BFCM and LGSSM-BFCM, and Eqs. (1) and (8) for the HMM-BFCM.

Third, we propose the ARMA-QFCM, HMM-QFCM, and LGSSM-QFCM. These optimization problems were constructed by replacing $d_{i,k}^{(\text{FCM})}$ with $d_{i,k}^{(\text{ARMA})}$, $d_{i,k}^{(\text{HMM})}$, and $d_{i,k}^{(\text{LGSSM})}$ in the QFCM optimization problem. The ARMA-QFCM, HMM-QFCM, and LGSSM-QFCM were constructed based on the following optimization problems:

$$\underset{u,\phi,\psi,\sigma^2,\pi}{\text{minimize}} \sum_{i=1}^{C}\sum_{k=1}^{N}(\pi_i)^{1-m}(u_{i,k})^m d_{i,k}^{(\text{ARMA})} + \frac{\lambda^{-1}}{m-1}\sum_{i=1}^{C}\sum_{k=1}^{N}(\pi_i)^{1-m}(u_{i,k})^m, \quad (19)$$

$$\underset{u,\omega,A,\nu,\Psi,\pi}{\text{minimize}} \sum_{i=1}^{C}\sum_{k=1}^{N}(\pi_i)^{1-m}(u_{i,k})^m d_{i,k}^{(\text{HMM})} + \frac{\lambda^{-1}}{m-1}\sum_{i=1}^{C}\sum_{k=1}^{N}(\pi_i)^{1-m}(u_{i,k})^m, \quad (20)$$

$$\underset{u,\mu,P,G,\Gamma,F,\Sigma,\pi}{\text{minimize}} \sum_{i=1}^{C}\sum_{k=1}^{N}(\pi_i)^{1-m}(u_{i,k})^m d_{i,k}^{(\text{LGSSM})} + \frac{\lambda^{-1}}{m-1}\sum_{i=1}^{C}\sum_{k=1}^{N}(\pi_i)^{1-m}(u_{i,k})^m.$$
$$(21)$$

subject to Eq. (1) for the ARMA-QFCM and LGSSM-QFCM, and Eqs. (1) and (8) for the HMM-QFCM.

3.2 Algorithm

The proposed clustering algorithms were derived by solving the optimization problems presented in Eqs. (13)–(21) under the constraints of Eq. (1) for the ARMA-KLFCM, ARMA-BFCM, ARMA-QFCM, LGSSM-KLFCM, LGSSM-BFCM, and LGSSM-QFCM, and Eqs. (1) and (8) for the HMM-KLFCM, HMM-BFCM, and HMM-QFCM. The necessary conditions for optimality are summarized in following algorithm, where certain details are omitted for the sake of brevity.

Algorithms 1 (ARMA-KLFCM,   ARMA-BFCM,   ARMA-QFCM, HMM-KLFCM, HMM-BFCM, HMM-QFCM, LGSSM-KLFCM, LGSSM-BFCM, LGSSM-QFCM).

STEP 1. Set the number of clusters C. Set the initial membership u and the initial variables controlling the cluster size π. Set the fuzzification parameters (m, λ). Set the autoregressive order p and moving average order q for the ARMA-KLFCM, ARMA-BFCM, and ARMA-QFCM, the number of states L for the HMM-KLFCM, HMM-BFCM, and HMM-QFCM, and the dimension of state variables M for the LGSSM-KLFCM, LGSSM-BFCM, and LGSSM-QFCM. Set the initial series models parameters $\{\phi, \psi, \sigma^2\}$ for the ARMA-KLFCM, ARMA-BFCM, and ARMA-QFCM, $\{\omega, A, \nu, \Psi\}$ for the HMM-KLFCM, HMM-BFCM, and HMM-QFCM, $\{\mu, P, G, \Gamma, F, \Sigma\}$ for the LGSSM-KLFCM, LGSSM-BFCM, and LGSSM-QFCM.

STEP 2. Calculate α, β, γ and ξ using the forward-backward algorithm for the HMM-KLFCM, HMM-BFCM, and HMM-QFCM. Calculate $\hat{\mu}$ and $\hat{V}$ using the Kalman filter and smoother for the LGSSM-KLFCM, LGSSM-BFCM, and LGSSM-QFCM.

STEP 3. Calculate the intermediate variables $\hat{Z}_{i,k,t}$, $\check{Z}_{i,k,t}$ and $J_{i,k,t}$ as

$$\hat{Z}_{i,k,t} = \hat{V}_{i,k,t|T} J_{i,k,t-1}^{\top} - \hat{\mu}_{i,k,t|T} \hat{\mu}_{i,k,t-1\,T}^{\top}, \tag{22}$$

$$\check{Z}_{i,k,t} = \hat{V}_{i,k,t|T} - \hat{\mu}_{i,k,t|T} \hat{\mu}_{i,k,t|T}^{\top}, \tag{23}$$

$$J_{i,k,t} = \hat{V}_{i,k,t|t} G_i^{\top} \hat{V}_{i,k,t+1|t}^{-1}. \tag{24}$$

STEP 4. Calculate d as $d_{i,k} = d_{i,k}^{(\mathrm{ARMA})}$ for the ARMA-KLFCM, ARMA-BFCM, and ARMA-QFCM, $d_{i,k} = d_{i,k}^{(\mathrm{HMM})}$ for the HMM-KLFCM, HMM-BFCM, and HMM-QFCM, $d_{i,k} = d_{i,k}^{(\mathrm{LGSSM})}$ for the LGSSM-KLFCM, LGSSM-BFCM, and LGSSM-QFCM.

STEP 5. Calculate u as

$$u_{i,k} = \left[\sum_{i'=1}^{C} \frac{\pi_{i'}}{\pi_i} \exp\left(\lambda(d_{i,k} - d_{i',k}) \right) \right]^{-1} \tag{25}$$

for the ARMA-KLFCM, HMM-KLFCM, and LGSSM-KLFCM,

$$u_{i,k} = \left[\sum_{i'=1}^{C} \frac{\pi_{i'}}{\pi_i} \left(\frac{d_{i,k}}{d_{i',k}} \right)^{1/(m-1)} \right]^{-1} \tag{26}$$

for the ARMA-BFCM, HMM-BFCM, and LGSSM-BFCM,

$$u_{i,k} = \left[\sum_{i'=1}^{C} \frac{\pi_{i'}}{\pi_i} \left(\frac{1 - \lambda(1-m)d_{i',k}}{1 - \lambda(1-m)d_{i,k}} \right)^{1/(1-m)} \right]^{-1} \tag{27}$$

for the ARMA-QFCM, HMM-QFCM, and LGSSM-QFCM.

STEP 6. Calculate π as

$$\pi_i = \frac{\sum_{k=1}^{N} u_{i,k}}{N} \tag{28}$$

for the ARMA-KLFCM, HMM-KLFCM, and LGSSM-KLFCM,

$$\pi_i = \left[\sum_{i'=1}^{C} \left(\frac{\sum_{k=1}^{N} (u_{i',k})^m d_{i',k}}{\sum_{k=1}^{N} (u_{i,k})^m d_{i,k}} \right)^{1/m} \right]^{-1} \tag{29}$$

for the ARMA-BFCM, HMM-BFCM, and LGSSM-BFCM,

$$\pi_i = \left[\sum_{i'=1}^{C} \left(\frac{\sum_{k=1}^{N} (u_{i',k})^m \left(1 - \lambda(1-m)d_{i',k} \right)}{\sum_{k=1}^{N} (u_{i,k})^m \left(1 - \lambda(1-m)d_{i,k} \right)} \right)^{1/m} \right]^{-1} \tag{30}$$

for the ARMA-QFCM, HMM-QFCM, and LGSSM-QFCM.

Step 7. Calculate series models parameters as

$$\left(\sum_{k=1}^{N} u_{i,k} W_{i,k} \right) \eta_i = \sum_{k=1}^{N} u_{i,k} h_{i,k}, \tag{31}$$

$$\sigma_i^2 = \frac{\sum_{k=1}^{N} u_{i,k} \sum_{t=1}^{T} e_{i,k,t}^2}{T \sum_{k=1}^{N} u_{i,k}}, \tag{32}$$

where,

$$W_{i,k} = \begin{pmatrix} \langle 1,1 \rangle_{0,0} & \langle 1,x_k \rangle_{0,1} & \cdots & \langle 1,x_k \rangle_{0,p} & \langle 1,e_{i,k} \rangle_{0,1} & \cdots & \langle 1,e_{i,k} \rangle_{0,q} \\ \langle x_k,1 \rangle_{1,0} & \langle x_k,x_k \rangle_{1,1} & \cdots & \langle x_k,x_k \rangle_{1,p} & \langle x_k,e_{i,k} \rangle_{1,1} & \cdots & \langle x_k,e_{i,k} \rangle_{1,q} \\ \vdots & \vdots & \ddots & \vdots & \vdots & \ddots & \vdots \\ \langle x_k,1 \rangle_{p,0} & \langle x_k,x_k \rangle_{p,1} & \cdots & \langle x_k,x_k \rangle_{p,p} & \langle x_k,e_{i,k} \rangle_{p,1} & \cdots & \langle x_k,e_{i,k} \rangle_{p,q} \\ \langle e_{i,k},1 \rangle_{1,0} & \langle e_{i,k},x_k \rangle_{1,1} & \cdots & \langle e_{i,k},x_k \rangle_{1,p} & \langle e_{i,k},e_{i,k} \rangle_{1,1} & \cdots & \langle e_{i,k},e_{i,k} \rangle_{1,q} \\ \vdots & \vdots & \ddots & \vdots & \vdots & \ddots & \vdots \\ \langle e_{i,k},1 \rangle_{q,0} & \langle e_{i,k},x_k \rangle_{q,1} & \cdots & \langle e_{i,k},x_k \rangle_{q,p} & \langle e_{i,k},e_{i,k} \rangle_{q,1} & \cdots & \langle e_{i,k},e_{i,k} \rangle_{q,q} \end{pmatrix}, \tag{33}$$

$$\eta_i = (\phi_{i,0}, \phi_{i,1}, \ldots, \phi_{i,p}, \psi_{i,1}, \ldots, \psi_{i,q})^\top, \tag{34}$$

$$h_{i,k} = (\langle x_k,1 \rangle_{0,0}, \langle x_k,x_k \rangle_{0,1}, \ldots \langle x_k,x_k \rangle_{0,p}, \langle x_k,e_{i,k} \rangle_{0,1}, \ldots, \langle x_k,e_{i,k} \rangle_{0,q})^\top, \tag{35}$$

$$\langle f,g \rangle_{j,j'} = \sum_{t=1}^{T} f_{t-j} g_{t-j'}, \quad f_t = 0, g_t = 0 \text{ for } t \notin [1,T] \tag{36}$$

ϕ and ψ are obtained by solving Eq. (31) for the ARMA-KLFCM,

$$\left(\sum_{k=1}^{N} (u_{i,k})^m W_{i,k} \right) \eta_i = \sum_{k=1}^{N} (u_{i,k})^m h_{i,k}, \tag{37}$$

$$\sigma_i^2 = \frac{\sum_{k=1}^{N} (u_{i,k})^m \sum_{t=1}^{T} e_{i,k,t}^2}{T \sum_{k=1}^{N} (u_{i,k})^m} \tag{38}$$

where, $W_{i,k}, \eta_i$ and $h_{i,k}$ are calculated using Eqs. (33)–(36), and ϕ and ψ are obtained by solving Eq. (37) for the ARMA-BFCM and ARMA-QFCM,

$$\omega_{i,\ell} = \frac{\sum_{k=1}^{N} u_{i,k} \gamma_{i,k,1,\ell}}{\sum_{k=1}^{N} u_{i,k} \sum_{\ell'=1}^{L} \gamma_{i,k,1,\ell'}}, \tag{39}$$

$$A_{i,\ell,\ell'} = \frac{\sum_{k=1}^{N} u_{i,k} \sum_{t=2}^{T} \xi_{i,k,t,\ell,\ell'}}{\sum_{k=1}^{N} u_{i,k} \sum_{t=2}^{T} \sum_{\ell''=1}^{L} \xi_{i,k,t,\ell,\ell''}}, \tag{40}$$

$$\nu_{i,\ell} = \frac{\sum_{k=1}^{N} u_{i,k} \sum_{t=1}^{T} \gamma_{i,k,t,\ell} x_{k,t}}{\sum_{k=1}^{N} u_{i,k} \sum_{t=1}^{T} \gamma_{i,k,t,\ell}}, \tag{41}$$

$$\Psi_{i,\ell} = \frac{\sum_{k=1}^{N} u_{i,k} \sum_{t=1}^{T} \gamma_{i,k,t,\ell} (x_{k,t} - \nu_{i,\ell})(x_{k,t} - \nu_{i,\ell})^\top}{\sum_{k=1}^{N} u_{i,k} \sum_{t=1}^{T} \gamma_{i,k,t,\ell}} \tag{42}$$

for the HMM-KLFCM,

$$\omega_{i,\ell} = \frac{\sum_{k=1}^{N} (u_{i,k})^m \gamma_{i,k,1,\ell}}{\sum_{k=1}^{N} (u_{i,k})^m \sum_{\ell'=1}^{L} \gamma_{i,k,1,\ell'}}, \tag{43}$$

$$A_{i,\ell,\ell'} = \frac{\sum_{k=1}^{N} (u_{i,k})^m \sum_{t=2}^{T} \xi_{i,k,t,\ell,\ell'}}{\sum_{k=1}^{N} (u_{i,k})^m \sum_{t=2}^{T} \sum_{\ell''=1}^{L} \xi_{i,k,t,\ell,\ell''}}, \tag{44}$$

$$\nu_{i,\ell} = \frac{\sum_{k=1}^{N} (u_{i,k})^m \sum_{t=1}^{T} \gamma_{i,k,t,\ell} x_{k,t}}{\sum_{k=1}^{N} (u_{i,k})^m \sum_{t=1}^{T} \gamma_{i,k,t,\ell}}, \tag{45}$$

$$\Psi_{i,\ell} = \frac{\sum_{k=1}^{N} (u_{i,k})^m \sum_{t=1}^{T} \gamma_{i,k,t,\ell} \left(x_{k,t} - \nu_{i,\ell} \right) \left(x_{k,t} - \nu_{i,\ell} \right)^{\top}}{\sum_{k=1}^{N} (u_{i,k})^m \sum_{t=1}^{T} \gamma_{i,k,t,\ell}} \tag{46}$$

for the HMM-BFCM and HMM-QFCM,

$$\mu_i = \frac{\sum_{k=1}^{N} u_{i,k} \hat{\mu}_{i,k,1\,T}}{\sum_{k=1}^{N} u_{i,k}}, \tag{47}$$

$$P_i = \frac{\sum_{k=1}^{N} u_{i,k} \left(\check{Z}_{i,k,1} - \mu_i \hat{\mu}_{i,k,1\,T}^{\top} - \hat{\mu}_{i,k,1\,T} \mu_i^{\top} + \mu_i \mu_i^{\top} \right)}{\sum_{k=1}^{N} u_{i,k}}, \tag{48}$$

$$G_i = \left(\sum_{k=1}^{N} u_{i,k} \sum_{t=2}^{T} \hat{Z}_{i,k,t} \right) \cdot \left(\sum_{k=1}^{N} u_{i,k} \sum_{t=2}^{T} \check{Z}_{i,k,t-1} \right)^{-1}, \tag{49}$$

$$\Gamma_i = \frac{1}{(T-1) \sum_{k=1}^{N} u_{i,k}} \sum_{k=1}^{N} u_{i,k} \sum_{t=2}^{T} \left\{ \check{Z}_{i,k,t} - G_i \hat{Z}_{i,k,t}^{\top} \right.$$
$$\left. - \hat{Z}_{i,k,t} G_i^{\top} + G_i \check{Z}_{i,k,t-1} G_i^{\top} \right\}, \tag{50}$$

$$F_i = \left(\sum_{k=1}^{N} u_{i,k} \sum_{t=1}^{T} x_{k,t} \hat{\mu}_{i,k,t|T}^{\top} \right) \cdot \left(\sum_{k=1}^{N} u_{i,k} \sum_{t=1}^{T} \check{Z}_{i,k,t} \right)^{-1}, \tag{51}$$

$$\Sigma_i = \frac{1}{T \sum_{k=1}^{N} u_{i,k}} \sum_{k=1}^{N} u_{i,k} \sum_{t=1}^{T} \left\{ x_{k,t} x_{k,t}^{\top} - F_i \hat{\mu}_{i,k,t|T} x_{k,t}^{\top} \right.$$
$$\left. - x_{k,t} \hat{\mu}_{i,k,t|T}^{\top} F_i^{\top} + F_i \check{Z}_{i,k,t} F_i^{\top} \right\} \tag{52}$$

for the LGSSM-KLFCM,

$$\mu_i = \frac{\sum_{k=1}^{N}(u_{i,k})^m \hat{\mu}_{i,k,1\,T}}{\sum_{k=1}^{N}(u_{i,k})^m}, \tag{53}$$

$$P_i = \frac{\sum_{k=1}^{N}(u_{i,k})^m \left(\check{Z}_{i,k,1} - \mu_i \hat{\mu}_{i,k,1\,T}^\top - \hat{\mu}_{i,k,1\,T}\mu_i^\top + \mu_i\mu_i^\top \right)}{\sum_{k=1}^{N}(u_{i,k})^m}, \tag{54}$$

$$G_i = \left(\sum_{k=1}^{N}(u_{i,k})^m \sum_{t=2}^{T} \hat{Z}_{i,k,t} \right) \cdot \left(\sum_{k=1}^{N}(u_{i,k})^m \sum_{t=2}^{T} \check{Z}_{i,k,t-1} \right)^{-1}, \tag{55}$$

$$\Gamma_i = \frac{1}{(T-1)\sum_{k=1}^{N}(u_{i,k})^m} \sum_{k=1}^{N}(u_{i,k})^m \sum_{t=2}^{T} \left\{ \check{Z}_{i,k,t} - G_i \hat{Z}_{i,k,t}^\top \right.$$
$$\left. -\hat{Z}_{i,k,t}G_i^\top + G_i\check{Z}_{i,k,t-1}G_i^\top \right\}, \tag{56}$$

$$F_i = \left(\sum_{k=1}^{N}(u_{i,k})^m \sum_{t=1}^{T} x_{k,t}\hat{\mu}_{i,k,t|T}^\top \right) \cdot \left(\sum_{k=1}^{N}(u_{i,k})^m \sum_{t=1}^{T} \check{Z}_{i,k,t} \right)^{-1}, \tag{57}$$

$$\Sigma_i = \frac{1}{T\sum_{k=1}^{N}(u_{i,k})^m} \sum_{k=1}^{N}(u_{i,k})^m \sum_{t=1}^{T} \left\{ x_{k,t}x_{k,t}^\top - F_i\hat{\mu}_{i,k,t|T}x_{k,t}^\top \right.$$
$$\left. -x_{k,t}\hat{\mu}_{i,k,t|T}^\top F_i^\top + F_i\check{Z}_{i,k,t}F_i^\top \right\} \tag{58}$$

for the LGSSM-BFCM and LGSSM-QFCM.

STEP 8. For ARMA-KLFCM, ARMA-BFCM and ARMA-QFCM, if the variables $(u, \phi, \psi, \sigma^2, \pi)$ converge, terminate this algorithm. Otherwise, return to STEP 2. For the HMM-KLFCM, HMM-BFCM, and HMM-QFCM, if the variables $(u, \omega, A, \nu, \Psi, \pi)$ converge, terminate this algorithm. Otherwise, return to STEP 2. For the LGSSM-KLFCM, LGSSM-BFCM, and LGSSM-QFCM, if the variables $(u, \mu, P, G, \Gamma, F, \Sigma, \pi)$ converge, terminate this algorithm. Otherwise, return to STEP 2.

4   Numerical Experiments

This section describes numerical experiments conducted using artificial datasets. In the first experiment, we compared the clustering accuracy of the proposed methods (ARMA-KLFCM, ARMA-BFCM, and ARMA-QFCM) to that of a conventional method (ARMA mixtures), using the "Shapelet Sim" dataset from the University of California Riverside (UCR) time series classification archive [13]. The "Shapelet Sim" dataset contains 180 objects in two classes. In the second experiment, we compared the clustering accuracy of the proposed methods (HMM-KLFCM, HMM-BFCM, and HMM-QFCM) to that of a conventional method (mixture of HMMs), using the "Synthetic Control" dataset from UCR archive. The "Synthetic Control" dataset contains 300 objects in six classes.

In the third experiment, we compared the clustering accuracy of the proposed methods (LGSSM-KLFCM, LGSSM-BFCM, and LGSSM-QFCM) to that of a conventional method (mixtures of LGSSMs), using an artificial dataset based on the "Synthetic Control" dataset. The experimental results were evaluated using the Adjusted Rand Index (ARI) [14], As the ARI approached 1, the clustering accuracy increased. The autoregressive and moving average orders were set to $(p, q) = (3, 0)$ for the ARMA-KLFCM, ARMA-BFCM, ARMA-QFCM, and ARMA mixtures. The number of states was set to $L = 2$ for the HMM-KLFCM, HMM-BFCM, HMM-QFCM, and mixture of HMMs. The dimension of the state variables was set to $M = 1$ for the LGSSM-KLFCM, LGSSM-BFCM, LGSSM-QFCM and mixtures of LGSSMs. The initial values for membership and cluster size are set according to actual labels. The initial parameters $(\phi_i, \psi_i, \sigma_i^2)$ for the i-th cluster in the ARMA-KLFCM, ARMA-BFCM, ARMA-QFCM, and ARMA mixtures were set equal to the ARMA parameter obtained from the first object in the i-th class. The initial parameters for the HMM-based and LGSSM-based methods were set in a similar manner. We must note that these initial settings are not applicable in real situations, but these experimental results indicate whether each algorithm produce a good result from a good initial setting, or not. These experiments are preparatory before more comprehensive ones. The fuzzification parameters m, λ were set as follows: $m \in \{1 + 10^{-15}, 1 + 10^{-5}, 1 + 10^{-4}, 1 + 10^{-3}, 1 + 10^{-2}, 1.05, 1.1, 1.5, 2.0, 3.0\}$, $\lambda \in \{0.01, 0.1, 0.5, 1, 5, 10, 100, 1000, 10000, 1.79769 \times 10^{308}\}$. Table 1 summarizes the highest ARI values for each method. It can be observed that all the proposed methods exhibit the same or better clustering accuracy compared to the conventional methods, except for one proposed method: LGSSM-BFCM. Furthermore, the Q-type methods demonstrate the highest clustering accuracy overall. This is because the Q-type methods have two fuzzification parameters. Therefore, the Q-type methods can perform clustering flexibly based on adjustment of the two fuzzification parameters.

Table 1. Highest ARI

	mixture model	KL-type	B-type	Q-type
ARMA	0.674287	0.692733	0.692733	0.711487
HMM	0.701996	0.718215	0.718215	0.718215
LGSSM	0.517625	0.533866	0.503011	0.543717

5 Conclusion

In this work, we proposed nine fuzzy clustering algorithms based on ARMA model, HMM, and LGSSM. The numerical experiments indicated that the Q-type fuzzification methods demonstrated the highest clustering accuracy among the proposed methods. In the future, we plan to compare the clustering accuracy of the proposed methods and conventional methods under randomized initial settings.

References

1. Bezdek, J.C.: Pattern Recognition with Fuzzy Objective Function Algorithms. Plenum Press, New York (1981)
2. Miyamoto, S., Mukaidono, M.: Fuzzy c-means as a regularization and maximum entropy approach. In: Proceedings of IFSA 1997, vol. 2, pp. 86–92 (1997)
3. Yasuda, M.: Tsallis entropy based fuzzy C-means clustering with parameter adjustment. In: Proceedings of SCIS&ISIS2012, pp. 1534–1539 (2012)
4. Miyamoto, S., Kurosawa, N.: Controlling cluster volume sizes in fuzzy c-means clustering. In Proceedings of SCIS&ISIS2004, pp. 1–4 (2004)
5. Kanzawa, Y.: On fuzzy clustering based on Tsallis entropy-regularization. In: Proceedings of 30th Fuzzy System Symposium, pp. 452–457 (2014)
6. D'Urso, P., Maharaj, E.A.: Autocorrelation-based fuzzy clustering of time series. Fuzzy Sets Syst. **160**, 3565–3589 (2009)
7. Lopez-Oriona, A., Vilar, J.A., D'Urso, P.: Quantile-based fuzzy clustering of multivariate time series in the frequency domain. Fuzzy Sets Syst. **443**, 115–154 (2022)
8. Xiong, Y., Yeung, D.Y.: Mixtures of ARMA models for model-based time series clustering. In: Proceedings of ICDM2002, pp. 717–720 (2002)
9. Alon, J., Sclaroff, S., Kollios, G., Pavlovic, V.: Discovering clusters in motion time-series data. In: Proceedings of CVPR2003, vol. 1, pp. 375–381 (2003)
10. Umatani, R., Imai, T., Kawamoto, K., Kunimasa, S.: Time series clustering with an EM algorithm for mixtures of linear gaussian state space models. Pattern Recognit. **138**, 109375 (2023)
11. Leonald, E.B., Ted, P., George, S., Norman, W.: A maximization technique occurring in the statistical analysis of probabilistic functions of Markov chains. Ann. Math. Stat. **41**(1), 164–171 (1970)
12. Brian, D.O.A., Anderson, John, B.M., et al.: Optimal Filtering. Prentice-Hall, New Jersey (1979)
13. Hoang, A.D., et al.: The UCR time series classification archive (2018). https://www.cs.ucr.edu/~eamonn/time_series_data_2018/
14. Hubert, L., Arabie, P.: Comparing Partitions. J. Classif. **2**(1), 193–218 (1985)

On Some Fuzzy Clustering for Series Data

Yuto Suzuki and Yuchi Kanzawa[(✉)] [ID]

Shibaura Institute of Technology, 3-7-5 Toyosu, Koto, Tokyo 135-8548, Japan
{ma23101,kanzawa}@shibaura-it.ac.jp

Abstract. Various fuzzification techniques have been applied to clustering algorithms for vectorial data, such as Yang-type fuzzification and extended q-divergence-regularization, whereas only a few such techniques have been applied to fuzzy clustering algorithms for series data. In this regard, this study presents four fuzzy clustering algorithms for series data. The first two algorithms are obtained by penalizing each optimization problem in the two conventional algorithms: Bezdek-type fuzzy dynamic-time-warping (DTW) c-means and Bezdek-type fuzzy c-shape, with the cluster-size controller fixed. The other two algorithms are obtained from a conventional algorithm, q-divergence-based fuzzy DTW c-means or q-divergence-based fuzzy c-shape, by distinguishing two fuzzificators for membership from those for cluster-size controllers. Numerical experiments are conducted to evaluate the performance of the proposed algorithms.

Keywords: fuzzy clustering · series data · dynamic-time-warping · shape-based-distance · fuzzy c-means

1 Introduction

The fuzzy c-means (FCM) proposed by Bezdek [4] is a representative fuzzy clustering algorithm for vectorial data. To distinguish this algorithm from other variants such as entropy-regularized FCM (EFCM) [1] and Tsallis-entropy-based FCM (TFCM) [2,3], it is referred to as Bezdek-type FCM (BFCM) in this study. These FCM variants were extended by introducing a cluster-size controller, resulting in the modified BFCM (MBFCM) [5], KL-divergence-regularized FCM (KLFCM) [5], and q-divergence-based FCM (QFCM) [6] algorithms. In particular, it was shown that QFCM is an extension of both MBFCM and KLFCM, and it outperforms MBFCM and KLFCM in terms of clustering accuracy. The fuzzification technique used in QFCM is referred to as Q-type fuzzification. Penalized FCM proposed by Yang [7] is developed by penalizing BFCM with the logarithm of cluster-size controller; this method and this fuzzification technique are referred to as Yang-type FCM (YFCM) and Y-type fuzzification. Kanzawa proposed fuzzy clustering for vectorial data, referred to as extended QFCM (EQFCM) [8], and showed that EQFCM is an extension of both QFCM and YFCM, and outperforms QFCM and YFCM in terms of

K. Honda et al. (Eds.): IUKM 2023, LNAI 14376, pp. 205–217, 2024.
https://doi.org/10.1007/978-3-031-46781-3_18

clustering accuracy. The fuzzification technique used in EQFCM is referred to as EQ-type fuzzification.

The aforementioned algorithms use the squared Euclidean distance to measure the dissimilarity between the object and the cluster center, whereas dynamic-time-warping (DTW) and shape-based-distance (SBD) measure the representative dissimilarities with respect to the series data. Fujita et al. adopted DTW to measure object-cluster dissimilarities, and proposed three fuzzy clustering algorithms for series data: Bezdek-type fuzzy dynamic-time-warping c-means (BFDTWCM), KL-divergence-regularized fuzzy dynamic-time-warping c-means (KLFDTWCM), and q-divergence-based fuzzy dynamic-time-warping c-means (QFDTWCM) [9]. Furthermore, Fujita et al. adopted SBD to measure object-cluster dissimilarities, and proposed three fuzzy clustering algorithms for series data: Bezdek-type fuzzy c-shape (BFCS), KL-divergence-regularized fuzzy c-shape (KLFCS) and q-divergence-based fuzzy c-shape (QFCS) [10]. Although various fuzzification techniques are used in these algorithms, neither Y-type or EQ-type fuzzification has been introduced for fuzzy clustering algorithms for series data. However, adopting these techniques to fuzzy clustering for series data has the potential to produce higher clustering accuracy; this is similar to the fact that such the fuzzification in fuzzy clustering for vectorial data produces a higher clustering accuracy than others.

In this study, we propose four fuzzy clustering algorithms for series data. The first two algorithms were obtained by penalizing the optimization problems of BFDTWCM and BFCS by fixing the cluster-size controller, referred to as Yang-type fuzzy dynamic-time-warping c-means (YFDTWCM) and Yang-type fuzzy c-shape (YFCS), respectively. The second two algorithms are obtained from QFDTWCM and QFCS by distinguishing the fuzzificators m and λ for membership from those for the cluster-size controllers and are referred to as extended q-divergence-based fuzzy dynamic-time-warping c-means (EQFDTWCM) and extended q-divergence-based fuzzy c-shape (EQFCS), respectively.

The remainder of this paper is organized as follows: Sect. 2 introduces some conventional algorithms; we formulate four algorithms in Sect. 3; Sect. 4 presents the experimental results for evaluating the proposed algorithms in terms of clustering accuracy, and Sect. 5 concludes this paper.

2   Preliminaries

2.1   Clustering for Vectorial Data

In this subsection, we explain the partitioning of the dataset denoted by $x = \{x_k \in \mathbb{R}^D\}_{k=1}^N$. Let $v = \{v_i \in \mathbb{R}^D\}_{i=1}^C$ be the set of cluster centers. Let $u = \{u_{i,k}\}_{(i,k)=(1,1)}^{(C,N)}$ be the membership of object $\#k$ with respect to cluster $\#i$, with the constraint

$$\sum_{i=1}^{C} u_{i,k} = 1. \tag{1}$$

Let $\alpha = \{\alpha_i\}_{i=1}^{C}$ be the i-th cluster-size controller, which has the constraint

$$\sum_{i=1}^{C} \alpha_i = 1. \tag{2}$$

The BFCM, QFCM, YFCM, and EQFCM algorithms are based on the following optimization problems:

$$\underset{u,v,\alpha}{\text{minimize}} \sum_{i=1}^{C} \sum_{k=1}^{N} (u_{i,k})^m \|x_k - v_i\|_2^2, \tag{3}$$

$$\underset{u,v,\alpha}{\text{minimize}} \sum_{i=1}^{C} \sum_{k=1}^{N} (\alpha_i)^{1-m} (u_{i,k})^m \|x_k - v_i\|_2^2$$
$$+ \frac{\lambda^{-1}}{m-1} \sum_{i=1}^{C} \sum_{k=1}^{N} \left((\alpha_i)^{1-m} (u_{i,k})^m - u_{i,k} \right), \tag{4}$$

$$\underset{u,v,\alpha}{\text{minimize}} \sum_{i=1}^{C} \sum_{k=1}^{N} (u_{i,k})^m \|x_k - v_i\|_2^2 + \lambda^{-1} \sum_{i=1}^{C} \sum_{k=1}^{N} (u_{i,k})^m \log\left((\alpha_i)^{-1} \right), \tag{5}$$

$$\underset{u,v,\alpha}{\text{minimize}} \sum_{i=1}^{C} \sum_{k=1}^{N} (\alpha_i)^{1-m'} (u_{i,k})^m \|x_k - v_i\|_2^2$$
$$+ \frac{\lambda^{-1}}{m'-1} \sum_{i=1}^{C} \sum_{k=1}^{N} (u_{i,k})^m \left((\alpha_i)^{1-m'} - 1 \right)$$
$$+ \frac{\lambda'^{-1}}{m-1} \sum_{i=1}^{C} \sum_{k=1}^{N} \left((u_{i,k})^m - u_{i,k} \right), \tag{6}$$

where $m > 1$, $m' > 1$, $\lambda > 0$, and $\lambda' > 0$ are fuzzificators. Notably, the YFCM optimization problem was obtained by penalizing the BFCM optimization problem using $\sum_{i=1}^{C} \sum_{k=1}^{N} (u_{i,k})^m \log((\alpha_i)^{-1})$. Furthermore, note that the EQFCM optimization problem is obtained from QFCM by distinguishing the fuzzificators m and λ for membership from those for the cluster-size controllers. These notes form the basis for deriving the proposed methods.

2.2   Clustering for Series Data

In this section, we introduce some clustering algorithms for series data denoted by $x = \{x_k^{(M)}\}_{k=1}^{N}$, where $x_k^{(M)}$ be a series $(x_{k,1}, \ldots, x_{k,M})$, and $x_{k,\ell} \in \mathbb{R}$ be its ℓ-th element. Moreover, let $x_k^{(\ell)}$ be a subseries $(x_{k,1}, \ldots, x_{k,\ell})$ of $x_k^{(M)}$. The set of cluster centers is denoted by $v = \{v_i^{(M)}\}_{i=1}^{C}$, where $v_i^{(M)}$ be a series $(v_{i,1}, \ldots, v_{i,M})$, and $v_{i,\ell} \in \mathbb{R}$ be its ℓ-th element.

The BFDTWCM and QFDTWCM algorithms were obtained by solving the following optimization problems:

$$\underset{u,v,\alpha}{\text{minimize}} \sum_{i=1}^{C}\sum_{k=1}^{N}(\alpha_i)^{1-m}(u_{i,k})^m \mathsf{DTW}_{i,k}, \tag{7}$$

$$\underset{u,v,\alpha}{\text{minimize}} \sum_{i=1}^{C}\sum_{k=1}^{N}(\alpha_i)^{1-m}(u_{i,k})^m \mathsf{DTW}_{i,k}$$

$$+ \frac{\lambda^{-1}}{m-1}\sum_{i=1}^{C}\sum_{k=1}^{N}\left((\alpha_i)^{1-m}(u_{i,k})^m - u_{i,k}\right), \tag{8}$$

subject to Eqs. (1) and (2), where $\mathsf{DTW}_{i,k}$ is the Dynamic-Time-Warping dissimilarity between the object #k and cluster #i defined as

$$\mathsf{DTW}_{i,k} = \mathsf{DTW}\left(x_k^{(M)}, v_i^{(M)}\right), \tag{9}$$

$$\mathsf{DTW}\left(x_k^{(\ell)}, v_i^{(\ell')}\right) = (x_{k,\ell} - v_{i,\ell'})^2 + \min \begin{cases} \mathsf{DTW}\left(x_k^{(\ell-1)}, v_i^{(\ell'-1)}\right), \\ \mathsf{DTW}\left(x_k^{(\ell-1)}, v_i^{(\ell')}\right), \\ \mathsf{DTW}\left(x_k^{(\ell)}, v_i^{(\ell'-1)}\right), \end{cases}$$

$$(\ell \in \{2,\ldots,M\}, \ell' \in \{2,\ldots,M\}), \tag{10}$$

$$\mathsf{DTW}\left(x_k^{(1)}, v_i^{(\ell)}\right) = (x_{k,1} - v_{i,\ell})^2 + \mathsf{DTW}\left(x_k^{(1)}, v_i^{(\ell-1)}\right),$$

$$(\ell \in \{2,\ldots,M\}), \tag{11}$$

$$\mathsf{DTW}\left(x_k^{(\ell)}, v_i^{(1)}\right) = (x_{k,\ell} - v_{i,1})^2 + \mathsf{DTW}\left(x_k^{(\ell-1)}, v_i^{(1)}\right),$$

$$(\ell \in \{2,\ldots,M\}), \tag{12}$$

$$\mathsf{DTW}\left(x_k^{(1)}, v_i^{(1)}\right) = (x_{k,1} - v_{i,1})^2. \tag{13}$$

Calculating $\mathsf{DTW}_{i,k}$ produces a sequence of pairs (ℓ, ℓ') ($\ell \in \{1,\cdots,M\}, \ell' \in \{1,\cdots,M\}$) for each element in the series, which is known as the warping path. Here, we introduce the matrices $\Omega^{(i,k)} \in \{0,1\}^{M \times M}$ ($i \in \{1,\ldots,C\}, k \in \{1,\ldots,N\}$) as

$$\Omega_{\ell,\ell'}^{(i,k)} = \begin{cases} 1 & (\ (\ell,\ell') \in \text{warping path}), \\ 0 & (\text{otherwise}). \end{cases} \tag{14}$$

The BFCS and QFCS algorithms are based on the following optimization problems:

$$\underset{u,v,\alpha}{\text{minimize}} \sum_{i=1}^{C} \sum_{k=1}^{N} (\alpha_i)^{1-m} (u_{i,k})^m \text{SBD}_{i,k}, \tag{15}$$

$$\underset{u,v,\alpha}{\text{minimize}} \sum_{i=1}^{C} \sum_{k=1}^{N} (\alpha_i)^{1-m} (u_{i,k})^m \text{SBD}_{i,k}$$

$$+ \frac{\lambda^{-1}}{m-1} \sum_{i=1}^{C} \sum_{k=1}^{N} \left((\alpha_i)^{1-m} (u_{i,k})^m - u_{i,k} \right), \tag{16}$$

subject to Eqs. (1) and (2), where $\text{SBD}_{i,k}$ is the Shape-Based-Distance dissimilarity between the object #k and cluster #i defined as

$$\text{SBD}_{i,k} = \left(1 - \text{NCC}_{i,k}(\omega_{i,k}^* - M) \right)^2, \tag{17}$$

$$\omega_{i,k}^* = \arg \max_{\omega \in \{1,2,\cdots,2M-1\}} \{\text{NCC}_{i,k}(\omega - M)\}, \tag{18}$$

$$\text{NCC}_{i,k}(\omega - M) = \frac{R_{\omega-M}\left(x_k^{(M)}, v_i^{(M)}\right)}{R_0(x_k^{(M)}, x_k^{(M)}) R_0(v_i^{(M)}, v_i^{(M)})}, \tag{19}$$

$$R_{\omega-M}(x_k^{(M)}, v_i^{(M)}) = \begin{cases} \text{Shift}(x_k^{(M)}; M - \omega)^{\mathsf{T}} v_i^{(M)} & (\omega - M \geq 0), \\ v_i^{(M)\mathsf{T}} \text{Shift}(x_k^{(M)}; M - \omega) & (\omega - M < 0). \end{cases} \tag{20}$$

Here, if $\omega \geq M$, $\text{Shift}(x_k^{(M)}; M - \omega)$ is shifted forward by $|M - \omega|$ with a zero set behind it. If $\omega < M$, $\text{Shift}(x_k^{(M)}; M - \omega)$ is shifted back by $|M - \omega|$ with a zero set in front of it.

The algorithms of BFDTWCM and BFCS are omitted because only these optimization problems are used to derive the proposed algorithms. The QFDTWCM and QFCS algorithms are summarized as follows:

Algorithms 1 (QFDTWCM, QFCS).

STEP 1. Set the number of clusters C. Set the fuzzificator m and λ. Set the initial cluster center v, and initial membership u.

STEP 2. Calculate v as
(a) Calculate DTW from Eqs. (9)–(13).
(b) Calculate v as

$$v_i^{(M)} = \left(\sum_{k=1}^{N} (u_{i,k})^m \Omega^{(i,k)} x_k^{(M)} \right) \oslash \left(\sum_{k=1}^{N} (u_{i,k})^m \Omega^{(i,k)} \mathbf{1} \right), \tag{21}$$

where $\oslash$ describes element-wise division, and $\mathbf{1}$ is the M-dimensional vector with all elements equal to one.

(c) If DTW does not converge, go to (a).
for QFDTWCM, and
(a) Calculate SBD from Eqs. (17)–(20).
(b) Obtain a maximizer $\tilde{v}_i^{(M)}$ as

$$\tilde{v}_i^{(M)} = \arg\max_a \frac{a^\mathsf{T} Q_i a}{a^\mathsf{T} a}, \tag{22}$$

$$Q_i = \left(E - \frac{1}{M}\mathbb{1} \right)$$

$$\times \left(\sum_{k=1}^{N} (u_{i,k})^m \mathrm{Shift}(x_k^{(M)}; \omega_{i,k}^* - M) \mathrm{Shift}(x_k^{(M)}; \omega_{i,k}^* - M)^\mathsf{T} \right)$$

$$\times \left(E - \frac{1}{M}\mathbb{1} \right), \tag{23}$$

where $\mathbb{1}$ is the M-dimensional square matrix with all elements equal to one.
(c) If the value of

$$\sum_{k=1}^{N} (u_{i,k})^m \mathrm{SBD}_{i,k} \tag{24}$$

with $v_i^{(M)} = \tilde{v}_i^{(M)}$ is lower than that of $v_i^{(M)} = -\tilde{v}_i^{(M)}$, then update $v_i^{(M)}$ as $\tilde{v}_i^{(M)}$. Otherwise, update $v_i^{(M)}$ as $-\tilde{v}_i^{(M)}$.
for QFCS.

STEP 3. Calculate α as

$$\alpha_i = \left[\sum_{j=1}^{C} \left(\frac{\sum_{k=1}^{N}(1-\lambda(1-m)\mathrm{DTW}_{j,k})(u_{j,k})^m}{\sum_{k=1}^{N}(1-\lambda(1-m)\mathrm{DTW}_{i,k})(u_{i,k})^m} \right)^{1/m} \right]^{-1} \tag{25}$$

for QFDTWCM, and

$$\alpha_i = \left[\sum_{j=1}^{C} \left(\frac{\sum_{k=1}^{N}(1-\lambda(1-m)\mathrm{SBD}_{j,k})(u_{j,k})^m}{\sum_{k=1}^{N}(1-\lambda(1-m)\mathrm{SBD}_{i,k})(u_{i,k})^m} \right)^{1/m} \right]^{-1} \tag{26}$$

for QFCS.

STEP 4. Calculate u as

$$u_{i,k} = \left[\sum_{j=1}^{C} \frac{\alpha_j}{\alpha_i} \left(\frac{1-\lambda(1-m)\mathrm{DTW}_{j,k}}{1-\lambda(1-m)\mathrm{DTW}_{i,k}} \right)^{1/(1-m)} \right]^{-1} \tag{27}$$

for QFDTWCM, and

$$u_{i,k} = \left[\sum_{j=1}^{C} \frac{\alpha_j}{\alpha_i} \left(\frac{1 - \lambda(1-m)\mathsf{SBD}_{j,k}}{1 - \lambda(1-m)\mathsf{SBD}_{i,k}} \right)^{1/(1-m)} \right]^{-1} \quad (28)$$

for QFCS.

STEP 5. If (u, v, α) does not converge, go to STEP 2

3   Proposed Algorithm

3.1   Concept

In this study, we propose four fuzzy clustering algorithms for series data.

The first two algorithms were obtained by penalizing the optimization problems of BFDTWCM and BFCS by fixing the cluster-size controller using $\sum_{i=1}^{C} \sum_{k=1}^{N} (u_{i,k})^m \log((\alpha_i)^{-1})$ and are referred to as YFDTWCM and YFCS, respectively. This derivation is similar to that of YFCM by penalizing the BFCM optimization problem using $\sum_{i=1}^{C} \sum_{k=1}^{N} (u_{i,k})^m \log((\alpha_i)^{-1})$. The YFDTWCM and YFCS optimization problems are then given by

$$\underset{u,v,\alpha}{\text{minimize}} \sum_{i=1}^{C} \sum_{k=1}^{N} (u_{i,k})^m \mathsf{DTW}_{i,k} + \lambda^{-1} \sum_{i=1}^{C} \sum_{k=1}^{N} (u_{i,k})^m \log\left((\alpha_i)^{-1}\right), \quad (29)$$

$$\underset{u,v,\alpha}{\text{minimize}} \sum_{i=1}^{C} \sum_{k=1}^{N} (u_{i,k})^m \mathsf{SBD}_{i,k} + \lambda^{-1} \sum_{i=1}^{C} \sum_{k=1}^{N} (u_{i,k})^m \log\left((\alpha_i)^{-1}\right), \quad (30)$$

respectively, subject to Eqs. (1) and (2).

The second two algorithms were obtained from QFDTWCM and QFCS by distinguishing the fuzzificators m and λ for membership from those for the cluster-size controllers; the resulting algorithms are referred to as EQFDTWCM and EQFCS, respectively. This derivation is similar to that of EQFCM obtained from QFCM by distinguishing the fuzzificators m and λ for membership from those of the cluster-size controllers. The EQFDTWCM and EQFCS optimization problems are then given by

$$\underset{u,v,\alpha}{\text{minimize}} \sum_{i=1}^{C} \sum_{k=1}^{N} (\alpha_i)^{1-m'} (u_{i,k})^m \text{DTW}_{i,k}$$

$$+ \frac{\lambda^{-1}}{m'-1} \sum_{i=1}^{C} \sum_{k=1}^{N} (u_{i,k})^m \left((\alpha_i)^{1-m'} - 1 \right)$$

$$+ \frac{\lambda'^{-1}}{m-1} \sum_{i=1}^{C} \sum_{k=1}^{N} \left((u_{i,k})^m - u_{i,k} \right), \tag{31}$$

$$\underset{u,v,\alpha}{\text{minimize}} \sum_{i=1}^{C} \sum_{k=1}^{N} (\alpha_i)^{1-m'} (u_{i,k})^m \text{SBD}_{i,k}$$

$$+ \frac{\lambda^{-1}}{m'-1} \sum_{i=1}^{C} \sum_{k=1}^{N} (u_{i,k})^m \left((\alpha_i)^{1-m'} - 1 \right)$$

$$+ \frac{\lambda'^{-1}}{m-1} \sum_{i=1}^{C} \sum_{k=1}^{N} \left((u_{i,k})^m - u_{i,k} \right), \tag{32}$$

respectively, subject to Eqs. (1) and (2).

3.2 Algorithm

The proposed clustering algorithms are obtained by solving the optimization problems given in Eqs. (29), (30), (31), and (32) subject to the constraints in Eqs. (1) and (2). The analysis of the necessary conditions for optimality, although the details are omitted for brevity, is summarized by the following algorithm:

Algorithms 2 (YFDTWCM, YFCS, EQFDTWCM, EQFCS).

STEP 1. Set the cluster number C. Set the fuzzificator m and λ for YFDTWCM and YFCS. Set the fuzzificator m, m', λ and λ' for EQFDTWCM and EQFCS. Initialize cluster center and membership, v, u.

STEP 2. Calculate v as
  (a) Calculate DTW from Eqs. (9)–(13).
  (b) Calculate v using Eq. (21).
  (c) If v does not converge, go to (a).
  for YFDTWCM and EQFDTWCM, and
  (a) Calculate SBD from Eqs. (17)–(20).
  (b) Obtain a maximizer $\tilde{v}_i^{(M)}$ of Eqs. (22) and (23).
  (c) If the value of Eq. (24) with $v_i^{(M)} = \tilde{v}_i^{(M)}$ is lower than that of $v_i^{(M)} = -\tilde{v}_i^{(M)}$, then update $v_i^{(M)}$ as $\tilde{v}_i^{(M)}$. Otherwise, update $v_i^{(M)}$ as $-\tilde{v}_i^{(M)}$.
  for YFCS and EQFCS.

STEP 3. Calculate α as

$$\alpha_i = \frac{\sum_{k=1}^{N}(u_{i,k})^m}{\sum_{j=1}^{C}\sum_{k=1}^{N}(u_{j.k})^m} \qquad (33)$$

for YFDTWCM and YFCS,

$$\alpha_i = \frac{\left(\sum_{k=1}^{N}(u_{i,k})^m(1-\lambda(1-m')\text{DTW}_{i,k})\right)^{1/m'}}{\sum_{j=1}^{C}\left(\sum_{k=1}^{N}(u_{j,k})^m(1-\lambda(1-m')\text{DTW}_{j,k})\right)^{1/m'}} \qquad (34)$$

for EQFDTWCM, and

$$\alpha_i = \frac{\left(\sum_{k=1}^{N}(u_{i,k})^m(1-\lambda(1-m')\text{SBD}_{i,k})\right)^{1/m'}}{\sum_{j=1}^{C}\left(\sum_{k=1}^{N}(u_{j,k})^m(1-\lambda(1-m')\text{SBD}_{j,k})\right)^{1/m'}} \qquad (35)$$

for EQFCS.

STEP 4. Calculate u as

$$u_{i,k} = \left[\sum_{j=1}^{C}\left(\frac{\text{DTW}_{j,k}-\lambda^{-1}\log(\alpha_j)}{\text{DTW}_{i,k}-\lambda^{-1}\log(\alpha_i)}\right)^{1/(1-m)}\right]^{-1} \qquad (36)$$

for YFDTWCM,

$$u_{i,k} = \left[\sum_{j=1}^{C}\left(\frac{\text{SBD}_{j,k}-\lambda^{-1}\log(\alpha_j)}{\text{SBD}_{i,k}-\lambda^{-1}\log(\alpha_i)}\right)^{1/(1-m)}\right]^{-1} \qquad (37)$$

for YFCS,

$$u_{i,k} = \left[\sum_{j=1}^{C}\left(\frac{\lambda'(m-1)\left((\alpha_j)^{1-m'}(1-\lambda(1-m')\text{DTW}_{j,k})-1\right)+\lambda(m'-1)}{\lambda'(m-1)\left((\alpha_i)^{1-m'}(1-\lambda(1-m')\text{DTW}_{i,k})-1\right)+\lambda(m'-1)}\right)^{1/(1-m)}\right]^{-1} \qquad (38)$$

for EQFDTWCM, and

$$u_{i,k} = \left[\sum_{j=1}^{C}\left(\frac{\lambda'(m-1)\left((\alpha_j)^{1-m'}(1-\lambda(1-m')\text{SBD}_{j,k})-1\right)+\lambda(m'-1)}{\lambda'(m-1)\left((\alpha_i)^{1-m'}(1-\lambda(1-m')\text{SBD}_{i,k})-1\right)+\lambda(m'-1)}\right)^{1/(1-m)}\right]^{-1} \qquad (39)$$

for EQFCS.

STEP 5. If (u,v,α) does not converge, go to STEP 2

4 Numerical Experiment

This section presents numerical experiments that illustrate the proposed methods based on four artificial datasets: "BME," "CBF," "SyntheticControl," and "UMD," which were obtained from the UCR time series classification archive [11]. The data number, class number, and dimensions of these datasets are summarized in Table 1. QFDTWCM, QFCS, YFDTWCM, YFCS, EQFDTWCM, and EQFCS are applied to these datasets with the initial setting using the actual label, where the fuzzificators are set to $m \in \{1.001, 1.01, 1.1, 1.3, 1.5, 2.0, 3.0\}$ and $\lambda \in \{0.1, 0.5, 1, 5, 10, 25, 100, 1000\}$ for YFDTWCM and YFCS, and $m, m' \in \{1.001, 1.01, 1.1, 1.3, 1.5, 2.0, 3.0\}$ and $\lambda, \lambda' \in \{0.1, 0.5, 1, 5, 10, 25, 100, 1000\}$ for EQFDTWCM and EQFCS. We use the Adjusted Rand Index (ARI) [12] to evaluate the clustering results. ARI value should be equal to or smaller than one, and higher values are preferred. The highest ARI value in each method and the parameter value at which the highest ARI value was achieved are listed in Tables 2, 3, 4 and 5, from which we observe that

- EQFDTWCM achieves higher ARI values than QFDTWCM and YFDTWCM, and

Table 1. Datasets used in the experiments

Name	Data-number	Class-number	Series-length
BME	150	3	128
CBF	900	3	128
SyntheticControl	300	6	60
UMD	144	3	150

Table 2. Highest ARI values for the BME dataset

Method	Highest ARI	Fuzzification parameter
QFDTWCM	0.228885	$(m, \lambda) = (3, 5)$
YFDTWCM	0.311871	$(m, \lambda) = (1.1, 1000)$
EQFDTWCM	0.336694	$(m, m', \lambda, \lambda') = \{(1.001, 1.001, 1, 0.1), (1.001, 1.001, 0.5, 0.1), (1.001, 1.01, 1, 0.1), (1.001, 1.01, 0.5, 0.1), (1.01, 1.001, 1, 0.1), (1.01, 1.001, 0.5, 0.1), (1.01, 1.01, 1, 0.1), (1.01, 1.01, 0.5, 0.1), (1.01, 1.1, 0.5, 0.1), (1.1, 1.001, 1, 0.1)\}$
QFCS	0.327499	$(m, \lambda) = \{(1.001, 1000), (1.01, 1000)\}$
YFCS	0.327499	$(m, \lambda) = \{(1.1, 1000), (1.3, 1000), (1.5, 1000), (2, 1000)\}$
EQFCS	0.382696	$(m, m', \lambda, \lambda') = (1.3, 1.001, 100, 1)$

Table 3. Highest ARI values for the CBF dataset

Method	Highest ARI	Fuzzification parameter
QFDTWCM	0.772908	$(m, \lambda) = \{(1.001, 10), (1.001, 25)\}$
YFDTWCM	0.775213	$(m, \lambda) = (1.001, 5)$
EQFDTWCM	0.775245	$(m, m', \lambda, \lambda') = (1.001, 1.001, 100, 1000)$
QFCS	0.727811	$(m, \lambda) = (1.001, 1000)$
YFCS	0.750414	$(m, \lambda) = (3, 100)$
EQFCS	0.757187	$(m, m', \lambda, \lambda') = (3, 1.1, 100, 100)$

Table 4. Highest ARI values for the SyntheticControl dataset

Method	Highest ARI	Fuzzification parameter
QFDTWCM	0.952862	$(m, \lambda) = \{(1.001, 1), (1.001, 5), (1.001, 10),$ $(1.001, 25), (1.001, 100), (1.001, 1000), (1.01, 1),$ $(1.01, 5), (1.01, 10), (1.01, 25), (1.01, 100),$ $(1.01, 1000), (1.1, 1), (1.1, 5), (1.1, 10),$ $(1.1, 25), (1.1, 100), (1.1, 1000), (1.3, 25)\}$
YFDTWCM	0.952862	$(m, \lambda) = \{(1.001, 1000), (1.001, 100), (1.001, 25),$ $(1.001, 10), (1.001, 5), (1.001, 1), (1.001, 0.5),$ $(1.01, 1000), (1.01, 100), (1.01, 25), (1.01, 10),$ $(1.01, 5), (1.01, 1), (1.01, 0.5), (1.1, 1000),$ $(1.1, 100), (1.1, 25), (1.1, 10), (1.1, 5),$ $(1.1, 1), (1.1, 0.5)\}$
EQFDTWCM	0.960406	$(m, m', \lambda, \lambda') = \{(1.1, 1.1, 10, 1),$ $(1.1, 1.1, 25, 1), (1.1, 1.1, 100, 1),$ $(1.1, 1.1, 1000, 1), (1.5, 1.5, 5, 25)\}$
QFCS	0.760055	$(m, \lambda) = (1.001, 1000)$
YFCS	0.773160	$(m, \lambda) = (1.01, 1000)$
EQFCS	0.810965	$(m, m', \lambda, \lambda') = (1.001, 1.3, 1000, 1000)$

- EQFCS achieves higher ARI values than QFCS and YFCS

Therefore, EQ-type fuzzification is superior to the other fuzzifications. EQFDTWCM achieves the highest ARI value for "CBF," "SyntheticControl," and "UMD" datasets, whereas EQFCS achieves the highest ARI value for "BME" dataset. To determine the performance of each method in terms of accuracy, further experiments using more datasets are necessary, and will be a topic for future work. Comparing YFDTWCM with QFDTWCM, YFDTWCM achieves higher or equal ARI values for "BME," "CBF," and "SyntheticControl" datasets, whereas QFDTWCM achieves higher ARI values for "UMD" dataset. Comparing YFCS with QFCS, YFCS achieves higher or equal ARI values for "BME," "CBF," and "SyntheticControl" datasets, whereas QFCS achieves higher ARI values for "UMD" dataset. Therefore, Y-type fuzzification

Table 5. Highest ARI values for the UMD dataset

Method	Highest ARI	Fuzzification parameter
QFDTWCM	0.572375	$(m, \lambda) = (2, 25)$
YFDTWCM	0.432572	$(m, \lambda) = \{(1.1, 1000), (1.1, 100), (1.1, 25),$ $(1.1, 10), (1.1, 5), (1.1, 1), (1.3, 1000),$ $(1.3, 100), (1.3, 25), (1.3, 10), (1.3, 5)\}$
EQFDTWCM	0.602259	$(m, m', \lambda, \lambda') = (1.5, 2, 5, 0.5)$
QFCS	0.216229	$(m, \lambda) = (3, 5)$
YFCS	0.141769	$(m, \lambda) = (3, 100)$
EQFCS	0.239858	$(m, m', \lambda, \lambda') = \{(1.001, 3, 5, 1), (1.01, 3, 5, 1)\}$

seems to be superior to Q-type fuzzification, but further experiments using more datasets are necessary, which is a future work.

5   Conclusion

This study proposed and implemented four algorithms for YFDTWCM, YFCS, EQFDTWCM, and EQFCS. The results of our evaluation using the four artificial datasets showed that the EQ-type fuzzified algorithms could achieve higher ARI values than the other fuzzified algorithms. The Y-type fuzzification is almost superior to the Q-type fuzzification, but further experiments using more datasets are required. Furthermore, experiments to investigate whether the DTW-based algorithms or the SBD-based ones is superior are also future work.

References

1. Miyamoto, S., Mukaidono, M.: Fuzzy c-means as a regularization and maximum entropy approach. In: Proceedings of the 7th International Fuzzy Systems Association World Congress (IFSA 1997), vol. 2, pp. 86–92 (1997)
2. Mènard, M., Courboulay, V., Dardignac, P.: Possibilistic and probablistic fuzzy clustering: unification within the framework of the nonextensive thermostatistics. Pattern Recogn. **2**, 86–92 (1997)
3. Yasuda, M.: Tsallis entropy based fuzzy cmeans clustering with parameter adjustment. In: Proceedings of the SCIS&ISIS2012, pp. 1534–1539 (2012)
4. Bezdek, J.C.: Pattern Recognition with Fuzzy Objective Function Algorithms. Plenum Press, New York (1981)
5. Miyamoto, S., Kurosawa, N.: Controlling cluster volume sizes in fuzzy c-means clustering. In: Proceedings of the SCIS&ISIS2004, pp. 1–4 (2004)
6. Kanzawa, Y.: On fuzzy clustering based on Tsallis entropy-regularization. In: Proceedings of the 30th Fuzzy System Symposium, pp. 452–457 (2014). (in Japanese)
7. Yang, M.S.: On a class of fuzzy classification maximum likelihood procedures. In: Fuzzy Sets and Systems, vol. 57, pp. 365–375 (1993)
8. Kanzawa, Y.: On a parameter extention for q-divergence-based fuzzy clustering. (in Japanese) (to be published in Proc. FSS2023)

9. Fujita, M., Kanzawa, Y.: On some fuzzy clustering algorithms for time-series data. In: Proceedings of the IUKM2022, pp. 168–181 (2022)
10. Fujita, M., Kanzawa, Y.: On some fuzzy clustering algorithms for time series data. In: Proceedings of the 37th Fuzzy System Symposium, pp. 626–631 (2021). (in Japanese)
11. Bagnall, A., Lines, J., Vickers, W., Keogh, E.: The UEA and UCR Time Series Classification Repository. www.timeseriesclassification.com (2018)
12. Lawrence, H., Phipps, A.: J. Classif. **2**(1), 193–218 (1985)

Collaborative Filtering Based on Rough C-Means Clustering with Missing Value Processing

Seiki Ubukata$^{(\boxtimes)}$ ⓘ and Kazushi Futakuchi

Osaka Metropolitan University, Sakai, Osaka 599-8531, Japan
`ubukata@omu.ac.jp`

Abstract. A significant problem in collaborative filtering (CF) tasks is that the data often contains many missing values. Conventional rough clustering-based CF employs single imputation methods, but this can introduce distortion into the data by substituting equable values that differ from the original ones, leading to potential biases in the results. Moreover, using the whole data strategy, in which rows containing missing values in the data matrix are deleted, to handle missing values leads to a problem where, because most of the data in CF tasks consists of missing values, nearly all rows are deleted, making the analysis challenging. Therefore, we propose a method, referred to as RCM-PDS, which introduces the partial distance strategy (PDS) into RCM. The goal of this approach is to perform clustering using the original data matrix with missing values without prior imputation. Additionally, we propose its application to CF as RCM-PDS-CF. Furthermore, we validate its recommendation performance through numerical experiments using a real-world dataset.

Keywords: Clustering · Rough set theory · Collaborative filtering

1 Introduction

Recommender systems, as found in e-commerce sites like Amazon and video streaming services like YouTube and Netflix, utilize collaborative filtering (CF) [1–3]. Enhancing the recommendation performance directly improves usability, purchase rate, viewing rate, and consequently, the company's profits. As big data is accumulated every day, clustering, a technique that classifies and summarizes data automatically without supervision, is gaining attention in today's world where the effective use of such data is in demand. In clustering-based CF [4–8], clusters of users with similar preferences are extracted, and contents with high preference within the cluster to which the user belongs are recommended. The data in CF tasks consists of a vast number of users and items, and in addition to the diverse preference patterns of users, the boundaries of preference patterns are not clear, and ambiguity and uncertainty are assumed to be inherent in the data.

This work was supported by JSPS KAKENHI Grant Number JP20K19886.

K. Honda et al. (Eds.): IUKM 2023, LNAI 14376, pp. 218–229, 2024.
https://doi.org/10.1007/978-3-031-46781-3_19

Hard C-means (HCM; k-means) is one of the most popular clustering algorithms [9]. In HCM-type clustering, each object is exclusively assigned to a unique cluster, making it unsuitable for clustering user preference information that includes ambiguity and uncertainty. Therefore, methods for handling ambiguity and uncertainty in data are required in clustering-based CF. There are soft computing approaches aimed at human-like flexible information processing capabilities, such as fuzzy theory [10] and rough set theory [11].

In this study, we focus on rough clustering [12], which is based on rough set theory and handles the uncertainty inherent in data. Rough clustering deals with the certainty, possibility, and uncertainty of each object's belonging to clusters. Various methods such as generalized rough C-means (GRCM) [13], rough set C-means (RSCM) [14], and rough membership C-means (RMCM) [15] have been proposed as rough clustering algorithms. Furthermore, rough clustering-based CF has been proposed, and its effectiveness has been reported [16].

In this study, we adopt GRCM with membership normalization (GRCM-MN) [17], which calculates the cluster center using normalized membership, and it will be referred to as RCM for simplicity. RCM allows for the belonging of each object to multiple clusters and can express cluster overlap by relaxing the condition of belonging based on a linear function based on the distance to the nearest cluster center.

One problem with data in CF tasks is that it contains many missing values. Conventional rough clustering-based CF uses an approach (single imputation method) to pre-fill missing values with constants or averages, but this can lead to bias in the results as equable values different from the original values are substituted, causing distortion in the data. One approach to missing values is the whole data strategy (WDS), which deletes rows containing missing values to create a data matrix without missing values for analysis, but in CF tasks, most of the data is missing values, and deleting most rows makes analysis difficult. On the other hand, the partial distance strategy (PDS) is an approach that utilizes partial distance, which ignores dimensions containing missing values when calculating distances [18]. With PDS, it is possible to analyze using information from dimensions that are not missing without deleting missing rows.

In this study, we propose RCM-PDS, which introduces PDS into RCM, aiming to handle missing values within the clustering algorithm without pre-filling missing values, and further propose its application to CF as RCM-PDS-CF. We also verify the recommendation performance of the proposed method through numerical experiments using a real-world dataset.

The structure of this paper is as follows. In Sect. 2, we give an overview of HCM, RCM, RCM-CF, single imputation method, whole data strategy, and partial distance strategy. In Sect. 3, we explain the proposed method, RCM-PDS, and its application to CF, RCM-PDS-CF. In Sect. 4, we show the setup and results of numerical experiments, and discuss the results. In Sect. 5, we conclude the study.

2   Preliminaries

2.1   Hard C-Means

HCM is one of the most popular clustering algorithms. We consider a problem of extracting C clusters that consist of objects with similar features from the data recording m observed features for n objects. The whole set of objects is denoted as $W = \{x_1, \ldots, x_i, \ldots, x_n\}$. Each object i is represented by an m-dimensional real-valued feature vector $x_i = (x_{i1}, \ldots, x_{ij}, \ldots, x_{im})^\top$, and each cluster c has a cluster center $b_c = (b_{c1}, \ldots, b_{cj}, \ldots, b_{cm})^\top$ as a representative point.

A sample algorithm of HCM is shown below.

Step 1 Set the number of clusters C.
Step 2 Select C initial cluster centers b_c randomly from W.
Step 3 Calculate the membership u_{ci} of object i to cluster c using nearest assignment.

$$d_i^{\min} = \min_{1 \leq l \leq C} d_{li}, \tag{1}$$

$$u_{ci} = \begin{cases} 1 & (d_{ci} \leq d_i^{\min}), \\ 0 & (\text{otherwise}). \end{cases} \tag{2}$$

where d_{ci} is the distance between cluster center b_c and object x_i.
Step 4 Calculate the cluster center b_c by using the following equation.

$$b_c = \frac{\sum_{i=1}^{n} u_{ci} x_i}{\sum_{i=1}^{n} u_{ci}}. \tag{3}$$

Step 5 Repeat **Step 3**–**4** until there is no change in u_{ci}.

In HCM, each object is exclusively assigned to a single cluster. Therefore, it cannot handle the ambiguity and uncertainty inherent in data, such as an object belonging to multiple clusters simultaneously.

2.2   Rough C-Means

RCM is a type of rough clustering that introduces the perspective of rough set theory into HCM. In RCM-type clustering, the concept of lower approximation, upper approximation, and boundary region in rough set theory is referred to, and for each cluster, the lower area, upper area, and boundary area are introduced. This allows for the handling of certainty, possibility, and uncertainty of object's belonging to clusters.

In RCM, the parameters α and β are used to adjust the degree of overlap between clusters during the assignment of each object to cluster. These parameters form a linear function with the distance to the nearest cluster center, which relaxes the condition of the nearest assignment in HCM. This allows for membership in multiple clusters, thereby achieving cluster overlap.

A sample algorithm of RCM is shown below:

Step 1 Set the number of clusters C, and the parameters α ($\alpha \geq 1$), β ($\beta \geq 0$) to adjust the overlap of clusters. After initializing the membership of the upper area $\overline{u}_{ci}$ of object i to cluster c under the constraints of Eqs. (4) and (5), calculate the normalized membership $\tilde{u}_{ci}$ using Eq. (6).

$$\overline{u}_{ci} \in \{0, 1\}, \forall c, i, \tag{4}$$

$$\sum_{l=1}^{C} \overline{u}_{li} \neq 0, \forall i, \tag{5}$$

$$\tilde{u}_{ci} = \frac{\overline{u}_{ci}}{\sum_{l=1}^{C} \overline{u}_{li}}. \tag{6}$$

Step 2 Calculate the cluster center $\boldsymbol{b}_c$ using the following equation:

$$\boldsymbol{b}_c = \frac{\sum_{i=1}^{n} \tilde{u}_{ci} \boldsymbol{x}_i}{\sum_{i=1}^{n} \tilde{u}_{ci}}. \tag{7}$$

Step 3 Calculate the distance from object i to the nearest cluster center, $d_i^{\min}$, the upper area membership $\overline{u}_{ci}$, and the normalized membership $\tilde{u}_{ci}$ in sequence using Eqs. (1), (8), and (6).

$$\overline{u}_{ci} = \begin{cases} 1 & \left(d_{ci} \leq \alpha d_i^{\min} + \beta\right), \\ 0 & \text{(otherwise)}. \end{cases} \tag{8}$$

Step 4 Repeat **Steps 2–3** until there are no change in $\overline{u}_{ci}$.

RCM yields the same result as HCM when $\alpha = 1$ and $\beta = 0$. As α and β increase, the upper area expands and cluster overlap increases.

2.3   RCM-CF

RCM-CF extracts clusters of users with similar preferences using RCM and recommends contents with high preference within the cluster to which each user belongs.

A sample procedure of RCM-CF is shown below:

Step 1 Apply RCM to the $n \times m$ rating matrix $R = \{r_{ij}\}$ to obtain the normalized membership $\tilde{u}_{ci}$ and the cluster center $\boldsymbol{b}_c$, where n is the number of users and m is the number of items.

Step 2 Calculate the recommendation degree $\hat{r}_{ij}$ of item j to user i using the following equation:

$$\hat{r}_{ij} = \sum_{c=1}^{C} \tilde{u}_{ci} b_{cj}. \tag{9}$$

Step 3 Set a threshold $\eta \in [\min\{\hat{r}_{ij}\}, \max\{\hat{r}_{ij}\}]$. If the recommendation degree $\hat{r}_{ij}$ is equal to or greater than η, recommend item j to user i.

$$\tilde{r}_{ij} = \begin{cases} 1 & (\hat{r}_{ij} \geq \eta), \\ 0 & \text{(otherwise)}. \end{cases} \tag{10}$$

2.4   Approaches to Missing Values in Clustering

Single Imputation Method. In conventional clustering methods, if the data matrix contains missing values, it is impossible to perform calculations as it is, hence the need for missing value processing. The single imputation method is a common approach to dealing with missing values in conventional clustering-based CF, which involves substituting missing values in the data matrix with uniform values such as constants or averages. While the single imputation method has advantages such as easy imputation of missing values, it has disadvantages like potential distortions in data due to substitution with values different from the original ones.

Whole Data Strategy (WDS). In WDS, rows containing missing values in the data matrix are deleted, and clustering is performed on the resulting data matrix that does not contain missing values. This strategy is easy to implement as it does not require special processing in the clustering algorithm. However, it results in loss of information as the valid values in the rows containing missing values are also deleted.

Partial Distance Strategy (PDS). PDS enables clustering of data with missing values by introducing a partial distance that calculates distance while ignoring missing dimensions, thus utilizing the valid values contained in missing rows without deleting them. This method has the advantage of lower information loss than WDS as it is not necessary to delete missing rows.

As an example, consider two $m = 4$ dimensional vectors, the object $x_i = (1, ?, 3, ?)^\top$, and the cluster center $b_c = (2, 4, 5, 1)^\top$, where "?" represents a missing value. Ignoring the dimensions containing missing values, the squared distance d_{ci}^2 between the cluster center b_c and the object x_i is calculated as follows:

$$d_{ci}^2 = (1 - 2)^2 + (3 - 5)^2. \tag{11}$$

Since the squared distance decreases according to the number of missing dimensions, we scale by multiplying by (total number of dimensions)/(number of used dimensions), and compute the partial distance by taking the square root as follows:

$$d_{ci} = \sqrt{\left(\frac{4}{4 - 2}\right)\{(1 - 2)^2 + (3 - 5)^2\}}. \tag{12}$$

3   Collaborative Filtering Based on Rough C-Means Clustering with Missing Value Processing

A significant problem in CF tasks is that the data often contains many missing values. Conventional rough clustering-based CF employs single imputation methods, but this can introduce distortion into the data by substituting equable values that differ from the original ones, leading to potential biases in the results.

Moreover, using the whole data strategy (WDS) to handle missing values leads to a problem where, because most of the data in CF tasks consists of missing values, nearly all rows are deleted, making the analysis challenging. Therefore, we propose a method, referred to as RCM-PDS, which introduces the partial distance strategy (PDS) into RCM. The goal of this approach is to perform clustering using the original data matrix with missing values without prior imputation. Additionally, we propose its application to CF as RCM-PDS-CF.

3.1   RCM-PDS

RCM-PDS introduces PDS into RCM, enabling clustering of a data matrix with missing values by utilizing partial distances that ignore dimensions with missing values.

A sample algorithm of RCM-PDS is as follows:

Step 1 Set the number of clusters C, and parameters α ($\alpha \geq 1$), β ($\beta \geq 0$) that adjust the overlap degree of the clusters. After randomly initializing the upper area membership $\bar{u}_{ci}$ under the constraints of Eqs. (4) and (5), calculate the normalized membership $\tilde{u}_{ci}$ using Eq. (6).

Step 2 Calculate the cluster center b_c using the following equation:

$$b_{cj} = \frac{\sum_{i=1}^{n} \tilde{u}_{ci} I_{ij} x_{ij}}{\sum_{i=1}^{n} \tilde{u}_{ci} I_{ij}}, \tag{13}$$

$$I_{ij} = \begin{cases} 0 & (x_{ij} \text{ is missing}), \\ 1 & (x_{ij} \text{ is observed}). \end{cases} \tag{14}$$

Step 3 Introduce PDS and calculate the partial distance d_{ci} between cluster center b_c and object x_i using the following equation:

$$d_{ci} = \sqrt{\frac{m}{I_i} \sum_{j=1}^{m} I_{ij}(x_{ij} - b_{cj})^2}, \tag{15}$$

$$I_i = \sum_{j=1}^{m} I_{ij}. \tag{16}$$

Step 4 Sequentially calculate the distance to the nearest cluster center $d_i^{\min}$, the upper area membership $\bar{u}_{ci}$, and the normalized membership $\tilde{u}_{ci}$ using Eqs. (1), (8), and (6), respectively. Here, the distance d_{ci} represents a partial distance.

Step 5 Repeat **Step 2–4** until there is no change in $\bar{u}_{ci}$.

3.2   RCM-PDS-CF

RCM-PDS-CF extracts clusters of users with similar preferences using RCM-PDS and recommends items with high preferences within the cluster each user belongs to.

A sample procedure of RCM-PDS-CF is as follows:

Step 1 Apply RCM-PDS to the $n \times m$ rating matrix $R = \{r_{ij}\}$, and calculate the normalized membership $\tilde{u}_{ci}$ and the cluster center $\boldsymbol{b}_c$, where n is the number of users and m is the number of items.

Step 2 Calculate the recommendation degree $\hat{r}_{ij}$ for item j for user i using Eq. (9).

Step 3 Set a threshold $\eta \in [\min\{\hat{r}_{ij}\}, \max\{\hat{r}_{ij}\}]$, and if the recommendation degree $\hat{r}_{ij}$ is greater than or equal to η, recommend item j to user i (Eq. (10)).

4   Numerical Experiments

4.1   Experimental Overview

Numerical experiments were conducted using a real-world dataset (MovieLens-100k dataset) to verify the recommendation performance of the proposed method. We applied both the conventional method (RCM based on single imputation) and the proposed method (RCM-PDS) to the MovieLens-100k dataset. The comparison and verification of changes in recommendation performance were made when parameters such as the degree of cluster overlap α, β, and the number of clusters C were varied.

In RCM with single imputation, the following four types of imputations were examined:

RCM_user Substituting missing values with each user's average rating
RCM_item Substituting missing values with each item's average rating
RCM_all Substituting missing values with the overall average rating
RCM_const(3) Substituting missing values with the constant 3

In the proposed method, in addition to the standard method of adjusting the scale, we also tested a method without scale adjustment (RCM-PDS-noscale method). The ROC-AUC indicator was used as an indicator of recommendation performance.

4.2   Dataset

The MovieLens-100k data is composed of $100,000$ ratings from 943 users on $1,682$ movies, collected by GroupLens Research (https://grouplens.org/). This dataset is sparse with many unrated values. For clustering and evaluation, in this experiment, we extracted $n = 690$ users who rated 30 or more movies and $m = 583$ movies that were rated by more than 50 users. We created a data matrix of 690×583 using data containing $77,201$ ratings. Approximately 10%, i.e., $7,721$ ratings, were used as test data from the total set. The training data was derived from the data matrix, in which elements that were adopted for the test data were replaced with unrated values. Unrated values are treated as missing values.

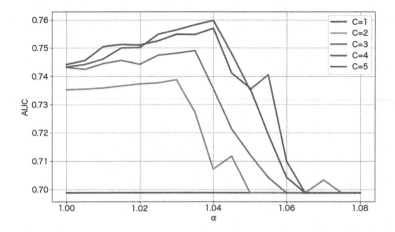

Fig. 1. Change in AUC by α at each C in RCM-PDS

4.3 Evaluation Indicator

In this study, the ROC-AUC indicator was adopted as the evaluation indicator for recommendation performance. From here on, we will simply refer to ROC-AUC as AUC. AUC is the area under the curve of the ROC curve (receiver operating characteristic curve), which is obtained by plotting the true positive rate (TPR) against the false positive rate (FPR) as the threshold changes. In random recommendations, the AUC is about 0.5, and the closer the AUC is to 1, the better the recommendation performance.

4.4 Experimental Results

Influence of Parameters in the Proposed Method. Firstly, Fig. 1 shows the changes in AUC by α when the number of cluster C is set to $\{1, 2, 3, 4, 5\}$ in RCM-PDS. Here, β was fixed at 0.0 and α was changed in increments of 0.005 in the range $[1.00, 1.08]$. The AUC was calculated as the average of 10 trials with random initial values. If no result could be obtained due to zero division, etc., the result was excluded. The average was calculated if results from at least three trials were obtained. When $\alpha = 1.0$, the results are identical to HCM. When $C = 1$, no clustering is performed, and the average rating of each movie becomes the recommendation score, resulting in an AUC of 0.6987. When $C = 2$ or more, as α increases, the AUC also rises to a certain extent. However, if it increases further, the AUC tends to decrease and eventually matches to the value when $C = 1$. Regarding the maximum values at each C, they increase as C ranges from 1 to 5. Specifically, the maximum value of 0.7597 is recorded at $\alpha = 1.04$ when $C = 5$.

Next, Fig. 2 shows the changes in AUC by β under the same conditions. Here, α was fixed at 1.0 and β was changed in increments of 0.1 in the range $[0.0, 2.0]$. When $\beta = 0.0$, the results are identical to HCM. Comparing with the

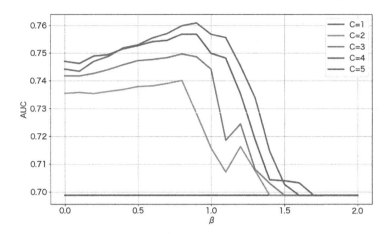

Fig. 2. Change in AUC by β at each C in RCM-PDS

case of changing α, there was no significant difference in the trend of changes in recommendation performance when changing β. When $C = 5$ and $\beta = 0.9$, it recorded the maximum value of 0.7608.

Comparison Between the Proposed Method and Conventional Methods. Figure 3 shows the changes in maximum AUC when α and β are changed at each $C \in \{1, 4, \ldots, 28, 30\}$ in the proposed method and four types of conventional methods. At each C, $\alpha \in [1.000, 1.080]$ was changed in increments of 0.05, and $\beta \in [0.0, 2.0]$ was changed in increments of 0.1. In each method, when $C = 1$, the average value of each item's rating is used as the recommendation degree. Among the conventional methods, RCM_item method showed the highest AUC at all C. The maximum AUC among the conventional methods was when $C = 16, \alpha = 1.01, \beta = 0.0$, recording a maximum value of 0.739. For the proposed method, the AUC increased as C increased up to around $C = 10$, and after that, although there were some ups and downs, the result was almost flat. The maximum AUC in the proposed method was when $C = 30, \alpha = 1.015, \beta = 1.0$, recording a maximum value of 0.764. It was confirmed that the proposed method has higher recommendation performance than any conventional method regardless of the number of clusters.

Conventional methods introduce distortion to the data and cause bias in the results due to the use of a single imputation method. On the other hand, the proposed method improves recommendation performance. It extracts user preference patterns without distorting data, thanks to the partial distance strategy. The results suggest that introducing a partial distance strategy is effective in CF tasks based on rough clustering.

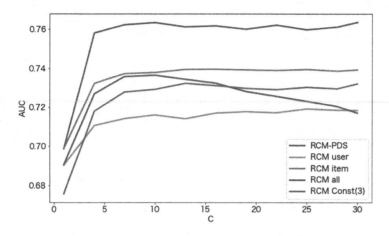

Fig. 3. Change in maximum AUC by C of each method

Effects of Scale Adjustment in the Proposed Method. We present the comparative results of the recommendation performance between the method with scale adjustment and the method without scale adjustment (RCM-PDS-noscale). Figure 4 shows the change in the maximum AUC of RCM-PDS and RCM-PDS-noscale when α and β were varied for each $C \in \{1, 4, \ldots, 28, 30\}$. For each C, $\alpha \in [1.000, 1.080]$ was varied in steps of 0.05 and $\beta \in [0.0, 2.0]$ was varied in steps of 0.1. Until $C = 10$, both methods showed similar values, with the AUC increasing as C increased. As C further increased, both methods remained almost flat, but RCM-PDS-noscale demonstrated a better ability to prevent a performance decrease, indicating a higher AUC.

In the method with scale adjustment, it was observed that when C increased, almost all or all clusters merged, resulting in some instances where clustering could not be performed effectively. This scenario could be the cause of the decrease in the average value of the maximum AUC. There is a difference in behavior when calculating the upper area membership in Eq. (8) between the method with scale adjustment and the one without. The scale adjustment method used in this study depends only on the data dimensionality and the number of missing dimensions of the object, so when focusing on a certain object, the ratio of distances to each cluster center does not change. Therefore, the multiplication-based α is not affected by scale adjustment, while the addition-based β is affected by scale adjustment. Without scale adjustment, if the number of missing dimensions varies between different objects, the scale difference becomes relatively large. Therefore, it is considered that users with fewer ratings are more likely to belong to many clusters, and users with many ratings tend to belong to fewer clusters. As a result, users with fewer ratings are likely to belong to more clusters and refer to the preference information of many users, while users with many ratings are likely to belong to relatively fewer clusters and refer to the information of a small number of users with similar preferences.

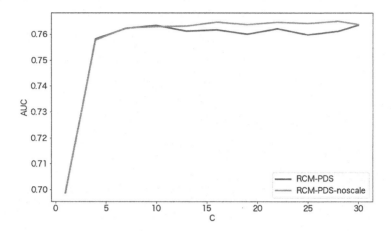

Fig. 4. Changes in AUC by C in RCM-PDS and RCM-PDS-noscale methods

In the present data, this type of recommendation was suitable, which may have led to these results. Furthermore, in the absence of scale adjustment, the fact that users with many ratings tend to belong to fewer clusters could prevent the clusters from merging, leading to stable performance.

5   Conclusion

In this study, we proposed RCM-PDS as a rough clustering approach for data including missing values, which introduces a partial distance strategy to rough C-means. Furthermore, we proposed RCM-PDS-CF as an application of RCM-PDS to collaborative filtering. We verified its recommendation performance through numerical experiments using the real-world dataset, MovieLens-100k. The results confirmed that, in the MovieLens-100k dataset, introducing a partial distance strategy to RCM, regardless of the number of clusters, demonstrated a higher recommendation performance compared to the conventional single imputation methods. This implies that introducing a partial distance strategy is effective for dealing with missing values in rough clustering-based collaborative filtering tasks. In addition, we compared the presence or absence of scale adjustment in the calculation of partial distances, confirming that when the number of clusters is large, not performing scale adjustment provides a higher recommendation performance.

Future work includes the introduction of handling missing values in the cluster centers and exploring more appropriate scale adjustment methods.

References

1. Linden, G., Smith, B., York, J.: Amazon.com recommendations: item-to-item collaborative filtering. IEEE Internet Comput. **7**(1), 76–80 (2003)
2. Smith, B., Linden, G.: Two decades of recommender systems at Amazon.com. IEEE Internet Comput. **21**(3), 12–18 (2017)
3. Su, X., Khoshgoftaar, T.M.: A survey of collaborative filtering techniques. Adv. Artif. Intell. **2009**, 1–19 (2009)
4. Ungar, L.H., Foster, D.P.: Clustering methods for collaborative filtering. In: AAAI Workshop on Recommendation Systems, vol. 1, pp. 114–129 (1998)
5. Chee, S.H.S., Han, J., Wang, K.: RecTree: an efficient collaborative filtering method. In: International Conference on Data Warehousing and Knowledge Discovery, pp. 141–151 (2001)
6. O'Connor, M., Herlocker, J.: Clustering items for collaborative filtering. In: Proceedings of the ACM SIGIR Workshop on Recommender Systems (SIGIR 1999), vol. 128 (1999)
7. Sarwar, B.M., Karypis, G., Konstan, J.A., Riedl, J.: Recommender systems for large-scale E-commerce: scalable neighborhood formation using clustering. In: Proceedings of the 5th International Conference on Computer and Information Technology (ICCIT 2002), vol. 1, pp. 291–324 (2002)
8. Xue, G.-R., Lin, C., Yang, Q., et al.: Scalable collaborative filtering using cluster-based smoothing. In: Proceedings of the ACM SIGIR Conference, pp. 114–121 (2005)
9. MacQueen, J.: Some methods of classification and analysis of multivariate observations. In: Proceedings of 5th Berkeley Symposium on Mathematical Statistics and Probability, pp. 281–297 (1967)
10. Zadeh, L.A.: Fuzzy sets. Inf. Control **8**, 338–353 (1965)
11. Pawlak, Z.: Rough sets. Int. J. Comput. Inf. Sci. **11**(5), 341–356 (1982)
12. Ubukata, S.: Development of rough set-based C-means clustering. J. Jpn. Soc. Fuzzy Theory Intell. Inf. **32**(4), 121–127 (2020)
13. Ubukata, S., Notsu, A., Honda, K.: General formulation of rough C-means clustering. Int. J. Comput. Sci. Netw. Secur. **17**(9), 29–38 (2017)
14. Ubukata, S., Umado, K., Notsu, A., Honda, K.: Characteristics of rough set C-means clustering. J. Adv. Comput. Intell. Intell. Inf. **22**(4), 551–564 (2018)
15. Ubukata, S., Notsu, A., Honda, K.: The rough membership k-means clustering. In: Proceedings of the 5th International Symposium on Integrated Uncertainty in Knowledge Modelling and Decision Making, pp. 207–216 (2016)
16. Ubukata, S., Takahashi, S., Notsu, A., Honda, K.: Basic consideration of collaborative filtering based on rough C-means clustering. In: Proceedings of Joint 11th International Conference on Soft Computing and Intelligent Systems and 21st International Symposium on Advanced Intelligent Systems, pp. 256–261 (2020)
17. Ubukata, S., Kato, H., Notsu, A., Honda, K.: Rough set-based clustering utilizing probabilistic membership. J. Adv. Comput. Intell. Intell. Inf. **22**(6), 956–964 (2018)
18. Hathaway, R.J., Bezdek, J.C.: Fuzzy c-means clustering of incomplete data. IEEE Trans. Syst. Man Cybern., Part B (Cybern.) **31**(5), 735–744 (2001)

Collaborative Filtering Based on Probabilistic Rough Set C-Means Clustering

Seiki Ubukata$^{(\boxtimes)}$ and Kazuma Ehara

Osaka Metropolitan University, Saka, Osaka 599-8531, Japan
`ubukata@omu.ac.jp`

Abstract. Collaborative filtering (CF) is a technique for realizing recommender systems found in e-commerce sites and video streaming sites. Appropriate content recommendations to individual users will improve usability, purchase rates, viewing rates, and corporate profits. Clustering is a technique for automatically classifying and summarizing the data by extracting clusters composed of similar objects. Clustering-based CF extracts clusters of users with similar interests and preferences, and recommends highly preferred contents in the cluster to each user. Rough set C-means (RSCM) is one of the rough clustering methods based on rough set theory that can deal with the uncertainty of belonging of object to clusters considering the granularity of the object space. Probabilistic rough set C-means (PRSCM) is an extension of RSCM based on a probabilistic rough set model. In this study, we propose a collaborative filtering approach based on probabilistic rough set C-means clustering (PRSCM-CF). Furthermore, we verify the recommendation performance of the proposed method through numerical experiments using real-world datasets.

Keywords: Clustering · Rough set theory · Collaborative filtering

1 Introduction

Collaborative filtering (CF) is utilized in recommender systems seen on e-commerce sites, such as Amazon, and video streaming services like YouTube and Netflix [1–3]. Improving the performance of content recommendations leads to improvements in usability, purchase rates, viewing rates, and corporate profits. In today's age, where the effective use of accumulated big data is essential, clustering - a technique for automatically classifying and summarizing data without supervision - is garnering attention. In clustering-based CF, clusters of users with similar preferences are extracted, and contents with high preference degree within the cluster are recommended.

This work was supported by JSPS KAKENHI Grant Number JP20K19886.

K. Honda et al. (Eds.): IUKM 2023, LNAI 14376, pp. 230–242, 2024.
https://doi.org/10.1007/978-3-031-46781-3_20

Hard C-means (HCM; k-means) [4] is one of the most popular clustering algorithms, and the effectiveness of HCM-based CF (HCM-CF) has been reported [5]. In CF tasks, the data consist of a massive number of users and items, and user preference patterns are diverse. In addition, the boundaries of users' preference patterns are not clear, and thus ambiguity and uncertainty are inherent in the data. Therefore, conventional HCM-type clustering, which clearly divides clusters, cannot adequately extract the cluster structure of users' preference patterns. Therefore, a method for handling the ambiguity and uncertainty of data is required in clustering-based CF.

There are soft computing approaches aiming for human-like flexible information processing, such as fuzzy theory and rough set theory. Rough clustering is a method in which the viewpoint of rough set theory [6,7] is introduced. Rough set C-means (RSCM) has been proposed as a rough clustering considering granularity [9], and the effectiveness of CF based on RSCM has been reported [10]. In RSCM, lower approximation, upper approximation, and boundary regions that indicate certainty, possibility, and uncertainty in each cluster, respectively, are detected.

In the probabilistic rough set model (PRSM) [11], the definitions of lower and upper approximations in rough set theory are relaxed by the α-cut of the rough membership value, representing the proportion of the cluster to the object's neighborhood. This allows for gradual changes in the lower and upper approximations, thereby achieving intermediate approximations. Probabilistic approximation based on the α-cut allows for flexible approximation. Probabilistic RSCM (PRSCM), which introduces the concept of a probabilistic rough set model to RSCM, has been proposed [12].

In this study, we propose a collaborative filtering approach based on probabilistic rough set C-means clustering (PRSCM-CF). Furthermore, we verify the recommendation performance of the proposed method through numerical experiments using real-world datasets. We verify the impact of the flexible approximation based on the smooth transition between lower and upper approximations by PRSCM on the CF task.

The structure of this paper is as follows. In Sect. 2, we outline RSCM, RSCM-based CF, and PRSCM as preparation. In Sect. 3, we explain the proposed CF based on PRSCM (PRSCM-CF). In Sect. 4, we present the setup and results of the numerical experiments, along with a discussion of the findings. In Sect. 5, we summarize this study.

2 Preliminaries

2.1 Hard C-Means

HCM is one of the most popular clustering algorithms. We consider the problem of extracting C clusters, each consisting of objects with similar features, from the data that records m observed features for n objects. The whole set of objects is denoted as $W = \{x_1, \ldots, x_i, \ldots, x_n\}$. Each object i is represented by an m-dimensional real-valued feature vector $x_i = (x_{i1}, \ldots, x_{ij}, \ldots, x_{im})^\top$, and each cluster c has a cluster center $\boldsymbol{b}_c = (b_{c1}, \ldots, b_{cj}, \ldots, b_{cm})^\top$ as a representative point.

A sample algorithm of HCM is shown below.

Step 1 Set the number of clusters C.
Step 2 Select C initial cluster centers $\boldsymbol{b}_c$ randomly from W.
Step 3 Calculate the membership u_{ci} of object i to cluster c using the nearest assignment.

$$d_i^{\min} = \min_{1 \le l \le C} d_{li}, \tag{1}$$

$$u_{ci} = \begin{cases} 1 & (d_{ci} \le d_i^{\min}), \\ 0 & (\text{otherwise}), \end{cases} \tag{2}$$

where d_{ci} is the distance between cluster center $\boldsymbol{b}_c$ and object $\boldsymbol{x}_i$.
Step 4 Calculate the cluster center $\boldsymbol{b}_c$ by using the following equation.

$$\boldsymbol{b}_c = \frac{\sum_{i=1}^n u_{ci} \boldsymbol{x}_i}{\sum_{i=1}^n u_{ci}}. \tag{3}$$

Step 5 Repeat **Step 3-4** until there is no change in u_{ci}.

In HCM, each object is exclusively assigned to a single cluster. Therefore, it cannot handle the ambiguity and uncertainty inherent in data, such as an object belonging to multiple clusters simultaneously.

2.2   Rough Set C-Means

RSCM is a type of rough clustering that introduces a perspective of rough set theory into HCM. RSCM is performed in the approximation space $\langle W, R \rangle$ consisting of the set of objects W and the binary relation $R \subseteq W \times W$. The neighborhood of the object defined by R is treated as a granule. By determining the object's belonging to the clusters on a granular basis, the algorithm can handle the certainty, possibility, and uncertainty of the object's belonging to the cluster while considering granularity. In this study, we adopt RSCM with membership normalization (RSCM-MN) [13], which simplifies the calculation of the cluster center, as RSCM.

A sample algorithm of RSCM is shown below.

Step 1 Set the number of clusters, C. Set the neighborhood radius δ and calculate the δ-neighborhood relation R_{it} based on the Euclidean distance D_{it} between objects i and t.

$$D_{it} = \|\boldsymbol{x}_t - \boldsymbol{x}_i\| \tag{4}$$

$$= \left(\sum_{j=1}^m (x_{tj} - x_{ij})^2 \right)^{\frac{1}{2}}. \tag{5}$$

$$R_{it} = \begin{cases} 1 & (D_{it} \leq \delta), \\ 0 & (\text{otherwise}). \end{cases} \tag{6}$$

Here, the value of δ can be determined by the τ-percentile of the inter-object distance distribution, denoted by $[D_{it}]$, and τ can be set within the range $\tau \in [0, 100]$.

Step 2 Determine the initial cluster center b_c randomly.

Step 3 Calculate the membership u_{ci} of object i to the temporary cluster c using the nearest assignment, as shown in Eq. (2). Calculate the rough membership μ_{ci} to cluster c of object i.

$$\mu_{ci} = \frac{\displaystyle\sum_{t=1}^{n} u_{ct} R_{it}}{\displaystyle\sum_{t=1}^{n} R_{it}}. \tag{7}$$

Calculate the membership $\overline{u}_{ci}$ to the upper approximation of cluster c of object i.

$$\overline{u}_{ci} = \begin{cases} 1 & (\mu_{ci} > 0), \\ 0 & (\text{otherwise}). \end{cases} \tag{8}$$

Normalize $\overline{u}_{ci}$ for each object so that the total sum becomes 1, and obtain the normalized membership $\tilde{u}_{ci}$.

$$\tilde{u}_{ci} = \frac{\overline{u}_{ci}}{\displaystyle\sum_{l=1}^{C} \overline{u}_{li}}. \tag{9}$$

Step 4 Calculate the new cluster center b_c based on the normalized membership.

$$b_c = \frac{\displaystyle\sum_{i=1}^{n} \tilde{u}_{ci} x_i}{\displaystyle\sum_{i=1}^{n} \tilde{u}_{ci}}. \tag{10}$$

Step 5 Repeat **Step 3-4** until there is no change in u_{ci}.

2.3 Collaborative Filtering Based on RSCM

The effectiveness of RSCM-based CF (RSCM-CF) has been reported [10]. A sample procedure for RSCM-CF is as follows.

Let $X = \{r_{ij}\}$ be an $n \times m$ matrix representing the preference (such as ownership and ratings) of m items by n users.

First, RSCM is applied to X to extract the cluster structure in X, and the normalized membership $\tilde{u}_{ci}$ and the cluster center $\boldsymbol{b}_c$ are calculated.

Next, the recommendation degree $\hat{r}_{ij}$ of item j for user i is calculated using the following equation.

$$\hat{r}_{ij} = \sum_{c=1}^{C} \tilde{u}_{ci} b_{cj}. \tag{11}$$

Finally, using a threshold $\eta \in [\min\{\hat{r}_{ij}\}, \max\{\hat{r}_{ij}\}]$, items with recommendation degree above the threshold are recommended. $\tilde{r}_{ij}$ is a variable that indicates whether or not to recommend item j to user i, and is calculated using the following equation.

$$\tilde{r}_{ij} = \begin{cases} 1 & (\hat{r}_{ij} \geq \eta), \\ 0 & (\text{otherwise}). \end{cases} \tag{12}$$

Item j is recommended to user i when $\tilde{r}_{ij} = 1$, and not recommended otherwise.

2.4 Probabilistic Rough Set C-Means

PRSCM is a type of rough clustering that introduces the concept of probabilistic approximation in probabilistic rough set models to RSCM.

In rough set theory, the lower and upper approximations are determined by the weak 1-cut ($\mu_{ci} \geq 1$) and the strong 0-cut ($\mu_{ci} > 0$) of rough membership, respectively. Considering a parameter $\alpha \in (0, 1]$, the weak α-cut ($\mu_{ci} \geq \alpha$) of rough membership is defined as a probabilistic lower approximation, also known as an α-lower approximation. The α-lower approximation is the same as the ordinary lower approximation when $\alpha = 1$ and is identical to the ordinary upper approximation as $\alpha \to 0$. By gradually changing α from 1 to 0, the approximation can gradually transition from the lower approximation to the upper approximation, enabling flexible approximation.

A sample algorithm for PRSCM is as follows.

Step 1 Set the number of clusters, C. Set the neighborhood radius δ and calculate the δ-neighborhood relation R_{it} based on the Euclidean distance D_{it} between object i and object t (Eqs. (4) and (6)). Set the parameter $\alpha \in (0, 1]$ for the α-lower approximation.

Step 2 Determine the initial cluster center $\boldsymbol{b}_c$ randomly.

Step 3 Calculate the membership u_{ci} to the temporary cluster c of object i by nearest assignment (Eq. (2)). Calculate the rough membership μ_{ci} to cluster c of object i (Eq. (7)). Calculate the membership $\underline{u}_{ci}^{\alpha}$ to the α-lower approximation of cluster c of object i.

$$\underline{u}_{ci}^{\alpha} = \begin{cases} 1 & (\mu_{ci} \geq \alpha), \\ 0 & (\text{otherwise}). \end{cases} \tag{13}$$

Step 4 Calculate the new cluster center $\boldsymbol{b}_c$ based on the α-lower approximation.

$$\boldsymbol{b}_c = \frac{\sum\limits_{i=1}^{n} \underline{u}_{ci}^{\alpha} \boldsymbol{x}_i}{\sum\limits_{i=1}^{n} \underline{u}_{ci}^{\alpha}}. \tag{14}$$

Step 5 Repeat **Steps 3-4** until there is no change in u_{ci}.

3 Collaborative Filtering Based on Probabilistic Rough Set C-Means Clustering

In this study, we propose collaborative filtering based on PRSCM (PRSCM-CF). This method aims to improve recommendation performance by extracting more suitable cluster structures through more flexible approximations using probabilistic lower approximations that relax the definition of rough approximations.

3.1 Procedure of PRSCM-CF

A sample procedure of PRSCM-CF is as follows.

We consider $X = \{r_{ij}\}$ as an $n \times m$ matrix representing the preference (ownership status, evaluation value, etc.) of m items by n users.

Initially, we apply PRSCM to X, extract the cluster structure in X, and calculate the α-lower approximate memberships $\underline{u}_{ci}^{\alpha}$ and the cluster centers $\boldsymbol{b}_c$.

Next, we calculate the recommendation degree $\hat{r}_{ij}$ of item j for user i using the following equation:

$$\hat{r}_{ij} = \sum_{c=1}^{C} \underline{u}_{ci}^{\alpha} b_{cj}. \tag{15}$$

Finally, we use a threshold $\eta \in [\min\{\hat{r}_{ij}\}, \max\{\hat{r}_{ij}\}]$ and recommend items with a recommendation degree greater than or equal to the threshold. We define $\tilde{r}_{ij}$ as a variable that indicates whether to recommend item j to user i and calculate it using Eq. (12). We recommend item j to user i when $\tilde{r}_{ij} = 1$, and do not recommend it otherwise.

3.2 Handling Noise Users

As the threshold parameter α of the probabilistic lower approximation approaches 0, the possible region is detected, causing significant overlap of clusters. As α approaches 1, more certain regions are detected, making each cluster smaller. In PRSCM, a pattern in which "a user belongs to multiple clusters" and a pattern in which "a user does not belong to any cluster" may appear.

In this study, we call "a user who does not belong to any cluster" a *noise user* because they do not show tendencies similar to the preference patterns of any clusters. From Eq. (15), for noise user i, $\underline{u}_{ci}^{\alpha} = 0$ for all c, so the recommendation degree $\tilde{r}_{ij}$ is 0 for all items j, and no effective recommendation can be made. Therefore, when evaluating recommendation performance, we need to make some adaptations such as "excluding noise users and evaluating only non-noise users," "making recommendations for noise users using the overall average," and "evenly distributing the membership of noise users ($\underline{u}_{ci}^{\alpha} = \frac{1}{C}$) before making recommendations."

4   Numerical Experiments

We apply the proposed method, probabilistic rough set C-means-based collaborative filtering (PRSCM-CF), to real-world datasets and evaluate its recommendation performance using the ROC-AUC indicator. Through the experiments, we confirm the impact of probabilistic approximation in the probabilistic rough set model on the recommendation performance.

4.1   Datasets

NEEDS-SCAN/PANEL. The NEEDS-SCAN/PANEL dataset was collected by Nikkei Inc. in 2000, and indicates whether 996 households own any of the 18 products, including consumer electronics. Each element r_{ij} of the data matrix $X = \{r_{ij}\}$ represents whether household i owns product j, with $r_{ij} = 1$ indicating ownership and $r_{ij} = 0$ indicating otherwise. From the 996 × 18 matrix elements, we randomly selected 1,000 elements as test data. The data matrix, with all 1,000 elements corresponding to the test data replaced with $r_{ij} = 0$, was used as training data for clustering and making recommendations.

The 18 surveyed products (with the number of owning households in parentheses) are {Automobile (825), Piano (340), VTR (933), Room Air Conditioner (911), Personal Computer (588), Word Processor (506), CD (844), VD (325), Motorcycle (294), Bicycle (893), Large Refrigerator (858), Medium/Small Refrigerator (206), Microwave Oven (962), Oven (347), Coffee Maker (617), Washing Machine (986), Clothes Dryer (226), Electric Dryer (242)}.

MovieLens-100k. The MovieLens-100k dataset, collected by GroupLens Research (https://grouplens.org/), consists of 100,000 ratings by 943 users for 1,682 movies. 1,000 ratings were randomly selected from the 100,000 ratings to serve as the test data, with the corresponding ratings in the training data being replaced with unrated values. The original ratings were graded on a scale of 1 to 5. If the original rating was 4 or higher, it was replaced with $r_{ij} = 1$; if it was 3 or lower, it was replaced with $r_{ij} = 0$. Unrated values were given a value of $r_{ij} = 0.5$.

4.2   Evaluation Metrics

The recommendation degree $\hat{r}_{ij}$ was calculated based on the α-lower approximation of the membership $\underline{u}_{ci}^{\alpha}$ and the cluster center b_c using PRSCM. The recommendation performance was measured using the ROC-AUC indicator. The ROC (receiver operating characteristic) curve is constructed by plotting the true positive rate against the false positive rate for various threshold settings. The true positive rate refers to the proportion of correct recommendations given to users, whereas the false positive rate refers to the proportion of incorrect recommendations. The AUC (area under the curve) is the area under the ROC curve, and a value around 0.5 would be expected for random recommendations. The higher the value (closer to 1), the better the recommendation performance.

4.3   NEEDS-SCAN/PANEL

In the following two subsections, the membership values for noise users i were distributed evenly. That is, for noise users, all values of c were replaced with $\underline{u}_{ci}^{\alpha} = \frac{1}{C}$. In the subsequent third subsection, the noise users were removed and the recommendation performance was evaluated only with non-noise users.

Changes in AUC by C for Each α at Various τ. Figures 1(a), 1(b), 1(c), and 1(d) show changes in AUC by the number of cluster C for each $\alpha = \{1.0, 0.75, 0.5, 0.25, 10^{-5}\}$ at $\tau = \{10, 30, 50, 70\}$, respectively. The conventional method, RSCM, is shown by the dashed line. In Fig. 1(a), while RSCM exhibits good recommendation performance around $AUC = 0.85$, PRSCM performs worse under all α. Particularly, performance is poor for $\alpha = 1.0$ lower approximation and $\alpha = 10^{-5}$ upper approximation, with some improvement seen in intermediate probabilistic approximations such as $\alpha = \{0.75, 0.5\}$. In Fig. 1(b), when the number of clusters is small, PRSCM($\alpha = \{0.5, 10^{-5}\}$) outperforms RSCM. However, other α show unstable behavior with performance degradation. Unlike the case for $\tau = 10$, the upper approximation with $\alpha = 10^{-5}$ displays stable performance, but overall, it still falls short of RSCM. In Fig. 1(c), while RSCM's performance drops with fewer clusters, PRSCM($\alpha = 10^{-5}$) exhibits stable performance. In Fig. 1(d), all methods show decreased performance. When the number of clusters is small, PRSCM($\alpha = \{0.25, 10^{-5}\}$) experiences a relatively smaller drop. However, the performance is worse than without clustering ($C = 1$), so the effect of clustering is not observed.

Changes in AUC by τ for Each α at Various C.   Figures 2(a), 2(b), and 2(c) show the changes in AUC by τ for $\alpha = \{1.0, 0.75, 0.5, 0.25, 10^{-5}\}$, respectively, for $C = \{2, 5, 10\}$. In the case of $C = 2$ in Fig. 2(a), the performance of RSCM deteriorated with increasing τ. On the other hand, for PRSCM, the performance improved with increasing τ when α was small, but the performance deteriorated with increasing τ when α was large. Similar trends were observed in Figs. 2(b) and 2(c), but the results were overall inferior to the performance of RSCM.

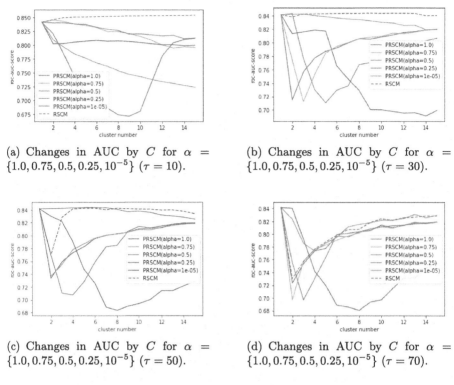

(a) Changes in AUC by C for $\alpha = \{1.0, 0.75, 0.5, 0.25, 10^{-5}\}$ ($\tau = 10$).

(b) Changes in AUC by C for $\alpha = \{1.0, 0.75, 0.5, 0.25, 10^{-5}\}$ ($\tau = 30$).

(c) Changes in AUC by C for $\alpha = \{1.0, 0.75, 0.5, 0.25, 10^{-5}\}$ ($\tau = 50$).

(d) Changes in AUC by C for $\alpha = \{1.0, 0.75, 0.5, 0.25, 10^{-5}\}$ ($\tau = 70$).

Fig. 1. Changes in AUC by C for different α values and τ parameters.

Effects of Removing Noise Users. In this section, we present the results of evaluating the recommendation performance using only non-noise users by removing noise users. Figures 3(a) and 3(b) show the changes in AUC by C for $\alpha = \{1.0, 0.75, 0.5, 0.25, 10^{-5}\}$ at $\tau = \{5, 10\}$, respectively. Figures 3(a) and 3(b) show almost the same trend. While RSCM shows stable performance, PRSCM with $\alpha = 1.0$ demonstrates better performance than RSCM when the number of clusters is small ($AUC = 0.86$). When α is large, each cluster becomes smaller, leading to the removal of more noise users. Therefore, we consider that the impact of noise was reduced, and more accurate recommendations could be made for non-noise users.

In the first and second subsections, it was difficult for PRSCM to outperform RSCM when the membership of noise users was evenly distributed. However, by removing noise users, it was suggested that more accurate recommendations than RSCM could be made.

4.4   MovieLens-100k

In Sect. 4.4, the membership values for noise user i were evenly distributed. In other words, for noise users, we replaced $\underline{u}_{ci}^{\alpha}$ with $\frac{1}{C}$ for all c. Figures 4(a), 4(b),

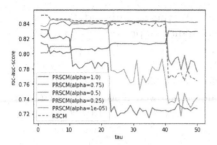
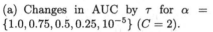
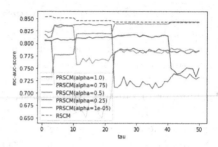

(a) Changes in AUC by τ for $\alpha = \{1.0, 0.75, 0.5, 0.25, 10^{-5}\}$ ($C = 2$).

(b) Changes in AUC by τ for $\alpha = \{1.0, 0.75, 0.5, 0.25, 10^{-5}\}$ ($C = 5$).

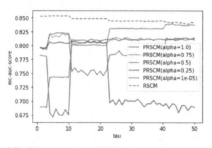

(c) Changes in AUC by τ for $\alpha = \{1.0, 0.75, 0.5, 0.25, 10^{-5}\}$ ($C = 10$).

Fig. 2. Changes in AUC by τ for different α values and C parameters.

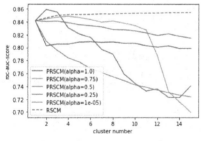
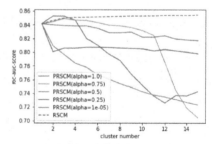

(a) Changes in AUC by C for $\alpha = \{1.0, 0.75, 0.5, 0.25, 10^{-5}\}$ ($\tau = 5$).

(b) Changes in AUC by C for $\alpha = \{1.0, 0.75, 0.5, 0.25, 10^{-5}\}$ ($\tau = 10$).

Fig. 3. Changes in AUC by C for different α values and τ parameters.

4(c), and 4(d) each show the changes in AUC due to the number of clusters C for each $\alpha = \{1.0, 0.75, 0.5, 0.25, 10^{-5}\}$ when $\tau = \{10, 30, 50, 70\}$. The conventional methods, RSCM and HCM, are shown in dotted lines. In Fig. 4(a), a performance drop was observed in all methods. Although the performance drop in PRSCM was relatively small compared to RSCM, the performance was worse than when $C = 1$ (no clustering), and thus the effect of clustering was not confirmed. In Fig. 4(b), for PRSCM($\alpha = 1.0$), the recommendation performance decreased up to $C = 3$, but the performance improved as C increased. On the other hand, for the other α, a decrease in recommendation performance was shown with an increase in C. In Fig. 4(c), PRSCM($\alpha = 1.0$) consistently demonstrated a higher recommendation performance than RSCM, regardless of the number of clusters. In contrast, the other α showed a lower performance than RSCM. In Fig. 4(d), similar to when $\tau = 50$, PRSCM($\alpha = 1.0$) consistently showed higher recommendation performance than RSCM. Furthermore, while a decline in recommendation performance was observed with RSCM when the number of clusters was large, the performance remained stable for probabilistic approximations in the middle range such as $\alpha = \{0.75, 0.5, 0.25\}$.

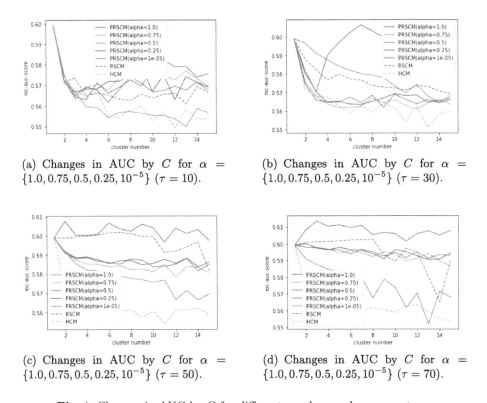

(a) Changes in AUC by C for $\alpha = \{1.0, 0.75, 0.5, 0.25, 10^{-5}\}$ ($\tau = 10$).

(b) Changes in AUC by C for $\alpha = \{1.0, 0.75, 0.5, 0.25, 10^{-5}\}$ ($\tau = 30$).

(c) Changes in AUC by C for $\alpha = \{1.0, 0.75, 0.5, 0.25, 10^{-5}\}$ ($\tau = 50$).

(d) Changes in AUC by C for $\alpha = \{1.0, 0.75, 0.5, 0.25, 10^{-5}\}$ ($\tau = 70$).

Fig. 4. Changes in AUC by C for different α values and τ parameters.

In the MovieLens-100k data, it was possible to outperform the recommendation performance of the conventional RSCM-CF under the lower approximation of $\alpha = 1.0$. In the NEEDS-SCAN/PANEL data, the performance was good in the case of upper approximation of $\alpha = 10^{-5}$, while in the MovieLens-100k data, the performance was good with a lower approximation of $\alpha = 1.0$. From this, it can be inferred that PRSCM closer to an upper approximation is suitable for dense data, and PRSCM closer to a lower approximation is more suitable for sparse data.

5   Conclusion

In this study, we proposed a collaborative filtering based on probabilistic rough set C-means clustering (PRSCM-CF), and validated its recommendation performance through numerical experiments using real-world datasets. In the case of NEEDS-SCAN/PANEL dataset, it was challenging for the proposed PRSCM-CF to outperform the conventional method, RSCM-CF, when the membership of noise users was distributed equally. However, it was suggested that by removing noise users, more accurate recommendations than RSCM-CF could be made. Using the MovieLens-100k dataset, PRSCM-CF was able to surpass the recommendation performance of RSCM-CF, depending on the threshold setting for the probabilistic lower approximation.

In the future, we plan to conduct experiments with finer parameter settings across various datasets, further examine the handling of noise users, and verify the scenarios where the proposed method can function effectively.

References

1. Linden, G., Smith, B., York, J.: Amazon.com recommendations: item-to-item collaborative filtering. IEEE Internet Comput. **7**(1), 76–80 (2003)
2. Smith, B., Linden, G.: Two decades of recommender systems at Amazon.com. IEEE Internet Comput. **21**(3), 12–18 (2017)
3. Su, X., Khoshgoftaar, T. M.: A survey of collaborative filtering techniques. In: Advances in Artificial Intelligence 2009 (2009)
4. MacQueen, J. B.: Some methods of classification and analysis of multivariate observations. In: Proceedings of 5th Berkeley Symposium on Mathematical Statistics and Probability, pp. 281–297 (1967)
5. Ubukata, S., Takahashi, S., Notsu, A., Honda, K.: Basic consideration of collaborative filtering based on rough C-means clustering. In: Proceedings of Joint 11th International Conference on Soft Computing and Intelligent Systems and 21st International Symposium on Advanced Intelligent Systems, pp. 256–261 (2020)
6. Pawlak, Z.: Rough sets. Int. J. Comput. Inf. Sci. **11**(5), 341–356 (1982)
7. Pawlak, Z.: Rough classification. Int. J. Man Mach. Stud. **20**(5), 469–483 (1984)
8. Ubukata, S.: Development of rough set-based C-means clustering. J. Jpn. Soc. Fuzzy Theory Intel. **32**(4), 121–127 (2020)
9. Ubukata, S., Umado, K., Notsu, A., Honda, K.: Characteristics of rough set C-means clustering. J. Adv. Comput. Intell. Intell. Inform. **22**(4), 551–564 (2018)
10. Ubukata, S., Murakami, Y., Notsu, A., Honda, K.: Basic consideration of collaborative filtering based on rough set C-means clustering. In: Proceedings of the 22nd International Symposium on Advanced Intelligent Systems, #OS19-4, pp. 1–6 (2021)
11. Yao, Y.Y.: Probabilistic rough set approximations. Int. J. Approx. Reason. **49**(2), 255–271 (2008)

12. Umado, K., Ubukata, S., Notsu, A., Honda, K.: A study on rough set C-means clustering based on probabilistic rough set. In: Proceedinngs of 28th Intelligent System Symposium, pp. 219–224 (2018)
13. Ubukata, S., Kato, H., Notsu, A., Honda, K.: Rough set-based clustering utilizing probabilistic membership. J. Adv. Comput. Intell. Intell. Inform. **22**(6), 956–964 (2018)

Identifying Topics on Social Impact from S&P1500 CSR/ESG Reports

Farukh Hamza[1], Ping-Yu Hsu[1], Wan-Yu Yang[1], Ming-Shien Cheng[2(✉)], and Yu-Chun Chen[1]

[1] Department of Business Administration, National Central University, No. 300, Jhongda Road, 32001 Jhongli City, Taoyuan County, Taiwan, Republic of China
984401019@cc.ncu.edu.tw

[2] Department of Industrial Engineering and Management, Ming Chi University of Technology, No. 84, Gongzhuan Road, 24301 Taishan District, New Taipei City, Taiwan, Republic of China
mscheng@mail.mcut.edu.tw

Abstract. The standard of measuring the social impact of an organization is actually not fully developed in theory and practice. Then what should be contained in social impact? As the social impact is subjective, and even hard to be defined and measured, there is no consensus on it to-date. Therefore, we expect to let the entrepreneur know what's the trending topic when mentioning the social impact in the business community. This study uses the CSR/ESG report of S&P 1500, and uses the Latent Dirichlet Allocation to analyze the topics. We confirmed the number of topics with the topic coherence score, and also improved the topic coherence score to 0.482. The five topics analyzed by the experimental results are "Safety and Health Initiatives in Education", "Brand Equity", "Water Economy", "Community-based supports and services", "Social Business". The process of naming the subject of this research not only refers to two books which can represent the social impact, and also invites other researchers to name it independently, both are different from previous researchers, and would be more objective as well.

Keywords: Social Impact · LDA · Topic Analysis · Corporate Social Responsibility · S&P 1500 · Topic Coherence · CSR · ESG · SDGs · Sustainability

1 Introduction

In 2015, the United Nations announced the "2030 Sustainable Development Goals (SDGs)," which contain 17 core goals that cover 169 detailed goals to guide citizens of the planet to work together toward sustainability. However, while there are some clear improvements from 2015 to now, such as child mortality and mortality from some diseases, there are still many things that we need to work on to achieve all of the SDGs by 2030, such as the continued rise in greenhouse gas emissions, the increasing global climate, and even the unabated rate of biodiversity loss…etc. As global citizens, we should pay attention to these hot issues together.

With the signing of the Glasgow Climate Pact, a new milestone in global sustainable development has been reached. Sustainable development is now the normative world-view and an issue that should be taken seriously as a resident of the planet, and it has established a goal that the world should strive for. However, when faced with the issue of sustainability, companies can not only invest in sustainability for long-term profit but also promote social responsibility.

Corporate social responsibility (CSR) is prominent in today's global corporate agenda, from community development and environmental protection to socially responsible business practices. More than ever before, companies are devoting significant resources to this path (Du, Bhattacharya, & Sen [1]). With the expansion of corporate social influence, corporate social responsibility (CSR) has been regarded as a key factor in corporate management (Kim, Yin, & Lee [2]). CSR measures outcomes beyond corporate performance, so do CSR programs provide the social benefits they promise? After decades of CSR research, we do not have an answer (Barnett, Henriques, & Husted [3]), and this has led to many companies not knowing how to demonstrate their efforts for the social good to their investors.

Some studies by García-Chiang [4] have pointed out that the implementation of CSR can contribute to the development of the region, but usually, most studies only focus on specific stakeholders rather than the broader social interests. However, as CSR has received more and more attention, many studies have measured CSR activities rather than their impact. Most of the studies related to "social impact" in the past have focused on the social impact of a single domain or multiple domains, such as the social impact of social media (Pulido, Ruiz-Eugenio, Redondo-Sama, & Villarejo-Carrballido [5]) and the social impact of political conflict during the COVID-19 epidemic (Blofield, Hoffmann, & Llanos [6]).

In the social responsibility report, major enterprises have written a lot about what activities they have done and whether these activities have any real effect, which is called "impact." Due to the subjective nature of the social impact, which is difficult to define and measure, there is no unanimous opinion so far, and because of this, we hope to let entrepreneurs understand what hot topics are being discussed in the social responsibility reports of the corporate sector today.

This study aims to provide a reference for companies to write CSR reports in the future by asking, "What are U.S. companies talking about when they mention social impact in their social responsibility reports?".

This study extracts the 2020 CSR reports of S&P 1500 companies, analyzes only the text related to the keyword "social impact," and then identifies the corresponding themes through the implied Dirichlet distribution. Most studies are very subjective in naming the final themes, so this study also proposes a new method for naming the themes. In sum, the research objectives that this study seeks to achieve are:1. Find out what U.S. companies are talking about regarding social impact; 2. Find the theme using the implied Dirichlet distribution. 3. Verify the results of the theme.

This paper organized as follow: (1) Introduction: Research background, motivation and purpose. (2) Related work: Review of scholars' researches on social impact, corporate social responsibility, topic coherence and naming topic. (3) Research methodology: Content of research process in this study. (4) Result analysis: Experimental results and the discussion of the test results. (5) Conclusion and future research: Contribution of the study, and possible future research direction is discussed.

2   Related Work

2.1   Social Impact

Most of the papers related to social impact are on social impact assessment (SIA), and they develop new principles for SIA or indicate what needs to be done. For example, Esteves et al. [7] argue that strengthening human rights impact assessment, social performance standards, supply chain management, and local economic development will increase the value of SIA to all stakeholders. Vanclay [8] argues that SIA needs to include analyzing, monitoring, managing, and planning relevant policies or programs and any social change processes resulting from these policies or programs.

Some studies have also examined what factors are associated with social impacts. Mancini & Sala [9] proposed that the social impact of the mining industry focuses on two indicators, human rights and the working environment. Das, Balakrishnan, & Vasudevan [10] studied the social impact of alcohol in India and hoped that health and safety stakeholders could develop effective prevention or treatment of alcohol-related diseases. This shows that many fields have been studying its social impacts, such as mining, healthcare, and even higher education (O'carroll, Harmon, & Farrell [11]), which is evident in the importance of social impacts.

In order to provide scholars with a clear understanding of the current state of social impact research, Rawhouser et al. [12] used top journals in the business field to determine whether the abstracts and keywords of papers were representative of social impact and used them as a sample to give companies some insights and suggestions. The difference is that this study used the entire CSR report with the keywords of social impact as the sample, which can better understand the hot topics related to social impact nowadays.

2.2   Corporate Social Responsibility

The first person to propose the concept of Corporate Social Responsibility (CSR) was a British scholar, Sheldon [13]. Although his concept was not taken seriously at the beginning, with the industrialization of Western countries, CSR has become one of the most important issues in the world, and it has been paid more and more attention to and importance by many countries.

The extent to which companies benefit or harm social welfare is a growing concern, and corporate behavior in this area is often referred to as Environmental, Social, and Governance (ESG) or CSR (Gillan, Koch, & Starks [14]). Since ESG is derived from CSR, the terms ESG and CSR are used interchangeably (Garcia, Mendes-Da-Silva, & Orsato [15]). In order to strengthen sustainable development in Taiwan, in August 2020, the FSC officially launched "Corporate Governance 3.0 - A Blueprint for Sustainable Development" and officially revised the CSR report to "Sustainability Report" (Online newspaper, 2021).

So, what else can CSR be measured besides donation figures, exposure, etc.? There are already many international metrics, such as the Global Reporting Initiative (GRI), the Dow Jones Sustainability Index (DJSI), and so on. Today, however, CSR is focused on the idea of creating "shared value," where a company's role is to create value for its shareholders and value for society, embodying a win-win proposition (Ali, Frynas, & Mahmood [16]). However, many companies have deliberately participated in or organized CSR activities to build and enhance their corporate reputation to attract customers Lai, Chiu, Yang, & Pai [17], which shows that companies have understood that CSR has become increasingly important to them Hsu [18].

Scholars Gutsche, Schulz, & Gratwohl [19] studied the impact of CSR disclosure and CSR performance on the firm value of S&P 500 companies from 2011 to 2014 and used two regressions to test the model and found that CSR performance related to environmental governance was positively associated with firm value. Despite all the CSR-related research mentioned above, no research has been conducted to examine what is included in the ESG reports written by companies. What exactly does the social impact issue include?

3   Research Methodology

This study aims to find out what companies in the S&P 15000 were talking about regarding "social impact" by using LDA with the theme of "CSR/ESG report." For this purpose, we first extracted the relevant text of the keyword "social impact" from the collected reports manually, built an LDA model through a series of data pre-processing to find the appropriate number of topics, adjusted the hyperparameters to improve the topic consistency scores, and finally named the topics based on the results (Fig. 1).

Fig. 1. The Research Process

3.1 Data Collection

The companies collected for this study were the S&P 1500 Index (S&P 15000), which combined three indices, which are the S&P 500, S&P MidCap 400, and S&P SmallCap 600, covering approximately 90% of the market capitalization in the United States. Since no regulation required all companies to write ESG reports, this study collected ESG reports from a total of 848 companies and focused on reports published in 2020. However, reports are usually generated in the year after the reporting year (Goloshchapova et al. [20]), so if the reports are published between January and June, we use the publication date minus one year so that reports from the first half of 2021 was adopted.

The source of ESG reports collected by the Institute is mainly from ResponsibilityReports.com. This website has the latest and most complete list of ESG reports on the Internet, such as sustainability reports, CSR reports, ESG reports, etc., and provides users with free access to view and download them. If you can't find a company's report on this website, you can go to each company's website to download it.

3.2 Topic Coherence

The concept of topic coherence was first introduced by Newman, Lau, Grieser, & Baldwin [21]. The topic coherence score is used to measure the semantic similarity between words in a topic and to solve the problem that the LDA model does not guarantee its interpretability in the output (Röder, Both, & Hinneburg [22]). Too low a k-value (total number of topics) for the LDA output will result in too few topics or too broad a meaning, while too high a value will result in topics that are uninterpretable or topics that are so similarly repeated that they should ideally be merged. Therefore, choosing the right number of topics is an important task.

There are many different measures of thematic coherence. This study provides a more formal description based on the Cv Coherence Score formula proposed by Syed & Spruit [23].

$W = \{W_1, \ldots, W_N\}$, W denotes the set of words, W' denotes one of the words and $W' \in W$, W^* denotes all the words in the set and $W^* \in W$ pair to $S_i = (W' = w_1), (W^* = w_1, w_2, w_3)$. If the probability of occurrence of two words is higher, the normalized pointwise mutual information (NPMI) will also be higher (Eq. 2). The consistency between w_i, w_j is calculated by NPMI, that is, the first word and all the words in the topic to calculate the NPMI to retain a vector (Eq. 1). If the meaning of the words is very similar, the vector and the vector will have a higher score than each other (Eq. 3).

$$\vec{v}(W') = \left\{ \sum_{w_i \in W'} \text{NPMI}(w_i, w_j)^{\gamma} \right\}_{j=1,\ldots,|W|} \tag{1}$$

$$\text{NPMI}(w_i, w_j)^{\gamma} = \left(\frac{\log \frac{P(w_i, w_j) + \epsilon}{P(w_i) \cdot P(w_j)}}{-\log(P(w_i, w_j) + \epsilon)} \right)^{\gamma} \tag{2}$$

$$\phi_{S_i}(\vec{u}, \vec{w}) = \frac{\sum_{i=1}^{|W|} u_i \cdot w_i}{\|\vec{u}_2\| \cdot \|\vec{w}_2\|} \tag{3}$$

In summary, Cv is divided into the following parts. Firstly, it splits the data into many word pairs, calculates the probability of words or word pairs, calculates the vector strength of words and word sets, and calculates the coherence score.

3.3 Naming Topic

The topic model is an unsupervised model that only classifies data through learning, so most studies are very subjective about the final "naming" of the topic model. For example, Guo, Barnes, & Jia [24] named the topic by the weight of keywords, or most of the studies did not directly state how to "name" the topic but were named directly by scholars (Alam, Ryu, & Lee [25]). Some studies did not name them but only labeled them with "Theme 1", "Theme 2," etc. Therefore, this study proposes a new naming method for the naming of theme models.

First of all, the keywords of each topic were distributed in two books about social impact. These two books were recommended to me by Amazon.com on March 2, 2022, after searching for "measure social impact", and they are Social Return on Investment Analysis: Measuring the Impact of Social Investment (Palgrave Studies in Impact Finance) published in 2018 and Social Return on Investment Analysis: Measuring the Impact of Social Investment (Palgrave Studies in Impact Finance) published in 2014. We then looked at the number of pages where the keywords were located and took the intersection to find the consecutive page numbers from each book that best represented each topic. After all, three researchers finished naming the topics, and we held a group meeting to explain the origin and reasons for naming each topic and had a three-way discussion.

4 Research Experiment

4.1 Results After Adjusting Parameters

The results are presented in the form of a table (Table 1) and a text cloud (Fig. 2). The five words with the highest theme weight are "health," "product," "organization," "community," and "impact." When the number of topics is 5, and the hyperparameters alpha and beta are 0.01 and 0.91, respectively, the score of topic consistency increases from 0.369 to 0.482. Since this study intercepted paragraphs rather than full articles, the score of thematic coherence would be lower than the general use of full text. From the study of Syed & Spruit [23], we can see that the topic consistency scores of the two datasets using article abstracts fall around $0.45 - 0.48$. Then, we identified the keywords representing each topic, compiled the number of pages representing each topic from the two books mentioned in Sect. 3.3, and presented the following tables (Tables 2 and 3).

Table 1. Keywords of each topic

Topics	Keywords
1	health + student + safety + education + training + learn + worker + high + initiative + stem
2	product + program + equity + industry + diversity + brand + inclusion + list + consumer + commitment
3	organization + year + water + use + life + grant + total + economy + donation + end
4	community + provide + work + support + employee + opportunity + effort + partner + focus + need
5	impact + environmental + social + economic + report + stakeholder + business + issue + company + topic

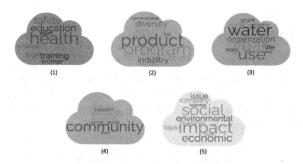

Fig. 2. Word Cloud of each topic

Table 2. Keywords of each topic within the page of 《Social Return on Investment Analysis: Measuring the Impact of Social Investment (Palgrave Studies in Impact Finance) 》

Keywords	Page
health + student + safety + education + training + learn + worker + high + initiative + stem	p. 333−345
product + program + equity + industry + diversity + brand + inclusion + list + consumer + commitment	p. 352−365
organization + year + water + use + life + grant + total + economy + donation + end	p. 77−83
community + provide + work + support + employee + opportunity + effort + partner + focus + need	p. 128−131
impact + environmental + social + economic + report + stakeholder + business + issue + company + topic	p. 19−25, p. 200−210

Table 3. Keywords of each topic within the page of 《Measuring and Improving Social Impacts: A Guide for Nonprofits, Companies, and Impact Investors》

Keywords	Page
health + student + safety + education + training + learn + worker + high + initiative + stem	p. 55–60
product + program + equity + industry + diversity + brand + inclusion + list + consumer + commitment	p. 35–43
organization + year + water + use + life + grant + total + economy + donation + end	p. 70–73
community + provide + work + support + employee + opportunity + effort + partner + focus + need	p. 195–199
impact + environmental + social + economic + report + stakeholder + business + issue + company + topic	p. 98–102

4.2 Results of Topic Naming

After compiling the page numbers according to Tables 2 and 3, respectively, against the text in the two books, the keywords were collated together with the keywords to the researchers (Fig. 3). Table 4 shows the results of naming the five themes by the three researchers, and the overall agreement rate of the naming results was 86.7%. The final naming of each theme is described below.

Keywords	Texts
"health" + "student" + "safety" + "education" + "training" + "learn" + "worker" + "high" + "initiative" + "stem"	"Helping people to help themselves" constitutes one of the main principles of Menschen für Menschen population. In addition to a lack of clean drinking water, health issues (for instance, the widespread eye inf challenges. By encouraging capacity-building within the local population, the measures implemented as pa facilities and hygiene measures), initiating and executing awareness campaigns about harmful traditional p various aspects from agriculture, education, health and earnings. WiD can be categorised as a sub-program the Ginde Beret project area, comprising more than 70 square kilometres and a population of approximatel
"community" + "provide" + "work" + "support" + "employee" + "opportunity" + "effort" + "partner" + "focus" + "need"	Application Example: SONG Multigeneration Co-housing Developments (1) German welfare organisations have joined together in a network called SONG (Soziales Neu Gestalten: introducing multigeneration co-housing developments in the 1990s in which social workers assist informa that invest in such models place financing and personnel at the local neigh- bourhood's disposal in order simultaneously: the housing market, the mar- ket for care and the "market" of communication and net community by providing infrastructures that help to reinforce local networks, con- tacts and structures of s services and voluntary involvement—which, in turn, can be interpreted as a political response to the questi that positive eco- nomic effects emerge in the process, which can be traced back to fortified social capital a (Netzwerk Soziales neu gestalten 2009; see also Kehl and Then 2013).
"product" + "program" + "equity" + "industry" + "diversity" + "brand" + "inclusion" + "list" + "consumer" + "commitment"	NODE is a network of companies in the oil and gas industry in the district of Agder, Norway, established The analysis thus did not look at the effects of all NODE companies on the region of Agder—but rather o cluster and all member firms on the region of Agder is a very different thing—far beyond the effect of the As the roles and number of participants in NODE are numerous, the establishment of a value chain was n different value for the different stakeholders—and for a continuous and potentially unlimited period of tim These key impact dimensions—Employer Branding, Public standing and Visibility, Reputation, Network

Fig. 3. Organized reference with keywords and inner texts

Table 4. Results of naming the five themes by the three researchers

Topics	Researcher 1	Researcher 2	Researcher 3
1	Promote Health Education	Helps Education	Talent Cultivation
2	Brand Equity	Brand Equity	Brand's Commitment to Products
3	Water Economy	Water–Energy–Food–Ecosystem Nexus	Life Cycle Pollution to Water Resources
4	Community Advocate	Elderly Community Support Service	Human Rights and Community Functions
5	Social Business	Impact Investing Management	Major themes in the report

Topic 1 is named "Safety and Health Initiatives in Education," which is named for education for the disadvantaged but not for education in general. More and more companies are now actively promoting education in rural areas. Education is the key to fostering a culture of safety and health and the strength of vulnerable children to stay safe and healthy in their homes, schools, communities, and even in the workplace in the future. However, raising children in remote areas is a difficult task. In some backward countries, child labor is still a problem, so some programs focus on protecting the safety and health of working children (Carothers, Breslin, Denomy, & Foad [26]) so that their working conditions and quality of life can be improved. The idea that children need to learn to stay safe and healthy can be overwhelming, so companies can teach children basic health and safety concepts and good eating and activity habits when implementing corporate social responsibility activities.

Topic 2 is named "Brand Equity." One of a company's most valuable assets is its brand, which has the ability to influence consumer behavior and provide its owners with the assurance of future sustainable revenue. Customer perception often plays an important role in a company because of its ability to attract new customers and retain existing ones. Relatively speaking, companies have the ability to control and build factors that are relevant to customers' perceptions of their brands. In 2014, Nielsen global corporate social responsibility report conducted a global survey on CSR. It is clear from the survey that consumers clearly and positively state that branding is one of the factors that influence purchasing decisions and that it not only helps to increase brand share but also has the potential to have a positive impact on our society. A study by Iqbal, Qureshi, Shahid, & Khalid [27] also found a strong correlation between CSR and brand equity and that companies that invest in CSR activities tend to have good consumer loyalty.

Topic 3 is named "Water Economy." Kofi A. Annan, Secretary General of the United Nations, has said, "The fierce competition for freshwater could become a source of conflict or war in the future." Water is closely related to many sources of energy, even agricultural irrigation, industrial and ecological development, etc. Water is also needed, and the quality of water has a great impact on human health. Water is also essential to maintain livelihoods, whether for direct consumption or as an input for production; it is used in almost all aspects of life. Therefore, the management, supply, and distribution of water resources is already an important global issue. Some companies may discharge industrial wastewater into neighboring rivers for their own convenience. Even some backward countries do not take special care of their rivers, so the whole river is dirty. However, rivers are the most important source of fresh water and have great significance for regulating the environment. A study by Kabir [28] has indicated that improving water pollution in rivers has a positive impact on humans, the environment, and the ecosystem.

Topic 4 is named "Community-based Supports and Services (CBSS)." CBSS are designed to help older adults living in the community stay safely at home and delay or prevent hospitalization (Siegler, Lama, Knight, Laureano, & Reid [29]). CBSS provides specific resources for older adults, including health programs, educational programs on health and aging, and housing, financial, and home security, among others, so that older adults are not neglected and forgotten by society. Yusriadi [30] also studied that the implementation of social services in the community can increase the income of families and thus improve their quality of life. It can be seen that these services have a positive impact on society, and there are now many companies, both domestic and foreign, that invest in long-term or short-term community services or create projects. Some even recruit colleagues and relatives to serve as volunteers, so those in need in the community can receive appropriate care.

Topic 5 is named "Social Business." There is no single definition of "Social Business," but Nobel Peace Prize Winner Muhammad Yunus proposes the most prominent one. He defined social business as a type of enterprise whose goal is to solve social, economic, and environmental problems affecting human beings, such as disease, hunger, pollution, etc. It is a non-profit organization. Social business has now become increasingly international, and the main reason for driving the corporate world into the international arena is the desire to create social impact (Misbauddin & Nabi [31]). Social businesses can help the poor provide funds and services they would otherwise not be able to afford and help them get a job. In recent years, companies have also realized that in the long run, investing in social enterprises can be effective in enhancing their image and even attracting more loyal customers.

5   Conclusion and Future Research

First, in order to understand the hot topics of social impact in U.S. companies, this study identified five themes through LDA. The five key themes are "Safety and Health Initiatives in Education," "Brand Equity," "Water Economy," "Community-based Supports and Services," and "Social Business." The score of coherence also increased from 0.369 to 0.482.

According to Corporate Knights' list of the top 100 sustainable companies in the world by 2021, only one company from Taiwan is included, and the top 10 are mostly companies from Europe and the US. We can see that Taiwan still has much room for improvement in this area. We hope to use the results of the five themes analyzed above as a reference for companies writing their ESG reports..

Nowadays, companies are looking at CSR/ESG as more than just an obligation or a responsibility; they are increasingly considering and even integrating it with their business strategies. Although ESG reporting is not entirely voluntary, and a large part of it is due to government regulations, it is a good start to familiarize listed companies with sustainability information.

As social impact issues develop rapidly, the topics discussed by companies will likely change in the next five years. In addition, since this study is based on companies in the United States, companies in different countries may need to pay more attention if they want to use the data. Therefore, the data collected in this study are subject to regional and time constraints.

The data used for this study is the S&P 1500 ESG report to find the impact issues. It may be possible to use some of the more representative social impact-related websites for this study, which may yield other findings that can be used as future recommendations.

Acknowledgment. We would like to thank National Science and Technology Council, Taiwan for generously supporting this research through project #112–2410-H-008–017-MY2.

References

1. Du, S., Bhattacharya, C.B., Sen, S.: Maximizing business returns to corporate social responsibility (CSR): The role of CSR communication. Int. J. Manag. Rev. **12**(1), 8–19 (2010)
2. Kim, M., Yin, X., Lee, G.: The effect of CSR on corporate image, customer citizenship behaviors, and customers' long-term relationship orientation. Int. J. Hosp. Manag. **88**, 102520 (2020)
3. Barnett, M.L., Henriques, I., Husted, B.W.: Beyond good intentions: designing CSR initiatives for greater social impact. J. Manag. **46**(6), 937–964 (2020)
4. García-Chiang, A.: Corporate social responsibility in the Mexican oil industry: social impact assessment as a tool for local development. Int. J. Corp. Soc. Responsib. **3**(1), 1–8 (2018). https://doi.org/10.1186/s40991-018-0038-z
5. Pulido, C.M., Ruiz-Eugenio, L., Redondo-Sama, G., Villarejo-Carballido, B.: A new application of social impact in social media for overcoming fake news in health. Int. J. Environ. Res. Public Health **17**(7), 2430 (2020)
6. Blofield, M., Hoffmann, B., Llanos, M.: Assessing the political and social impact of the COVID-19 crisis in Latin America (2020)
7. Esteves, A.M., Franks, D., Vanclay, F.: Social impact assessment: the state of the art. Impact Assess. Proj. Appraisal **30**(1), 34–42 (2012)
8. Vanclay, F.: International principles for social impact assessment. Impact Assess. Project Appraisal **21**(1), 5–12 (2003)
9. Mancini, L., Sala, S.: Social impact assessment in the mining sector: review and comparison of indicators frameworks. Resour. Policy **57**, 98–111 (2018)
10. Das, S.K., Balakrishnan, V., Vasudevan, D.: Alcohol: its health and social impact in India. Natl Med. J. India **19**(2), 94 (2006)
11. O'carroll, C., Harmon, C., Farrell, L.: The economic and social impact of higher education. Irish Univ. Assoc. **1**, 1–32 (2006)
12. Rawhouser, H., Cummings, M., Newbert, S.L.: Social impact measurement: current approaches and future directions for social entrepreneurship research. Entrep. Theory Pract. **43**(1), 82–115 (2019)
13. Sheldon, O.: The Philosophy of Management. Pitman (1924)
14. Gillan, S.L., Koch, A., Starks, L.T.: Firms and social responsibility: a review of ESG and CSR research in corporate finance. J. Corp. Finan. **66**, 101889 (2021)

15. Garcia, A.S., Mendes-Da-Silva, W., Orsato, R.J.: Sensitive industries produce better ESG performance: evidence from emerging markets. J. Clean. Prod. **150**, 135–147 (2017)
16. Ali, W., Frynas, J.G., Mahmood, Z.: Determinants of corporate social responsibility (CSR) disclosure in developed and developing countries: a literature review. Corp. Soc. Responsib. Environ. Manag. **24**(4), 273–294 (2017)
17. Lai, C.-S., Chiu, C.-J., Yang, C.-F., Pai, D.-C.: The effects of corporate social responsibility on brand performance: the mediating effect of industrial brand equity and corporate reputation. J. Bus. Ethics **95**(3), 457–469 (2010)
18. Hsu, K.-T.: The advertising effects of corporate social responsibility on corporate reputation and brand equity: evidence from the life insurance industry in Taiwan. J. Bus. Ethics **109**(2), 189–201 (2012)
19. Gutsche, R., Schulz, J.-F., Gratwohl, M.: Firm-value effects of CSR disclosure and CSR performance. In: EFMA-Conference Proceedings, pp. 1–31 (2017)
20. Goloshchapova, I., Poon, S.-H., Pritchard, M., Reed, P.: Corporate social responsibility reports: topic analysis and big data approach. Euro. J. Finance **25**(17), 1637–1654 (2019)
21. Newman, D., Lau, J.H., Grieser, K., Baldwin, T.: Automatic evaluation of topic coherence. Human language technologies: The 2010 Annual Conference of the North American Chapter of the Association for Computational Linguistics, pp. 100–108 (2010)
22. Röder, M., Both, A., Hinneburg, A.: Exploring the space of topic coherence measures. In: Proceedings of the Eighth ACM International Conference on Web Search and Data Mining, pp. 399–408 (2015)
23. Syed, S., Spruit, M.: Full-text or abstract? Examining topic coherence scores using latent dirichlet allocation. 2017 IEEE International Conference on Data Science and Advanced Analytics (DSAA), pp. 165–174 (2017)
24. Guo, Y., Barnes, S.J., Jia, Q.: Mining meaning from online ratings and reviews: tourist satisfaction analysis using latent Dirichlet allocation. Tour. Manage. **59**, 467–483 (2017)
25. Alam, M.H., Ryu, W.-J., Lee, S.: Joint multi-grain topic sentiment: modeling semantic aspects for online reviews. Inf. Sci. **339**, 206–223 (2016)
26. Carothers, R., Breslin, C., Denomy, J., Foad, M.: Promoting occupational safety and health for working children through microfinance programming. Int. J. Occup. Environ. Health **16**(2), 164–174 (2010)
27. Iqbal, F., Qureshi, A., Shahid, N., Khalid, B.: Impact of Corporate Social Responsibility on Brand Equity Thesis for: MBA Marketing, University of Bradford] (2013)
28. Kabir, M.R.: Social impact assessment of water pollution: a case study on Bangshi River, Savar BRAC University] (2014)
29. Siegler, E.L., Lama, S.D., Knight, M.G., Laureano, E., Reid, M.C.: Community-based supports and services for older adults: a primer for clinicians. J. Geriatr. (2015)
30. Yusriadi, Y.: Economic and Social Impacts of Social Entrepreneurship Implementation Service to Community (2021)
31. Misbauddin, S., Nabi, M.N.U.: Internationalization of Social Business: Toward a Comprehensive Conceptual Understanding. Emerald Publishing Limited, In Societal Entrepreneurship and Competitiveness (2019)

Descriptor-Based Information Systems and Rule Learning from Different Types of Data Sets with Uncertainty

Hiroshi Sakai[1($\boxtimes$)] and Michinori Nakata[2]

[1] Graduate School of Engineering, Kyushu Institute of Technology,
Kitakyushu, Tobata 804-8550, Japan
`sakai@mns.kyutech.ac.jp`
[2] Faculty of Management and Information Science, Josai International University,
Gumyo, Togane, Chiba 283-0002, Japan
`nakatam@ieee.org`

Abstract. We have coped with rule generation from tables (or information systems) and extend this framework to that from different types of data, like heterogeneous, uncertain, time series, clustered, etc. For this new framework, we use the term 'rule learning' instead of 'rule generation.' We newly specify descriptors, which take the role of words in data sets, and we define a DbIS (Descriptor-based Information System) Ω as the unified format for different data types. We prove one DbIS Ω is convertible to one NIS (Non-deterministic Information System), where we realized the NIS-Apriori system. Thus, we can obtain rules via DbIS and the NIS-Apriori system. Furthermore, we merge two DbISs Ω_1 and Ω_2 to one $\Omega_{1,2}$ using missing values, and we can inductively merge any number of DbISs. We also estimate some missing values by the self-obtained certain rules and consider the application of DbIS to big data analysis.

Keywords: rule learning · different types of data · the Apriori algorithm · uncertain data

1 Introduction

At the IUKM conference, topics like the manipulation, management, the usage, etc. of uncertain information, are focused on. We also coped with rule generation from tables and combined the rough sets based concepts [11,17] with the Apriori algorithm [1] to realize the rule generator from several data sets with uncertainty. We simply call it NIS-Apriori-based Rule Generation [9,13,14]. We enumerate its property below:

1. Generally, a rule τ from a table is a logical implication that satisfies some constraints. To handle rules from a table with incomplete information, we newly defined a certain rule and a possible rule based on possible world semantics [6,8].

© The Author(s), under exclusive license to Springer Nature Switzerland AG 2024
K. Honda et al. (Eds.): IUKM 2023, LNAI 14376, pp. 255–266, 2024.
https://doi.org/10.1007/978-3-031-46781-3_22

2. The number of the possible world increases exponentially. Therefore the calculation of two types of rules seemed to be difficult. For this issue related to the exponential order problem, we proved some propositions and gave a solution. Thus, we can now handle certain rules and possible rules.
3. The NIS-Apriori algorithm preserves the logical property. This algorithm only generates every logical implication, which is defined as a rule (soundness). This algorithm does not miss any logical implication, which is defined as a rule (completeness). Namely, this algorithm is sound and complete for defined rules. We implemented this algorithm in Python [5] and SQL [14], respectively.

Let us consider Fig. 1 to clarify the purpose of this research. Our previous research (surrounded by dotted lines) handled rules from one table data set with uncertainty [16]. However, rules from different types of data sets (not in the form of a categorical table) will be helpful for decision support. We extend the previous rule generation framework and propose the new framework in Fig. 1. We use the term 'rule learning' [3,18,19] instead of 'rule generation'. The definition of a Descriptor-based Information System (DbIS) Ω in Fig 1 is a unified framework of different types of data with uncertainty. We advance the preliminary research [16] and cope with the merging process of different types of data.

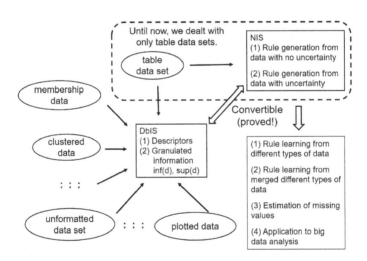

Fig. 1. The overview of rule learning from different types of data sets with uncertainty via DbIS. We handled a preliminary and similar figure in [16].

This paper is organized as follows. In Sect. 2, we briefly survey the previous research. In Sect. 3, we show a simple example and the definition of DbIS Ω, which is a unified format for handling rules from different types of data sets with uncertainty. We clarify the relation between DbIS, NIS, and NIS-Apriori-based rule generation. In Sect. 4, we discuss the merging procedure for two DbISs. In

Sect. 5, we show one method to estimate missing values in rule learning. The self-obtained certain rules may estimate such missing values. In Sect. 6, we consider an issue on discretization and descriptors for big data analysis. In Sect. 7, we summarize the framework of rule learning based on several different types of data.

2   Preliminary and Background of Research

Table 1 is a simple table. We term such a table as a Deterministic Information System (DIS) [11]. To handle information incompleteness, missing values '?' in Table 2 was considered [4]. If we replace each '?' with a set of all possible values, we have a Non-deterministic Information System (NIS) [8,10] in Table 3.

Table 1. An exemplary DIS ψ for five persons.

OB	P	Q	R	S	Dec
p_1	3	1	2	2	a
p_2	2	2	2	1	a
p_3	1	2	2	1	b
p_4	1	3	3	2	b
p_5	3	2	3	1	c

Table 2. An exemplary NIS Φ with missing value '?'. '?' is one of 1, 2, 3.

OB	P	Q	R	S	Dec
p_1	3	?	2	2	a
p_2	2	{2, 3}	2	?	a
p_3	?	2	2	{1, 2}	b
p_4	1	3	3	2	b
p_5	3	2	3	?	c

Table 3. An exemplary NIS Φ. Each '?' is replaced with a set {1, 2, 3}.

OB	P	Q	R	S	Dec
p_1	3	{1, 2, 3}	2	2	a
p_2	2	{2, 3}	2	{1, 2, 3}	a
p_3	{1, 2, 3}	2	2	{1, 2}	b
p_4	1	3	3	2	b
p_5	3	2	3	{1, 2, 3}	c

In Table 1, we see logical implications $[P, 3] \Rightarrow [Dec, c]$ from p_5 and $[R, 3] \wedge [S, 2] \Rightarrow [Dec, b]$ from p_4. A pair of the attribute and its value is termed a 'descriptor'. For example, $[P, 3]$ and $[Dec, c]$ are descriptors. Each logical implication consists of descriptors. Generally, a rule is defined as follows [1,11,17].

1. For $\tau : \wedge_i [A_i, v_i] \Rightarrow [Dec, v]$, if τ satisfies below for given threshold values α and β $(0 \leq \alpha, \beta \leq 1.0)$, we see this logical implication τ as a rule.
   $support(\tau) = | \wedge_i Obj([A_i, v_i]) \cap Obj([Dec, v]) | / |OB| \geq \alpha$,
   $accuracy(\tau) = | \wedge_i Obj([A_i, v_i]) \cap Obj([Dec, v]) | / | \wedge_i Obj([A_i, v_i]) | \geq \beta$.
   Here, $Obj([A, v])$ for a descriptor $[A, v]$ and $|S|$ for a set S mean a set of objects satisfying $[A, v]$ and the cardinality of S, respectively. For example, $support(\tau) = 2/5$ and $accuracy(\tau) = 2/3$ for $\tau : [R, 2] \Rightarrow [Dec, a]$ from p_1. If $| \wedge_i Obj([A_i, v_i]) | = 0$, we define $accuracy(\tau) = 0$.

2. If we fix threshold values in DIS, all rules are internally fixed. Rule generation means to obtain all rules. (We often handle a logical implication with the minimal condition part.)

In NIS Φ, we have one DIS by replacing a set of attribute values with an element of the set. We term such a DIS a 'derived DIS' from NIS. We define $DD(\Phi)=\{\ \psi \mid \psi$ is a derived DIS from $\Phi\}$ and rules below:

1. A certain rule τ, which is a rule in each of $\psi \in DD(\Phi)$.
2. A possible rule τ, which is a rule in at least one $\psi \in DD(\Phi)$.

In the above Φ, the number of elements in $DD(\Phi)$ increases in the exponential order. In the Mammographic data set [2], the number exceeds more than 10^{100}. Due to the number of derived DIS, realizing a rule generation system from NIS seemed to be hard. However, we gave one solution to this problem [13,14] and implemented some software tools. Some execution videos are in [15].

3   An Example and the Mathematical Framework

Let us consider the following example for defining DbIS Ω.

Example 1. In a table data set, like Table 1, we implicitly employ the pair [attribute,value] as a descriptor. If we employ such descriptors in Table 4, the *support* value will be low. We need to specify descriptors for generating rules.

Table 4. An additional table with *weight* and *height* for eight persons.

OB	p_1	p_2	p_3	p_4	p_5	p_6	p_7	p_8
weight	50	60	65	65	70	75	80	85
height	165	170	172	175	170	172	183	187

1. Case 1: We specify descriptors below:
    [$weight, light$] (supported by p_1),
    [$weight, normal$] (supported by p_2, p_3, p_4, p_5),
    [$weight, heavy$] (supported by p_6, p_7, p_8).
   Due to this specification, we can have one DIS from Table 1 and Table 4. In this case, the implemented software tool generated two additional rules for the constraint $accuracy \geq 1$.
    [$weight, light$] $\Rightarrow$ [Dec, a] ($supp(ort)=0.2$, $acc(uracy)=1.0$),
    [$P, 3$] $\wedge$ [$weight, normal$] $\Rightarrow$ [Dec, c] ($supp=0.2$, $acc=1.0$).

2. Case 2: We specify descriptors below:

$[weight, light]$ (supported by p_1, p_2, p_3, p_4),

$[weight, heavy]$ (supported by p_6, p_7, p_8).

This binary specification is simple, but the assignment of p_5 becomes uncertain. We think that p_5 possibly supports $[weight, light]$ and p_5 possibly supports $[weight, heavy]$. Due to this specification, we can have one NIS Φ with an attribute $weight$. In this case, $|DD(\Phi)|=2$, and the implemented software tool generated two additional possible rules for the constraint $accuracy \geq 1$.

$[weight, heavy] \Rightarrow [Dec, c]$ (max_supp=0.2, max_acc=1.0),

$[P, 3] \wedge [weight, light] \Rightarrow [Dec, a]$ (max_supp=0.2, max_acc=1.0).

Here, max_supp and max_acc are the maximum values of $support$ and $accuracy$ for two derived DISs in $DD(\Phi)$.

3. Two cases explained the relation between the specification of descriptors, DIS and NIS. In Case 1, there is no information incompleteness, and we have an equivalence class over a set $OB=\{p_1, p_2, p_3, p_4, p_5\}$ for each descriptor. On the other hand, there is information incompleteness in Case 2. We uniformly handle such information by two sets inf (certain information) and sup (possible information). Such two sets are variations of the equivalence classes (or granulated information in [7, 12]).

In Case 1,

$inf([weight, light])=\{p_1\}$, $sup([weight, light])=\{p_1\}$.

$inf([weight, normal])=\{p_2, p_3, p_4, p_5\}$,

$sup([weight, normal])=\{p_2, p_3, p_4, p_5\}$.

If there is no information incompleteness, $inf=sup$ holds.

In Case 2,

$inf([weight, light])=\{p_1, p_2, p_3, p_4\}$, $sup([weight, light])=\{p_1, p_2, p_3, p_4, p_5\}$.

$inf([weight, heavy])=\{\}$, $sup([weight, heavy])=\{p_5\}$.

4. In table data sets, we implicitly handled every descriptor $[A, val_A]$. By specifying descriptors and inf, sup, we handle information from different types of data sets with uncertainty.

Remark 1. We have to pay attention to two sets inf and sup.

1. Generally, $inf([A, val_A]) \subset sup([A, val_A])$ should hold for every descriptor $[A, val_A]$. The equation $inf([A, val_A])=sup([A, val_A])$ should also hold if there is no information incompleteness.

2. $\cup_{val_A} inf([A, val_A])=OB$ may not hold, and there may be $(sup([A, val_A]) \cap sup([A, val'_A])) \neq \emptyset$. Namely, $\{inf([A, val_A])\}$ for the attribute A may not be a set of equivalence classes. The same holds for $\{sup([A, val_A])\}$.

3. We may have contradiction for inappropriate $inf([A, val_A])$ and $sup([A, val_A])$. For example, we have contradiction, if $x \in inf([A, val_A]) \cap inf([A, val'_A])$ (the value of x is certainly val_A and certainly val'_A). We also have contradiction, if $x \notin \cup_{val_A} sup([A, val_A])$ (there is no value of x). To keep consistency in Fig. 1 and DbIS, we consider the constraints for $inf([A, val_A])$ and $sup([A, val_A])$. The constraints are specified in Definition 1.

We move to the mathematical framework of DbIS Ω. To handle rules from every data set, we at first clarify descriptors. Therefore, we propose a DbIS (Descriptor based Information System) in addition to DIS [11] and NIS [8,10].

Definition 1. *A Descriptor-based Information System DbIS Ω*

$$\Omega = (OB, AT \cup \{Dec\}, \{DESC(A)|A \in AT \cup \{Dec\}\}, Mappings(inf, sup))$$

consists of the following:

1. *A finite set OB of objects.*
2. *A finite set AT of attributes. Here, Dec is a decision attribute.*
3. *A finite set $DESC(A)=\{[A, val_A]|A : attribute\}$ of descriptors for the attribute A. The cardinality $|DESC(A)|$ is more than 2. A finite set VAL_A of attribute values for A is defined as $\cup_{[A,val_A]\in DESC(A)}\{val_A\}$.*
4. *Two mappings $inf : DESC(A) \to 2^{OB}$ and $sup : DESC(A) \to 2^{OB}$ below:*
   (4-a) $inf([A, val_A]) \subseteq sup([A, val_A])$, or $sup([A, val_A])=inf([A, val_A]) \cup diff([A, val_A])$ for $diff([A, val_A])=sup([A, val_A]) \setminus inf([A, val_A])$.
   (4-b) $inf([A, val_A]) \cap sup([A, val'_A])=\emptyset$ for any val'_A $(val_A \neq val'_A)$.
   (4-c) $\cup_{[A,val_A]\in DESC(A)}sup([A, val_A])=OB$ for every $A \in AT \cup \{Dec\}$.
   (4-d) For each $x \in diff([A, val_A])$, at least a distinct descriptor $[A, val'_A]$ satisfies $x \in diff([A, val_A]) \cap diff([A, val'_A])$.

The preliminary research on Definition 1 is in [16]. From now, we will prove that we can convert DbIS Ω to NIS Φ. We can easily show the converse also holds. Namely, DbIS and NIS are convertible. This property is stated in Fig. 1.

Lemma 1. *Let us consider DbIS Ω and a set*
$ATVal(x, A)=\{val_A \mid [A, val_A] \in DESC(A), x \in sup([A, val_A])\}$.
Then we have the following assertions.
(1) $ATVal(x, A)=\{val_A\}$, if $x \in inf([A, val_A])$ for one $[A, val_A] \in DESC(A)$.
(2) $ATVal(x, A)$ consists of more than two elements, if $x \notin inf([A, val_A])$ for any $[A, val_A] \in DESC(A)$.
(Proof)
(1) Due to (4-a), $inf([A, val_A]) \subseteq sup([A, val_A])$ holds. Therefore, $x \in inf([A, val_A])$ concludes $x \in sup([A, val_A])$. This $x \notin sup([A, val'_A])$ for any val'_A due to (4-b). Thus, this x is an element of only $sup([A, val_A])$, and we conclude $ATVal(x, A)=\{val_A\}$.
(2) Each object x is an element of at least one $sup([A, val_A])$ due to (4-c). Since $x \notin inf([A, val_A])$ for any $[A, val_A]$, $x \in diff([A, val_A])$. Due to (4-d), $x \in diff([A, val'_A])$ holds for one val'_A. Namely, $x \in sup([A, val'_A])$ is derived. Thus, we conclude $\{val_A, val'_A\} \subseteq ATVal(x, A)$, and $ATVal(x, A)$ consists of more than two elements.

Proposition 1. *The following holds.*
(1) Every DbIS Ω can be converted to one NIS.
(2) The converse also holds, namely, NIS Φ can be converted to one DbIS.
(Proof)

(1) Two sets OB and AT are given in DbIS Ω. $VAL_A = \cup_{[A,val_A] \in DESC(A)}\{val_A\}$ is a set of attribute values of A. For each descriptor $[A, val_A] \in DESC(A)$, we have a set $ATVal(x, A)$ of attribute values by Lemma 1. Namely, we can assign a set of attribute values to each pair $x \in OB$ and $A \in AT$. We can see that this assignment defines a table of NIS.

(2) For NIS, like Tables 2 and 3, two sets OB and AT are given. A descriptor $[A, val_A]$ is naturally defined by each attribute $A \in AT$ and $val_A \in VAL_A$. Two mappings, inf and sup, are defined by certain information and possible information, like Example 1. We can easily prove the constraints (4-a) to (4-d) based on the definition of inf and sup. Thus, we can convert one NIS into one DbIS.

Remark 2. The following is a summary of Sect. 3.

1. DbIS Ω is a unified format for rule generation from different types of data sets with uncertainty. The constraints (4-a) to (4-d) on *inf* and *sup* are necessary for keeping consistency of information.
2. Due to Proposition 1, we can convert DbIS to NIS, and the converse also holds.
3. In NIS, we can apply NIS-Apriori-based rule generation. Therefore, we can handle rules from different types of data sets with uncertainty via DbIS and NIS-Apriori-based rule generation.

4 Rule Learning from Two Different Types of Data Sets with Uncertainty

This section considers the merging procedure for two data sets. We repeatedly apply this procedure and have one DbIS combining several different types of data sets.

Example 2. In Example 1, let us consider the next case.
Case 3: In Case 1, we suppose Table 4 consists of p_1, p_2, p_3. Then, we have the following.
$ATVal(p_1, weight)=\{light\}$, $ATVal(p_2, weight)=\{normal\}$,
$ATVal(p_3, weight)=\{normal\}$.
However, there is no information for p_4 and p_5. In this case, we employ non-deterministic information (a kind of missing value) and have
$ATVal(p_4, weight)=\{light, normal, heavy\}$,
$ATVal(p_5, weight)=\{light, normal, heavy\}$.
Table 5 shows one merged NIS from Table 1 and Case 3. Non-deterministic information and missing values play an important role in merging data sets.

Definition 2. *For DbIS $\Omega_1 = (OB_1, AT_1 \cup \{Dec\}, \{DESC(A)\}, (inf_1, sup_1))$, and another DbIS $\Omega_2 = (OB_2, AT_2, \{DESC(A)\}, (inf_2, sup_2))$, we add information in Ω_2 to Ω_1 and have $\Omega_{1,2}$. We do not change the set OB_1 here.*
(1) If $OB_1 = OB_2$, we have the following DbIS $\Omega_{1,2}$.

$\Omega_{1,2}=(OB_1, AT_1 \cup AT_2 \cup \{Dec\}, \{DESC(A)\}, (inf_1 \cup inf_2, sup_1 \cup sup_2))$.
(2) If $OB_1 \neq OB_2$, we assert additional information to each $x \in OB_1 \cap OB_2$ and add a missing value ? or non-deterministic information to each $x \in OB_1 \setminus OB_2$, like Table 5.

Table 5. One merged NIS from Table 1 and Case 3 using non-deterministic information.

OB	P	Q	R	S	Dec	weight
p_1	3	1	2	2	a	{light}
p_2	2	2	2	1	a	{normal}
p_3	1	2	2	1	b	{normal}
p_4	1	3	3	2	b	{light, normal, heavy}
p_5	3	2	3	1	c	{light, normal, heavy}

For the merged data set $\Omega_{1,2}$ in Definition 2, we apply Proposition 1 and have one NIS via DbIS. Table 6 is the corresponding NIS from Table 2 and Case 2 in Table 4. Using the NIS-Apriori rule generator, we can have rules in Fig. 2 from Table 6. This execution is one example of rule learning in Fig. 1.

Table 6. One NIS from Table 2 and Case 2 in Table 4.

OB	P	Q	R	S	Dec	weight
p_1	3	?	2	2	a	{light}
p_2	2	{2,3}	2	?	a	{light}
p_3	?	2	2	{1,2}	b	{light}
p_4	1	3	3	2	b	{light}
p_5	3	2	3	?	c	{light, heavy}

In Fig. 2, the attribute value for p_3 and P affects $\tau_1 : [P,1] \Rightarrow [Dec, b]$. There are two cases. The values are 1 and others. If the value is 1, τ_1 occurs twice with no contradiction. If the value is other, τ_1 occurs one time and no contradiction. Namely, $min\_supp=1/5$ and $min\_acc=1.0$. The NIS-Apriori-based software does not handle all possible cases, but it calculates the same value as the exhaustive calculation [13,14].

The attribute value for p_5 and $weight$ affects $\tau_2 : [weight, heavy] \Rightarrow [Dec, c]$. There are two cases. The values are $light$ and $heavy$. If the value is $light$, τ_2 does not occur. If the value is $heavy$, τ_2 occurs one time and no contradiction. Namely, $max\_supp=1/5$ and $max\_acc=1.0$.

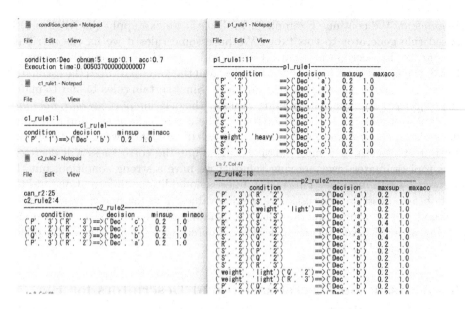

Fig. 2. The obtained certain rules (left side, $min\_supp \geq 0.1$ and $min\_acc \geq 0.7$) and possible rules (right side, $max\_supp \geq 0.1$ and $max\_acc \geq 0.7$) from Table 6.

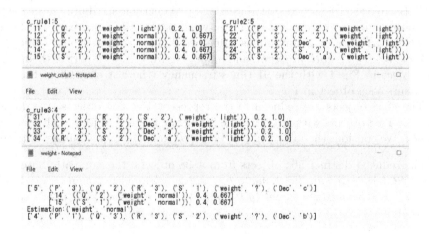

Fig. 3. The obtained certain rules (upper side, $min\_supp \geq 0.1$ and $min\_acc \geq 0.6$) and estimation of the attribute value (lower side) of p_4 and p_5.

5   Missing Value Estimation in Rule Learning

In Example 2, we used non-deterministic information for merging two data sets and had Table 5. In this case, we may impute missing values. We show this idea using an example.

Example 3. We continue Example 2. In Table 5, we can apply the NIS-Apriori-based rule generator to this table. We have some rules if we fix the attribute *weight* as the decision attribute. Certain rules are the most credible rules. In Fig. 3, the upper part shows obtained rules, and the lower part does the estimated values. No certain rule matches the object p_5. Since certain rules 14 and 15 match the object p_4, we impute its missing value by the decision part $[weight, normal]$. The total process took about 0.004(sec). The correctness of the imputed value depends on the *minsupp* and *minacc* values. If we have a weak condition, our software tool estimates most missing values, but the correctness will decrease. Our tool estimates fewer missing values if we have a strong condition, but the correctness will increase. This method needs no additional information, and the self-obtained certain rules estimate missing values. This missing value estimation method will be a new functionality for handling several data sets. We are also coping with this issue because rules from different types of data sets and missing values are closely related.

6   An Issue on Discretization and Descriptors for Big Data Analysis

This section focus on the Wine Quality data set [2]. The number of objects is 4898. There are 11 condition attributes and one decision attribute *quality*. The attribute value of *quality* is a score between 0 and 10. Since each of the 11 condition attributes is a continuous value, we divided each attribute value into three classes, i.e., *small*, *medium*, and *large*. Figure 4 shows a part of them. The first object in Fig. 4 with the 91 times frequency (the last element in the list) represents each object in Fig. 5.

The 4898 objects are reduced to the representative 564 objects. We do not consider 4898 objects with the one-time frequency but 564 objects with each frequency. We consider the following DbIS derived from the Wine Quality data set.

1. The reduced distinct 564 objects from 4898 objects, the same attribute.
2. $DESC(A)=\{[A, small], [A, medium], [A, large]\}$ for each condition attribute, and $DESC(quality)=\{[quality, 0], [quality, 1], \cdots, [quality, 10]\}$ for decision attribute.

```
[(('fixed', 's'), ('volatile', 's'), ('citric', 's'), ('residual', 's'), ('chlorides', 's'), ('free', 's'), ('total', 's'),
('density', 's'), ('pH', 'm'), ('sulphates', 's'), ('alcohol', 's'), ('quality', '6')), 91]
[(('fixed', 's'), ('volatile', 's'), ('citric', 's'), ('residual', 's'), ('chlorides', 's'), ('free', 's'), ('total', 's'),
('density', 's'), ('pH', 'm'), ('sulphates', 's'), ('alcohol', 'm'), ('quality', '7')), 93]
[(('fixed', 's'), ('volatile', 's'), ('citric', 's'), ('residual', 's'), ('chlorides', 's'), ('free', 's'), ('total', 's'),
('density', 's'), ('pH', 'm'), ('sulphates', 'm'), ('alcohol', 'm'), ('quality', '6')), 146]
[(('fixed', 'm'), ('volatile', 's'), ('citric', 's'), ('residual', 's'), ('chlorides', 's'), ('free', 's'), ('total', 'm'),
('density', 's'), ('pH', 'm'), ('sulphates', 'm'), ('alcohol', 's'), ('quality', '4')), 1]
```

Fig. 4. Four tuples of discretized objects. The first tuple represents the 91 objects (the last value in the list). The third tuple represents 146 objects.

object	fixed	volatile	citric	residual	chloride	free	total	density	pH	sulphat	alcohol	quality
2	6.3	0.3	0.34	1.6	0.049	14	132	0.994	3.3	0.49	9.5	6
7	6.2	0.32	0.16	7	0.045	30	136	0.9949	3.18	0.47	9.6	6
9	6.3	0.3	0.34	1.6	0.049	14	132	0.994	3.3	0.49	9.5	6
17	6.3	0.48	0.04	1.1	0.046	30	99	0.9928	3.24	0.36	9.6	6
58	6	0.19	0.26	12.4	0.048	50	147	0.9972	3.3	0.36	8.9	6
62	6	0.19	0.26	12.4	0.048	50	147	0.9972	3.3	0.36	8.9	6
124	6.9	0.19	0.28	5	0.058	14	146	0.9952	3.29	0.36	9.1	6

wineq   ⊕

Fig. 5. Each of 91 objects are reduced to the first tuple in Fig. 4.

3. A pair of two sets $(inf(d), sup(d))$ $(d \in DESC(A))$ is defined by the discretization. Here, $sup(d)=inf(d)$ holds for every descriptor d. We consider new 564 objects with each frequency, like in Fig. 4.

We compared the execution times for 4898 objects and 564 objects. The former took 13.366 (sec), and the latter took 10.963 (sec). The latter is about 82% of that of the former. Of course, the two programs generated the same rules.

In the Htru2 data set [2], 17898 objects are reduced to 144 objects. One discretized object represents 4431 objects. We do not need to handle 4431 objects sequentially. It is enough for us to slightly change the definition of *support* and *accuracy* with a frequency. For the condition $support \geq 0.001$ and $accuracy \geq 0.7$, the execution time of 1.720 (sec) is reduced to 0.198 (sec). The latter is about 10% of that of the former. This property will help reduce the objects in big data sets. Like this, the DbIS framework will also be helpful for big data analysis.

7   Concluding Remarks

We previously coped with rule generation from tables with uncertainty. Each rule consists of descriptors. In tables, we implicitly used a pair $[A, val_A]$ of an attribute and its attribute value as a descriptor. To handle different types of data sets, we should specify descriptors for rules, and then we can consider rules from different types of data sets. DbIS Ω in Definition 1 proposed this consideration. Fortunately, we can convert DbIS to NIS, where we implemented several software tools. Thus, we can obtain rules from different types of data sets via DbIS and NIS.

As shown at the lower-right part of Fig. 1, we handle rule generation, the merge process, missing value estimation, and big data analysis. We will extend our previous research to learning from different types of data sets.

Acknowledgment. The authors thank the reviewers for their helpful comments. This work is supported by JSPS (Japan Society for the Promotion of Science) KAKENHI Grant Number JP20K11954.

References

1. Agrawal, R., Srikant, R.: Fast algorithms for mining association rules in large databases. In: Bocca, J.B., Jarke, M., Zaniolo, C. (Eds.) Proceedings of VLDB 1994, Morgan Kaufmann, pp. 487–499 (1994)
2. Frank, A., Asuncion, A.: UCI machine learning repository. University of California, School of Information and Computer Science, Irvine, CA (2020). http://mlearn.ics.uci.edu/MLRepository.html. Accessed 7 Jan 2020
3. Fürnkranz, J., Kliegr, T.: A brief overview of rule learning. In: Bassiliades, N., Gottlob, G., Sadri, F., Paschke, A., Roman, D. (eds.) RuleML 2015. LNCS, vol. 9202, pp. 54–69. Springer, Cham (2015). https://doi.org/10.1007/978-3-319-21542-6_4
4. Grzymała-Busse, J.W., Werbrouck, P.: On the best search method in the LEM1 and LEM2 algorithms. In: Orłowska, E. (Ed.) Incomplete Information: Rough Set Analysis, Studies in Fuzziness and Soft Computing, vol. 13, pp. 75–91 (1998)
5. Jian, Z., Sakai, H., Watada, J., Roy, A., Hassan, M.: An A priori-based data analysis on suspicious network event recognition. In: Roger, B., Carlo, Z., (Eds.) Proceedings of IEEE Big Data, pp. 5888–5896 (2019)
6. Kripke, S.A.: Semantical considerations on modal logic. Acta Philosophica Fennica **16**, 83–94 (1963)
7. Lin, T.Y.: Granular Computing: Practices, Theories, and Future Directions. In: Meyers, R. (eds.) Encyclopedia of Complexity and Systems Science, pp. 4339–4355. Springer, New York (2009). https://doi.org/10.1007/978-0-387-30440-3_256
8. Lipski, W.: On semantic issues connected with incomplete information databases. ACM Trans. Database Syst. **4**(3), 262–296 (1979)
9. Nakata, M., Sakai, H.: Twofold rough approximations under incomplete information. Int. J. Gener. Syst. **42**(6), 546–571 (2013)
10. Orłowska, E., Pawlak, Z.: Representation of nondeterministic information. Theoret. Comput. Sci. **29**(1–2), 27–39 (1984)
11. Pawlak, Z.: Rough sets. Int. J. Comput. Inf. Sci. **11**(5), 341–356 (1982)
12. Pedrycz, W., Skowron, A., Kreinovich, V.: Handbook of Granular Computing, p. 1116. Wiley, New York (2008)
13. Sakai, H., Nakata, M.: Rough set-based rule generation and Apriori-based rule generation from table data sets: a survey and a combination. CAAI Trans. Intell. Technol. **4**(4), 203–213 (2019)
14. Sakai, H., Nakata, M., Watada, J.: NIS-Apriori-based rule generation with three-way decisions and its application system in SQL. Inf. Sci. **507**, 755–771 (2020)
15. Sakai, H.: Execution logs by RNIA software tools. http://www.mns.kyutech.ac.jp/~akai/RNIA
16. Sakai, H., Nakata, M.: Apriori-based rule generation with three-way decisions for heterogeneous and uncertain data. In: Proceedings of SCIS&ISIS, pp. 1–6 (2022). https://doi.org/10.1109/SCISISIS55246.2022.10001979
17. Skowron, A., Rauszer, C.: The discernibility matrices and functions in information systems. In: Słowiński, R. (Ed.) Intelligent Decision Support - Handbook of Advances and Applications of the Rough Set Theory, Kluwer Academic, pp. 331–62 (1992)
18. Association rule learning, Wikipedia
19. Wojtusiak, J.: Rule Learning. In: Seel, N.M. (eds.) Encyclopedia of the Sciences of Learning, pp. 2909–2911. Springer, Boston, MA (2012). https://doi.org/10.1007/978-1-4419-1428-6_1347

MFG-HUI: An Efficient Algorithm for Mining Frequent Generators of High Utility Itemsets

Hai Duong[1]([⊠]), Thong Tran[2,3], Tin Truong[1], and Bac Le[2,3]

[1] Department of Mathematics and Computer Science, Dalat University, Dalat, Vietnam
{haidv,tintc}@dlu.edu.vn
[2] Department of Computer Science, Faculty of Information Technology, University of Science, Ho Chi Minh City, Vietnam
thongt@dlu.edu.vn, lhbac@fit.hcmus.edu.vn
[3] Vietnam National University, Ho Chi Minh City, Vietnam

Abstract. The discovery of frequent generators of high utility itemsets (FGHUIs) holds great importance as they provide concise representations of frequent high utility itemsets (FHUIs). FGHUIs are crucial for generating nonredundant high utility association rules, which are highly valuable for decision-makers. However, mining FGHUIs poses challenges in terms of scalability, memory usage, and runtime, especially when dealing with dense and large datasets. To overcome these challenges, this paper proposes an efficient approach for mining FGHUIs using a novel lower bound called *lbu* on the utility. The approach includes effective pruning strategies that eliminate non-generator high utility branches early in the prefix search tree based on *lbu*, resulting in faster execution and reduced memory usage. Furthermore, the paper introduces a novel algorithm, MFG-HUI, which efficiently discovers FGHUIs. Experimental results demonstrate that the proposed algorithm outperforms state-of-the-art approaches in terms of efficiency and effectiveness.

Keywords: Frequent high utility itemset · Generators · Upper bound · Weak upper bound · Lower bound · Pruning strategy

1 Introduction

High utility itemset mining (HUIM) from quantitative datasets plays a crucial role in data mining. It aims to discover itemsets with high utility (HUI) and finds applications in diverse areas, such as analyzing user behavior in online stores, studying biomedical data, and cross-marketing [9, 15]. Developing efficient algorithms for HUIM is a non-trivial task because the utility function u does not adhere to the anti-monotonic ($\mathcal{AM}$) property, which is a powerful characteristic for reducing the search space in frequent itemset mining. To address this issue, researchers have developed upper bounds (UBs) on u that satisfy the $\mathcal{AM}$ and are greater than or equal to the actual utility values. These UBs are valuable for pruning low utility itemsets (LUIs) efficiently and reducing the search space in HUIM. The first UB, known as TWU (transaction weight utility), was proposed for the Two-Phase algorithm [10] and has been adopted by various HUIM algorithms such as GPA [7] and UP-Growth + [17]. The second UB, called *feub*, was introduced in [9] and is tighter than TWU. Other algorithms [6,12,1] have also utilized *feub*. . More recently, a weak upper bound (WUB) named *fwub* was presented in [2], which is even tighter than *feub* and enables early pruning of additional low utility itemsets.

A key limitation of HUIM is the potential for a large result set, $\mathcal{HUI}$, consisting of all high utility itemsets. This is particularly true for dense datasets, low minimum utility thresholds [18], or large-scale data, leading to computational complexity and significant memory requirements. Moreover, the extensive set of HUIs can complicate decision-making analysis for users. To address this, the paper focuses on discovering the set $\mathcal{FGHUI}$ of frequent generators of HUIs (FGHUI) [14,13], which serves as a concise representation of $\mathcal{HUI}$. The $\mathcal{FGHUI}$ set has several advantages: it has a smaller size compared to $\mathcal{FHUI}$ (frequent HUIs), making them easier to discover, requiring less storage space, and simplifying analysis. $\mathcal{FGHUI}$ aligns with the Minimum Description Length principle [8], representing minimal patterns in equivalence classes, where FHUIs share the same support and appear in the same transactions. Additionally, combining FGHUIs with frequent closed HUIs allows for generating non-redundant high utility association rules, which aids decision-making [14,11].

Despite the desirability of discovering FGHUIs, only a few algorithms have been proposed for this task [12, 13] [20]. However, these algorithms exhibit poor performance, especially when dealing with large databases or low minimum utility thresholds. Additionally, the concept of generator patterns utilized in HUG-Miner [4] is impractical because it can result in an empty set of generators for HUIs, rendering it impossible to generate high utility association rules by combining them with closed HUIs. To tackle these challenges, this study introduces the concept of FGHUIs, which is more practical and valuable. Furthermore, an efficient algorithm is developed to directly mine FGHUIs from the dataset, demonstrating superior runtime, memory consumption, and scalability.

Contributions. The key contributions of this paper are as follows. Firstly, an efficient method is proposed for constructing the $\mathcal{FGHUI}$ set accurately and completely, reducing the computational burden of checking itemset inclusion relations. Secondly, a novel lower bound (LB) named *lbu* is introduced, along with a general pruning strategy for non-generator FHUI branches. Thirdly, two novel pruning strategies, *P-NonGenHUI* and *LP-NonGenHUI*, are developed to early prune itemset branches in the prefix tree that do not contain FGHUIs. A novel algorithm, MFG-HUI, is also presented to efficiently discover $\mathcal{FGHUI}$ using these strategies. Experimental evaluations on real-life and synthetic datasets demonstrate that the proposed algorithm outperforms state-of-the-art approaches in terms of execution time, memory usage, and scalability.

The rest of this paper is structured as follows. Section 2 introduces the concepts and notations necessary for discovering FGHUIs. Section 3 details new theoretical results and algorithm. The results of the experiments conducted are discussed in Sect. 4. The paper concludes with Sect. 5.

2   Preliminaries and Problem Definition

This section presents fundamental concepts and notations related to the task of mining frequent and high utility itemsets.

Definition 1. (*Quantitative database*). Let $\mathcal{A} \overset{\text{def}}{=} \{a_j, j \in J \overset{\text{def}}{=} \{1, 2, \ldots, M\}\}$ be a finite set consisting of distinct items and $A \subseteq \mathcal{A}$ be a subset of $\mathcal{A}$. If A contains exactly k items, then it is referred to as a *k-itemset*. Without loss of generality, assume that the items within each itemset are sorted in a total order $\prec$ (e.g., lexicographical). The profit vector $\mathcal{P} \overset{\text{def}}{=} (p(a_j), j \in J$ and $p(a_j) > 0)$ represents the external utility of all items in $\mathcal{A}$, where $p(a_j)$ denotes the relative importance of item a_j, such as its unit profit, price, weight, or other factors. A *q-item* is a pair (a, q), where a is an item from $\mathcal{A}$ and q is a non-negative integer representing the purchase quantity or *internal* utility of item a. .. A quantitative database (QDB) $\mathcal{D}$ is a collection of transactions, $\mathcal{D} \overset{\text{def}}{=} \{T_i, i \in I \overset{\text{def}}{=} \{1, 2, \ldots, N\}\}$, where each transaction T_i, defined as $T_i \overset{\text{def}}{=} \{(a_j, q_{ij}), j \in J_i\}$ including a set of q-items, is uniquely identified by *TID* (e.g., *TID* $= i$) and J_i is an index subset of J, $J_i \subseteq J$. The *utility* of a q-item (a, q), denoted as $u((a, q))$, is defined as $u((a, q)) \overset{\text{def}}{=} q \times p(a)$. Intuitively, it is the amount of profit yielded by the sale of item a in the transaction where the q-item (a, q) appears. Table 1.a presents an example of a QDB, where $\mathcal{P} = (p(a) = 15, p(b) = 2, p(c) = 1, p(d) = 3, p(e) = 2, p(f) = 50)$, and Table 1.b shows an *integrated* QDB $\mathcal{D}'$ resulting from combining $\mathcal{D}$ and $\mathcal{P}$, i.e. $\mathcal{D}' \overset{\text{def}}{=} \{T'_i \overset{\text{def}}{=} \{(a_j, q'_{ij}), j \in J_i\}, J_i \subseteq J, i \in I\}$, where $q'_{ij} = q_{ij} * p(a_j)$ if $(a_j, q_{ij}) \in T_i$ (otherwise, $q'_{ij} = 0$), $\forall i \in I, j \in J$.

Table 1a. A QDB $\mathcal{D}$

TID	a	b	c	d	e	f
T_1	0	0	0	0	12	0
T_2	0	4	10	30	0	100
T_3	45	4	2	6	8	50
T_4	0	10	3	12	0	0
T_5	0	0	0	0	0	100

Table 1b. The integrated QDB $\mathcal{D}'$

TID	a	b	c	d	e	f
T_1	0	0	0	0	6	0
T_2	0	2	10	10	0	2
T_3	3	2	2	2	4	1
T_4	0	5	3	4	0	0
T_5	0	0	0	0	0	2

Running Example. Hereafter, the *integrated* QDB $\mathcal{D}'$ will serve as a running example. For brevity, the paper uses bd as a shorthand notation for the itemset $\{b, d\}$.

Table 2. Characteristics of datasets.

Dataset	#Trans.(D)	#Items (N)	Avg trans. Len (T)	Density (T*100/N %)
Chess	3,196	75	37	49.33
Connect	67,557	129	43	33.33
Mushroom	8,124	119	23	19.33
Pumsb	49,046	2113	74	3.50
BMS	77,512	3,340	4.62	0.14
T10I4N900D100K	100,000	900	9.1	1.01

Definition 2 *(Itemset support).* The support of an itemset A, denoted as $supp(A)$, is given by the count of transactions in the QDB $\mathcal{D}'$ that contain A, i.e., $supp(A) = |\rho(A)|$, where $\rho(A) \text{def} = \{T_i \in \mathcal{D}' | q'_{ij} > 0, \forall a_j \in A\}$.

Definition 3 *(Itemset utilities).* The utility of an item a_j ($j \in J$) in a transaction $T_i \in \mathcal{D}'$ is denoted and defined as $u(a_j, T_i) \overset{\text{def}}{=} q'_{ij}$. Then, the utilities of an itemset A in T_i containing A and in $\mathcal{D}'$ are denoted and calculated as $u(A, T_i) \overset{\text{def}}{=} \sum_{a_j \in A} u(a_j, T_i)$ and $u(A) \overset{\text{def}}{=} \sum_{T_i \in \rho(A)} u(A, T_i)$, respectively.

Definition 4 *(Transaction utility and total utility).* The transaction utility of a transaction T_i is denoted and defined as $tu(T_i) \overset{\text{def}}{=} \sum_{q'_{ij} > 0} q'_{ij}$. Then, the total utility TU of $\mathcal{D}'$ is defined as $TU \overset{\text{def}}{=} \sum_{T_i \in \mathcal{D}} tu(T_i)$.

Example 1. For the itemset $A = bf$, then $\rho(A) = \{T_2, T_3\}$, $u(A, T_2) = 4 + 100 = 104$ and $u(A, T_3) = 4 + 50 = 54$. Then, $u(A) = u(A, T_2) + u(A, T_3) = 104 + 54 = 158$. We have, $tu(T_4) = 10 + 3 + 12 = 25$, $TU = 12 + 144 + 115 + 25 + 100 = 396$.

Definition 5 *(Frequent and high utility itemset).* An itemset A is called a high utility itemset (HUI) if $u(A) \geq mu$ and an HUI A is called a frequent HUI (FHUI) if $supp(A) \geq ms$, where mu and ms are two user-defined positive *minimum utility* and *support thresholds*, respectively. The problem of FHUI mining ($\mathcal{FHUIM}$) is to exploit the set $\mathcal{FHUI} \overset{\text{def}}{=} \{A \in \mathcal{IS} | u(A) \geq mu \wedge supp(A) \geq ms\}$ of all FHUIs. Note that $\mathcal{FHUIM}$ is more general than HUIM with a special value of ms, $ms = 1$.

Definition 6 *(Forward extension).* For any two itemsets A and B such that $A \cap B = \varnothing$ and $A \prec B$ (i.e. $x \prec y$, $\forall x \in A$, $\forall y \in B$), let the itemset $S = A \oplus B$ denote a *forward extension* (FE) of A with B. If $B \neq \varnothing$, then S is said to be a *proper FE* of A. For the sake of brevity, let us denote the *branch Brh(A)* as the set including A and all its FEs in the prefix tree [3], and the *proper branch FBrh(A)* as $Brh(A)$ without itemset A.

Definition 7 *(Equivalence relation and equivalence class).* An equivalence relation on all itemsets of $\mathcal{IS}$ is denoted and defined as: $\forall A, B \in \mathcal{IS}, A \sim B \iff \rho(A) = \rho(B)$. Then, the equivalence class of A is denoted and defined as $[A] \overset{\text{def}}{=} \{B \in \mathcal{IS} | \rho(B) = \rho(A)\}$.

Definition 8 *(Sets $\mathcal{FGHUI}$ and $\mathcal{FHUGI}$).* Let A be an FHUI, i.e., $A \in \mathcal{FHUI}$. Then,

a) A is said to be a frequent generator of HUIs (FGHUI) if there is no FHUI that is a proper sub-itemset of A and has the same support. The set containing all FGHUIs is denoted and defined as.

$$\mathcal{FGHUI} \stackrel{\text{def}}{=} \{A \in \mathcal{FHUI} | \nexists B \in \mathcal{FHUI} : B \subset A \wedge supp(B) = supp(A)\}.$$

b) A is said to be a frequent *high utility generator itemset* (FHUGI) if there is no itemset that is a *proper* sub-itemset of A and has the same support. The set of all FHUGIs is denoted and defined as

$$\mathcal{FHUGI} \stackrel{\text{def}}{=} \{A \in \mathcal{FHUI} | \nexists B \in \mathcal{IS} : B \subset A \wedge supp(B) = supp(A)\}.$$

Discussions. Let $\mathcal{GI} \stackrel{\text{def}}{=} \{A \in \mathcal{IS} | \nexists R \in \mathcal{IS} : R \subset A \wedge supp(R) = supp(A)\}$, which is the set of all generator itemsets. It is clear that $\mathcal{FHUGI} = \mathcal{FHUI} \cap \mathcal{GI}$. Besides, note that $\mathcal{FGHUI}$ is different from $\mathcal{FHUGI}$. Since the u function is not *anti-monotonic* on each equivalence class, so for any $B \in [A]$ and $B \subset A$, if $u(A) \geq mu$, it does not ensure that $u(B) \geq mu$. Obviously, if $\mathcal{FHUI} \neq \varnothing$, $\mathcal{FHUGI} \subseteq \mathcal{FGHUI} \neq \varnothing$. Moreover, $\mathcal{FHUGI}$ can become empty, i.e., $\varnothing = \mathcal{FHUGI} \subset \mathcal{FGHUI}$. Thus, although any itemset A in $\mathcal{FHUI}$ is always recovered efficiently from $\mathcal{FGHUI}$ and closed HUIs (CHUI), it may not be recovered from the empty $\mathcal{FHUGI}$ set and CHUIs as shown in Example 2.

Example 2. Consider the integrated QDB $\mathcal{D}\prime = \{T_1 : (a,3)(b,6); T_2 : (b,2); T_3 : (a,15)(b,1)(c,5)\}$ with $mu = 19$ and $ms = 1$. Then, $\mathcal{FHUI} = \{ac_{20}^1, ab_{25}^2, abc_{21}^1\}$, where the notation ac_{20}^1 means that $u(ac) = 20$, $supp(ac) = 1$. It is found that although $\mathcal{FHUI} \neq \varnothing$, and $\mathcal{FGHUI} = \{ac_{20}^1, ab_{25}^2\} \neq \varnothing$, but $\mathcal{GI} = \{a_{18}^2, b_9^3, c_5^1\}$. Therefore, $\mathcal{FHUGI} = \mathcal{FHUI} \cap \mathcal{GI} = \varnothing$ and $\mathcal{FHUGI} \subset \mathcal{FGHUI}$.

Problem Statement. Given a QDB $\mathcal{D}\prime$ and two positive minimum utility and support thresholds mu and ms, respectively, defined by the user, the objective is to discover the set $\mathcal{FGHUI}$ of all FGHUIs.

Example 3. For $mu = 21$ and $ms = 1$, consider the itemset $A = bcd$. Then, $u(A) = 81 \geq mu$, $\rho(A) = \{T_2, T_3, T_4\}$, and $supp(A) = 3$. A is not a FGHUI, since there exists an FHUI $B = bc \in \mathcal{FHUI}$ with $u(B) = 33$, $A \supset B$, and $supp(B) = supp(A) = 3$. However, B is a FGHUI since $\nexists C \in \mathcal{FHUI} : B \supset C$ and $supp(C) = supp(B)$. We have $\mathcal{FGHUI} = \{a, f, d, bf, cf, df, ef, bc\}$, $|\mathcal{FGHUI}| = 8$, while $|\mathcal{FHUI}| = 53$.

3 Mining frequent generators of HUIs

3.1 Pruning Infrequent and Low Utility Itemsets

A function F on an itemset is considered *anti-monotonic* ($\mathcal{AM}$) if the condition $F(D) \leq F(C)$ holds for any two itemsets C and D such that $D \supseteq C$. In the context of the HUIM problem, the utility function u does not adhere to the $\mathcal{AM}$ property. To address this, upper bounds (UBs) and weak upper bounds (WUBs) on u were developed to enable early pruning of low utility itemsets (LUIs), while satisfying $\mathcal{AM}$ or $\mathcal{AM}$-like properties [16]. A function ub of itemsets is said to be an UB on u if $u(C) \leq ub(C), \forall C \in \mathcal{IS}$. A function wub is said to be a WUB on u if $u(S) \leq wub(C)$ for any itemset C and its *proper* FE $S, S = C \oplus D \supset C$ with $D \neq \varnothing$. For any two UBs ub_1 and ub_2, ub_1 is called tighter than ub_2 if $ub_1(C) \leq ub_2(C), \forall C \in \mathcal{IS}$.

Definiton 9 (*UB TWU* [10] *and WUB fwub* [2]). For any itemset A.

a) The transaction-weighted utilization (*TWU*) of A in $\mathcal{D}'$ is denoted and defined as
$TWU(A) \stackrel{\text{def}}{=} \sum_{T \in \rho(A)} tu(T)$

b) A WUB on u is denoted and defined as $fwub(A) \stackrel{\text{def}}{=} \sum_{T \in TC(A)}(u(A, T) + ru(A, T))$,
where $TC(A) \stackrel{\text{def}}{=} \{T \in \rho(A) | rem(A, T) \neq \varnothing\}$, $ru(A, T) \stackrel{\text{def}}{=} u(rem(A, T))$ is the
remaining utility of A in transaction T, and $rem(A, T)$ is the rest of T after the
last item of A.

Proposition 1 (*Properties of (W)UBs* [10, 2]).

a) *TWU* is an UB on u and *fwub* is a WUB on u, and *fwub* tighter than *TWU*, i.e.,
   $TWU(A) \geq fwub(A) \geq u(A)$ for any itemset A.
b) The $\mathcal{AM}$ property holds for the *TWU* UB, i.e., $TWU(B) \leq TWU(A), \forall B \supseteq A$.
c) For any proper FE S of $B \in \mathcal{IS}$, $S = B \oplus D$, and $D \neq \varnothing$, then $u(S) \leq fwub(B)$.

Strategy 1 (*Pruning LUIs based on TWU and fwub* [10, 2]). For any itemset A, if
$TWU(A) < mu$, then A and all its supersets are not HUIs. Thus, they can be eliminated
early. Meanwhile, if $fwub(A) < mu$, none of the proper FEs of A is high utility. Thus,
$FBrh(A)$ is pruned early to reduce the search space.

Strategy 1 has been commonly used in HUIM algorithms to reduce the search space
and QDB $\mathcal{D}'$. Specifically, if the *TWU* of a 1- or 2-itemset A, which has 1 or 2 items, is
less than mu, then none of the supersets of A is a HUI.

Example 4. For $mu = 240$, consider the itemset $A = bd$. Then, $\rho(A) = \{T_2, T_3, T_4\}$
and $TWU(A) = tu(T_2) + tu(T_3) + tu(T_4) = 284 \geq u(A) = 66$. For the itemset
$D = bdf \supseteq A$ and $E = bcd \supseteq A$, we have $TWU(D) = 259 \leq TWU(A)$ and $TWU(E) =$
$284 = TWU(A)$.

From now on, it is assumed that items in $\mathcal{A}$ are sorted in the ascending order of their
TWU values, denoted as $\prec_{twu}$, that is a, e, b, c, d, f since $TWU(a) = 115 < TWU(e) =$
$127 < TWU(b) = TWU(c) = TWU(d) = 284 < TWU(f) = 359$. In the process of
mining, $\prec_{twu}$ serves as the order for extending itemsets.

We have $TC(A) = \{T_2, T_3\}$. Note that $T_4 \notin TC(A)$ since $rem(A, T_4) = \varnothing$. Then,
$fwub(A) = u(A, T_2) + ru(A, T_2) + u(A, T_3) + ru(A, T_3) = 194 \leq TWU(A) = 284$.
Consider the proper FE $S = A \oplus f = bdf$ of A and $mu = 200$. Then, $u(S) = 194 \leq$
$fwub(A)$. Since $fwub(A) < mu$, all proper FEs of A are low utility itemsets (LUI) and
thus can be pruned, i.e., $FBrh(A)$ is pruned early. Note that if using *TWU*, then $FBrh(B)$
cannot be pruned as $TWU(A) = 284 > mu$.

If $FBrh(B)$, which includes all *proper* FEs of B, has no FGHUIs, it is called
NonGenFHU branch.

3.2 Constructing the Complete and Correct Set of FGHUIs

When employing a depth-first search algorithm for mining, where candidate itemsets are
arranged in a prefix tree based on a processing order $\prec$ over items, there is a condition
to be met when considering an itemset $B = A \oplus y$. This condition mandates that all

backward extensions C of B, i.e., $C = D \oplus y \oplus E$ with $A \subset D$, must already exist in the current prefix tree and have been evaluated prior. For B, any its super-itemset C can be represented in the form of $C = D \oplus y \oplus E$, with $A \subseteq D \prec y$ and $(E = \varnothing$ or $E \succ y)$. If C is considered before B, then $D = A + X \supset A$ with $X \neq \varnothing$. Otherwise, i.e., $X = \varnothing$ and $E \succ y$, then the FE C will be considered after B.

It is found that the representation of the set $\mathcal{FGHUI} \stackrel{\text{def}}{=} \{A \in \mathcal{FHUI} | \nexists B \in \mathcal{FHUI} : B \subset A \wedge supp(B) = supp(A)\}$ ($\subseteq \mathcal{FHUI}$) depends on checking two conditions $A \in \mathcal{FHUI}$ and $(\nexists B \in \mathcal{FHUI} : B \subset A \wedge supp(B) = supp(A)$. To enhance the efficiency of mining FGHUIs, it is necessary not only to utilize the (W)UBs-based strategies for pruning LUIs as mentioned above but also to devise an efficient approach for verifying the second condition.

Remark 1. Let $\mathcal{FGHUI}$ be the set of considered candidates of FGHUIs. Assume that the FHUI $B = A \oplus y$ is being considered. Consider the two following cases.

1. Check if there exists an FHU sub-itemset R of B (e.g. $B = bcd$), $R \subset B$. Then, such R can only be
   a. An itemset (e.g., $R \in \{c, cd, d\}$) that will be considered in next steps of the mining process.
   b. A candidate already considered in $\mathcal{FGHUI}$. Then, R is a *proper successive subset* of B such that their first items are the same, i.e., $R = b_1 b_2 \dots b_i$ for $i : 1 \leq i < k$ and $B = b_1 b_2 \dots b_k$. For example, for $B = bcd$, then $R \in \{b, bc\}$. Additionally, if $supp(R) = supp(B)$, B cannot be a FGHUI, so it is not added to $\mathcal{FGHUI}$. Otherwise, it is added to $\mathcal{FGHUI}$.
2. Check if there exists an FHU super-itemset S of B (e.g. $B = bd$), $S = C \oplus y \oplus D \supset A \oplus y$ with $D \neq \varnothing$ or $C \supset A$. Then, such S can only be
   a. A *forward extension* of B, $S = A \oplus y \oplus D$ with $C = A$ and $D \neq \varnothing$ (e.g. $S = bde$), which can only be found in next steps of the mining process.
   b. A *backward extension* of B, $S = C \oplus y \oplus D$ with $C \supset A$ (e.g. $S \in \{bcd, bcde, abd, \dots\}$), which can only be found in $\mathcal{FGHUI}$. Then, if there exists such an S with $supp(B) = supp(S)$, S is not a FGHUI, so it is removed from $\mathcal{FGHUI}$.

Theorem 1 asserts that upon the completion of the mining process using Remark 1, the entire set of FGHUIs is comprised solely of the remaining itemsets in $\mathcal{FGHUI}$.

Theorem 1 (*The correctness and completeness of constructing the $\mathcal{FGHUI}$ set*). By employing the mining process described in Remark 1, the complete and correct set $\mathcal{FGHUI}$ of all FGHUIs is discovered.

3.3 The Novel Lower Bound

Note that the u function exhibits monotonicity within each equivalence class but lacks anti-monotonicity. Consequently, it is not possible to employ strategies solely based on the monotonic property of u to efficiently eliminate *NonGenFHU* branches, which do not contain any FGHUIs. To overcome this challenge, it is necessary to devise lower bounds on u, as presented below. For any itemset A, a function $lb(A)$ is said to be a *lower bound* (LB) on $u(A)$ if $lb(A) \leq u(A)$. The $lb(A)$ function is called *monotonic w.r.t. FEs*, if $lb(A) \leq lb(B)$ for any FE B of A.

Definition 10 (*Lower bound on u*). For any *frequent* itemset A, i.e. $supp(A) \geq ms$, after sorting the series $\{u(A, T_i)|T_i \in \rho(A), i = 1..n\}$ in ascending order, we obtain the new series $\mathcal{F} \overset{\text{def}}{=} \{u_{i'}, i = 1..n\}$, with $n = |\rho(A)| \geq ms$. Then, a novel lower bound (LB) on u is denoted and defined as $lbu(A) = \sum_{i=1..ms} u_{i'}$, which is the ms smallest first values of $\mathcal{F}$.

Proposition 2 (*Properties of LB lbu on u*). For any *frequent* itemset A.

a. *lbu* is a LB on u, i.e. $lbu(A) \leq u(A)$.
b. *lbu* is *monotonic w.r.t. frequent forward* extensions, i.e. $lbu(A) \leq lbu(B), \forall B = A \oplus D \supseteq A$ and $supp(B) \geq ms$.

Example 5. For $ms = 2$, consider itemset $A = bd$ with $\rho(A) = \{T_2, T_3, T_4\}$. Then $u(A, T_2) = 34, u(A, T_3) = 10$, and $u(A, T_4) = 22$. After sorting the series $\{u(A, T_i)|T_i \in \rho(A)\}$ in ascending order, we obtain $\mathcal{F} \overset{\text{def}}{=} \{v_1, v_2, v_3\} = \{10, 22, 34\}$. Then, $lbu(A) = v_1 + v_2 = 10 + 22 = 32$. For a FE B of $A, A \subset B = A \oplus f = bdf$, then, $lbu(A) = 32 < lbu(B) = 60 + 134 = 194$.

3.4 The Novel Pruning Strategies for *NonGenFHU* Branches

Based on the LB *lbu* on u, the paper proposes a novel strategy for pruning *NonGenFHU* branches in Theorem 2.

Theorem 2 (*P-NonGenHUI*). For the current itemset B being examined, if there exists a sub-itemset $R \in \mathcal{FGHUI}, R \subset B$, such that $supp(R) = supp(B)$ and $lbu(R) \geq mu$. Then, $Brh(B)$ that cannot contain any FGHUIs can be pruned early.

It is worth noting that without employing the pruning condition $lbu(R) \geq mu$ in Theorem 2, numerous FGHUIs within the $Brh(B)$ branch may be pruned. In other words, the absence of the $lbu(R) \geq mu$ condition can lead to the exclusion of many FGHUIs from the final $\mathcal{FGHUI}$ result set, resulting in an incomplete set of discovered FGHUIs.

To implement the pruning strategy *P-NonGenHUI* in Theorem 2 based on the current itemset B, it is necessary to examine its support and subset relationship solely with patterns previously generated in $\mathcal{FGHUI}$. However, such checks can still be computationally expensive. To overcome this, and enable earlier and efficient branch pruning, a novel local pruning strategy called *LP-NonGenHUI* is introduced in Theorem 3. This strategy combines depth-first search and breadth-first search to retain crucial information about nodes extended from the current node. By doing so, it becomes possible to identify branches pruned by *LP-NonGenHUI*, thereby enhancing efficiency.

Theorem 3 (*LP-NonGenHUI*). For any two nodes $A = P \oplus x$ and $B = P \oplus y$ that have the same prefix (or parent node) in the prefix search tree such that $y \succ x$, assume that there is a FE S of A with the last item y of the node $B, S = A \oplus y$. If $(supp(S) = supp(B)$ and $lbu(B) \geq mu)$ or $(supp(S) = supp(A)$ and $lbu(A) \geq mu)$, then the *NonGenFHU* $Brh(S)$ *branch* is pruned early.

The novel theoretical results in Theorems 1–3 and Proposition 2 have been strictly proven. However, the proofs are not given in this study because of the space limitation.

3.5   The Proposed MFG-HUI Algorithm

To implement the proposed algorithm in this study, we adopted the data structure MPUN-list introduced in [2]. However, due to the space limitation, details about the data structure are omitted.

MFG-HUI($\mathcal{D}'$, mu, ms): *Input*: the QDB $\mathcal{D}'$, mu, ms; *Output*: the set $\mathcal{FGHUI}$.
1. Scan $\mathcal{D}'$ to determine TWU and the MPUN-lists of all items, and find
    $E \stackrel{\text{def}}{=} \{x \in \mathcal{A} \mid TWU(x) \geq mu \wedge supp(x) \geq ms\}$;   //Strategy 1
2. Remove irrelevant items in $\mathcal{A} \backslash E$ from $\mathcal{D}'$;
3. $\mathcal{FGHUI} := \emptyset$;
4. **for each** $x \in E$ **do**
5.   **Find-FGenHUI**(x, mu, ms, $\mathcal{FGHUI}$);
6. **return** $\mathcal{FGHUI}$;

Fig. 1. The MFG-HUI algorithm.

Find-FGenHUI(A, E, mu, ms, $\mathcal{FGHUI}$)
Input: the itemset A, set E, mu and ms thresholds and $\mathcal{FGHUI}$.
Output: $\mathcal{FGHUI}$ is updated.

1. **if** ($supp(A) < ms$ **or** $A.Prune\_Brh$ **or checkSubItemsets**(A, mu, $\mathcal{FGHUI}$)) **then**
2.   **return**;   //Theorems 2-3
3. **if** ($u(A) \geq mu$) **then UpdateFGHUI**(A, mu, $\mathcal{FGHUI}$);
4. **if** ($fwub(A) < mu$)) **then return**;   // Strategy 1
5. $x = lastItem(A); P = prefix(A); newE := \emptyset$;
6. **for each** $y \in E$ such that $y \succ x$ **do**
7.   **if** ($supp(Ay) \geq ms$) **then** $newE := newE \cup \{y\}$;
8. $List\_Exts\_of\_A = \emptyset$;
9. **for each** $y \in newE$ **do** {
10.   **Add** $C = Ay$ to $List\_Exts\_of\_A$ **and** construct the structure MPUN-list of C;
11.   **if** (($supp(A) = supp(C)$ **and** $lbu(A) \geq mu$) **or**
             ($supp(Py) = supp(C)$ **and** $lbu(Py) \geq mu$)) **then**
12. $C.Prune\_Brh =$ true;   //Theorem 3
13. }
14. **for each** $C \in List\_Exts\_of\_A$ **do**
15.   **Find-FGenHUI** (C, $NewE$, mu, ms, $\mathcal{FGHUI}$);

Fig. 2. The Find-FGenHUI procedure.

The MFG-HUI algorithm in Fig. 1 is designed to discover the set $\mathcal{FGHUI}$. It requires input of a QDB $\mathcal{D}'$ and two threshold values of mu and ms. The algorithm starts by scanning QDB $\mathcal{D}'$ to determine the TWU value and the MPUN-list of each item in the

$\mathcal{A}$ set, and find the E set of relevant items (line 1). Then, irrelevant items in $\mathcal{A}\backslash E$ are removed from $\mathcal{D}\prime$ to reduce the number of extended items (line 2). For each item x in E, it calls the Find-FGenHUI procedure (line 5). Finally, the algorithm returns the discovered set $\mathcal{FGHUI}$ as the output (line 6).

The Find-FGenHUI procedure depicted in Fig. 2 checks whether any of the pruning conditions at Line 1 are satisfied for the A itemset, where the checkExistingSubItemsets procedure is shown in Fig. 3. If yes, then the entire $Brh(A)$ branch is pruned and the procedure terminates. Next, if $u(A)$ is greater than or equal to mu, the UpdateFGHUI procedure in Fig. 4 is called to update the $\mathcal{FGHUI}$ set (line 3). The procedure also terminates if the condition $fwub(A) < mu$ is satisfied based on Strategy 1 (line 4). For each item y in the set $newE$, the procedure constructs the MPUN-list of $C = Ay$ (line 10). If the conditions at line 11 hold, the C itemset is marked for locally pruning according to Theorem 3. Finally, for each C of $List\_Exts\_of\_A$, the procedure recursively calls itself to discover larger FGHUIs (lines 14–15) (Fig. 3).

---

checkSubItemsets(B, mu, $\mathcal{FGHUI}$)
Input: the itemset B, threshold mu and $\mathcal{FGHUI}$; *Output*: *True* or *False*.
1. $Branch\_B\_is\_pruned = false$;
2.   **for each** *proper successive subset* R ($\in \mathcal{FGHUI}$) of B **do**   // *Remark 1*
3.   **if** $(supp(R) = supp(B))$ **then** {
4.   **if** $(lbu(R) \geq mu)$ **then** {
5. $Branch\_B\_is\_pruned = true$;   //*Theorem 2*
6.   **break**;
7.   }
8.   }
9.   **return** $Branch\_B\_is\_pruned$;

Fig. 3. The checkExistingSubItemsets procedure.

UpdateFGHUI(B, $\mathcal{FGHUI}$): *Input*: B and $\mathcal{FGHUI}$; *Output*: $\mathcal{FGHUI}$ is updated.
1. $\mathcal{FGHUI} := \mathcal{FGHUI} \cup \{B\}$;   // *B is an FGHU candidate*
2.   **for each** $S \in \mathcal{FGHUI}$ and $S \supset B$ **do**
3.   **if** $(supp(S) = supp(B))$ **then**
4.   **Remove** S from $\mathcal{FGHUI}$;   // *S is not an FGHUI*

Fig. 4. The UpdateFGHUI procedure.

4 Experimental Evaluation

This section presents a comparison between the proposed MFG-HUI algorithm and state-of-the-art HUCI-Miner (Gen) algorithm [14, 13] for discovering the $\mathcal{FGHUI}$ set on both real-life and synthetic datasets in Table 2, where HUCI-Miner (Gen) generates $\mathcal{FGHUI}$ from the set of all HUIs. The experiments were carried out on a Windows 10 computer with an Intel(R) Core(TM) i5–2400 CPU and 16 GB of RAM. All the algorithms were implemented in Java 7 SE, and the source code of HUCI-Miner (Gen) was obtained from the SPMF library [5]. Synthetic datasets in Table 2 were generated employing the IBM Quest data generator (obtained from [5]), which allows for the control of various parameters such as the average transaction length (T), average size of maximal frequent itemsets (I), number of distinct items (N), and number of transactions (in thousands) in the dataset (D).

4.1 Performance Evaluation of the Proposed Algorithm

Runtime. Figure 5 illustrates the runtime performance of the algorithms. The results show that MFG-HUI outperforms HUCI-Miner (Gen) with significant speed-ups of up to two orders of magnitude, particularly for dense datasets. For example, on Mushroom and Connect, MFG-HUI is approximately 105.2 and 50.4 times faster than HUCI-Miner (Gen) on average. The reason is that the proposed pruning strategies effectively reduce the search space by frequently satisfying pruning conditions. Additionally, MFG-HUI minimizes inclusion relationship checks, which are computationally costly. It is worth noting that HUCI-Miner (Gen) failed to complete within 5,000 s for certain *mu* values: 3% on Mushroom, 28% on Connect, and 1.8% on BMS. Therefore, their runtime and memory consumption are not shown in Fig. 5 and Fig. 6.

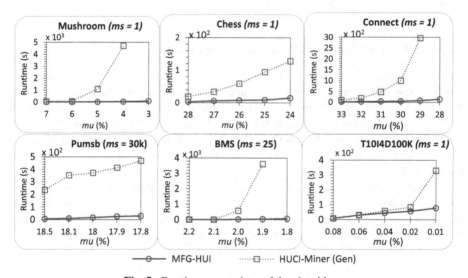

Fig. 5. Runtime comparison of the algorithms

Memory Usage. Figure 6 presents the memory usage evaluation of the algorithms in megabytes. The results show that MFG-HUI has lower peak memory consumption compared to HUCI-Miner (Gen) on all tested datasets. This can be attributed to the efficient pruning strategies employed by MFG-HUI, which utilize innovative techniques based on novel WUB and LB on the utility function and local pruning conditions.

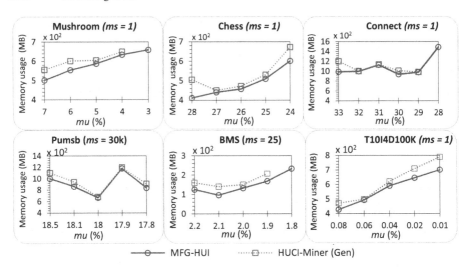

Fig. 6. Memory usage comparison of the algorithms.

These strategies allow for the quick elimination of non-generator high utility branches, resulting in the pruning of more unpromising candidates and smaller intermediate results compared to the HUCI-Miner (Gen) algorithm.

4.2 Scalability Evaluation of the Proposed Algorithm

To evaluate the scalability of the proposed algorithm, a second experiment was conducted involving the adjustment of parameters D and N for synthetic datasets. Figure 7 depicts the scalability of the algorithms in terms of runtime and memory usage, with the use of a (*) symbol indicating variations in the D and N parameters of the dataset. Synthetic datasets corresponding to different D and N values were generated using the IBM Quest Synthetic Data generator. Figure 7 reveals that when the values of the D and N parameters are increased, the proposed algorithm demonstrates superior scalability compared to HUCI-Miner (Gen). MFG-HUI displays efficient runtime and memory consumption, while HUCI-Miner (Gen) exhibits significant increases in both runtime and memory usage. Overall, MFG-HUI outperforms HUCI-Miner (Gen) in mining FGHUIs, making it a promising alternative for effectively discovering such itemsets.

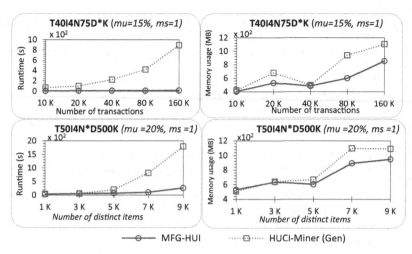

Fig. 7. Scalability of the algorithms for mining FGHUIs

5 Conclusions and Future Work

This paper presents an efficient method for discovering all FGHUIs and introduces a novel lower bound, *lbu*, for effective pruning of non-generator FHUIs. The P-NonGenHUI and LP-NonGenHUI pruning strategies are proposed to eliminate non-generator FHUI branches early in the mining process. Based on these contributions, a novel MFG-HUI algorithm is designed, which demonstrates faster mining, improved memory efficiency, and scalability compared to previous approaches in experimental evaluations. In the future, we plan to extend the theoretical results presented in this paper to address the problem of mining concise representations for patterns with different interestingness measures in quantitative databases.

Acknowledgment. This research is funded by Vietnam National Foundation for Science and Technology Development (NAFOSTED) under grant number 102.05–2021.52.

References

1. Dawar, S., et al.: A hybrid framework for mining high-utility itemsets in a sparse transaction database. Appl. Intell. **47**, 809–827 (2017)
2. Duong, H., et al.: Efficient algorithms for mining closed and maximal high utility itemsets. Knowl. Based Syst. **257**, 109921 (2022)
3. Fournier-Viger, P., et al.: EFIM-closed : fast and memory efficient discovery of closed high-utility itemsets. In: International Conference on Machine Learning and Data Mining in Pattern Recognition. pp. 199–213 (2016)
4. Fournier-Viger, P. et al.: Novel concise representations of high utility itemsets using generator patterns. In: International Conference on Advanced Data Mining and Applications. pp. 30–43 (2014)

5. Fournier-Viger, P., et al.: SPMF: a java open-source pattern mining library. J. Mach. Learn. Res. **15**, 3569–3573 (2014)
6. Krishnamoorthy, S.: HMiner: efficiently mining high utility itemsets. Expert Syst. Appl. **90**, 168–183 (2017)
7. Lan, G.C., et al.: An efficient projection-based indexing approach for mining high utility itemsets. Knowl. Inf. Syst. **38**(1), 85–107 (2014)
8. Li, J. et al.: Minimum description length principle: generators are preferable to closed patterns. In: Proceedings of the 21st National Conference on Artificial intelligence, AAAI 2006. pp. 409–414 (2006)
9. Liu, M., Qu, J.: Mining high utility itemsets without candidate generation. In: Proceedings of ACM International Conference on Information and Knowledge Management. pp. 55–64 (2012)
10. Liu, Y., et al.: Mining high utility itemsets based on pattern growth without candidate generation. Mathematics **9**(1), 1–22 (2021)
11. Mai, T. et al.: Efficient algorithm for mining non-redundant high-utility association rules. Sensors (Switzerland). **20**(4) (2020)
12. Ryang, H., Yun, U.: Indexed list-based high utility pattern mining with utility upper-bound reduction and pattern combination techniques. Knowl. Inf. Syst. **51**, 627–659 (2017)
13. Sahoo, J., et al.: An Algorithm for Mining High Utility Closed Itemsets and Generators. ArXiv. abs/1410.2, 1–18 (2014)
14. Sahoo, J., et al.: An efficient approach for mining association rules from high utility itemsets. Expert Syst. Appl. **42**(13), 5754–5778 (2015)
15. Shie, B.E., et al.: Mining interesting user behavior patterns in mobile commerce environments. Appl. Intell. **38**(3), 418–435 (2013)
16. Truong, T., et al.: Efficient vertical mining of high average-utility itemsets based on novel upper-bounds. IEEE Trans. Knowl. Data Eng. **31**(2), 301–314 (2018)
17. Tseng, V.S., et al.: Efficient algorithms for mining high utility itemsets from transactional databases. IEEE Trans. Knowl. Data Eng. **25**(8), 1772–1786 (2013)
18. Tseng, V.S., et al.: Efficient algorithms for mining the concise and lossless representation of high utility itemsets. IEEE Trans. Knowl. Data Eng. **27**(3), 726–739 (2015)

Negative Sentiments Make Review Sentences Longer: Evidence from Japanese Hotel Review Sites

Takumi Kato[(⊠)] [iD]

School of Commerce, Meiji University, Tokyo, Japan
takumi_kato@meiji.ac.jp

Abstract. Existing research on the psychology of the posters of customer reviews has mainly focused on their motivation for posting. However, there is little discussion about understanding the feelings of contributors based on review characteristics. This study used big data from a hotel reservation website Rakuten Travel, provided by Rakuten Group, Inc., a major Japanese IT company, and investigated the relationship between the number of characters in reviews and ratings. A multiple regression analysis showed that, the more words in the review, the lower the overall rating. Furthermore, the lower the rating of the individual items (location, room, meal, bath, service), the greater the negative effect of the number of characters in the review on the overall rating. Similarly, when only negative expressions were detected, the negative effect of review word count on overall rating was greater. Practitioners should recognize that customers are more likely to communicate negative than positive emotions. Consumers are less likely to express their emotional attitudes through writing than speaking. This is because, in the process of writing, there is more time to ponder on things to say and less emotion. Therefore, a strong negative feeling is associated with posting a long sentence with considerable effort. Practitioners should include both rating figures and review characteristics as variables in customer churn prediction models. It is effective for customer understanding to identify the generation mechanism for review features that cannot be comprehended at a glance.

Keywords: Customer Review · Sentiment · Word-of-Mouth · Hotel Industry · Big Data

1 Introduction

Reliability is the most important factor for customers when making online purchases, given that they cannot actually touch the products or services. An information source that customers trust is review or word-of-mouth (hereafter, review) [1]. Customer reviews reduce uncertainty about product or service quality and serve as a guide for customer attitudes and behavior [2, 3]. This is because they do not have commercial intentions, and cover both positive and negative aspects [4]. In particular, customers find negative reviews useful, and such posts influence their purchasing behavior [5].

© The Author(s), under exclusive license to Springer Nature Switzerland AG 2024
K. Honda et al. (Eds.): IUKM 2023, LNAI 14376, pp. 281–293, 2024.
https://doi.org/10.1007/978-3-031-46781-3_24

Based on the background, the success of web service platforms is highly dependent on customer reviews [6]. Practitioners are well aware of this, and use customer reviews to create purchasing experiences [7]. For example, product recommendations based on rating valence on a 5-star scale enhances the quality of the consumer's purchasing experience [8]. In other words, the source of competitiveness for a company is to understand customer reviews that are helpful for customers and utilize that knowledge in service development [9]. Hence, owing to the enormous influence of customer reviews, academic research on the topic has been given great importance over the past several decades [10].

Most existing research has focused on the impact and features of customer reviews on their readers, but understanding the posters of customer reviews is also important [11]. Understanding the underlying psychological mechanisms of people who post customer reviews can help companies empathize with their consumers and use their messages to co-create value [12, 13]. Existing research on the psychology of posters has mainly focused on their motivation for posting. However, there are few discussions about understanding the feelings of contributors from reviews. As it is difficult for customers to describe all of their experiences in writing, an approach that understands the psychology of the review characteristics is necessary. Therefore, this study expands knowledge of the psychology of review posters and clarifies the relationship between ratings and review length for budget hotel reservation websites in Japan.

2 Related Work and Hypothesis Development

2.1 Motivation for Posting Customer Reviews

The motivations for posting positive or negative reviews are different [11]. There are two main motivations for posting positive reviews: i) brand support: a high level of commitment to a brand motivates people to talk and interact with the brand [14, 15], and efforts to spread positive experiences are motivated by the customer's desire to help the brand [16]; and ii) self-enhancement: customers motivated by self-enhancement seek the opportunity to gain prestige by spreading positive reviews [17]. In other words, reviews are an effective means of advertising one's superiority over others [18]. From the perspective of the information receiver, the review of a trustworthy person is easily accepted as a sign of their expertise and usefulness of the information [19].

There are two main motivations for posting negative reviews: i) helping others: experience-based complaints strongly influence the writing of negative reviews [20] because communicating negative information can protect others [21]; and ii) revenge on the company: customers who have had a bad experience want revenge on the company and reduce their frustration and anxiety by sharing their bad consumption experiences [17].

2.2 Hypothesis Development

The features of reviews include rating valence [22], rating distribution [23], review length [24], and positive or negative expression [25]. The present study focuses on review length and positive or negative expression. Emotions embedded in reviews capture the reader's attention [26]. For example, for books on Amazon, the impact of a 1-star review is greater than that of a 5-star review [5]. In other words, negative information is attractive to customers; accordingly, the motivation of contributors is high. In addition, the motivation for negative reviews includes a strong feeling of revenge [17]. Accordingly, negative emotions are believed to lengthen review sentences, and the following hypotheses were derived:

H1: The more words in the review, the lower the overall rating.

H2: The lower the rating of the individual items (location, room, meal, bath, service), the greater the negative impact of the number of words in the review on the overall rating.

H3: The more negative expressions that are detected, the greater the negative impact of the number of words in the review on the overall rating.

3 Methodology

3.1 Data

This study used review data from a hotel reservation website Rakuten Travel [27] provided by Rakuten Group, Inc., a major Japanese IT company. Data of 82,824 reviews from approximately three years between January 1, 2017 and December 19, 2019 were adopted. The following six items were mainly included in the data: (a) overall rating (5-point Likert scale), (b) item rating (location, room, meal, bath, service; 5-point Likert scale), (c) purpose of use of the hotel (leisure or business), (d) companion, (e) hotel brand, and (f) review text. These data do not include the posters' personal information or attribute information. The purposes of hotel use were equally divided between leisure and business (Table 1). The target was 10 major budget hotel brands in Japan. Figure 1 shows the average overall rating (5-point Likert scale) for each brand.

Table 1. Breakdown of review data.

Item	Content	Number of Respondents	Composition Ratio
Purpose	Leisure	40,912	49.4%
	Business	41,912	50.6%
Fellow Travelers	Single	54,992	66.4%
	Coworker	4,793	5.8%
	Family	18,602	22.5%
	Friend	2,244	2.7%
	Partner	2,193	2.6%
Hotel Brand	Apa	13,297	16.1%
	Comfort	16,775	20.3%
	Daiwa_Roynet	5,512	6.7%
	Dormy_Inn	13,020	15.7%
	Keio_Presso	2,999	3.6%
	Mets	2,335	2.8%
	Richmond	6,404	7.7%
	Route_Inn	17,455	21.1%
	Sotetsu_Fresa	2,956	3.6%
	Sunroute	2,071	2.5%

3.2 Verification

The mean number of words in review text was 95.589 (Table 2). However, the distribution was skewed to the left, and very few reviews of 500 words or more were available (Fig. 2). Therefore, two types of evaluation were used: raw data (82,824 records) and screened data (27,185 records); only reviews with 100–500 words were extracted for the latter. This makes it possible to present the verification results in their original condition and as well-organized data.

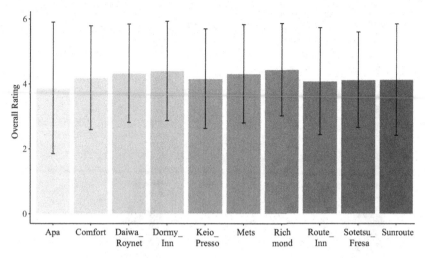

Fig. 1. Mean satisfaction rating for each hotel brand.

Table 2. Review word statistics.

	Sample Size	Mean	SD	Median	Min	Max
Raw Data	82,824	95.589	96.283	66	1	997
Screened Data	27,185	185.366	81.465	159	100	500

For H1–H3, overall rating (5-point Likert scale) was used as the objective variable. In the verification of H2, the mean value of item ratings (location, room, meal, bath, service) were used as conditional variables. Furthermore, H3 validation requires (f) extracting positive or negative expressions from the review text. Therefore, as shown in Table 3, eight expressions were detected via text mining using MeCab, an open-source Japanese morphological analysis engine; 22.2% of raw data and 32.8% of screened data contained positive expressions, and 10.9% of raw data and 20.7% of screened data contained negative expressions.

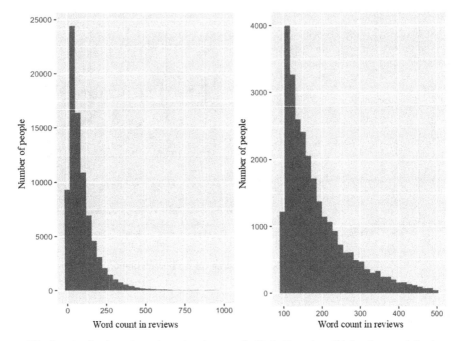

Fig. 2. Distribution of number of review words (Left: Raw data, Right: Screened data).

Table 3. Detection words for positive and negative expressions.

Word	Positive	Negative
1	Good	Bad
2	Satisfied	Dissatisfied
3	Excellent	Awful
4	Wonderful	Horrible
5	Nice	Terrible
6	Glad	Sad
7	Moving	Disappointing
8	Pleasant	Unpleasant

A multiple regression analysis including interactions using the variables shown in Table 4 was used to test the hypotheses. In the verification of H1, Model 1 was used with overall rating (No. 1) as the objective variable, word count in reviews (No. 28) as the explanatory variable, and control variables (No. 8–25). The word count in reviews had the large value of 1–997, so it was incorporated into the model in a standardized state. Owing to the large number of variables, variable selection was performed by a stepwise method based on AIC. In the verification of H2, overall rating (No. 1) was used

Table 4. Variable list.

No	Variable	Description	Data Type	Raw Data		Screened Data	
				Mean	SD	Mean	SD
1	Rating	Evaluation for overall rating	5-point Likert scale	4.179	0.845	4.125	0.969
2	R1_Location	Rating for hotel locations	5-point Likert scale	4.289	0.806	4.332	0.820
3	R2_Room	Rating for hotel rooms	5-point Likert scale	4.103	0.915	4.074	1.016
4	R3_Meal	Rating for hotel meals	5-point Likert scale	3.968	0.969	3.965	1.048
5	R4_Bath	Rating for hotel baths	5-point Likert scale	3.894	0.956	3.881	1.029
6	R5_Service	Rating for hotel services	5-point Likert scale	4.039	0.888	4.049	0.997
7	Mean_Rating	Mean of each item rating	Mean of Nos.2–6	4.059	0.681	4.060	0.725
8	Leisure	Dummy of the leisure purpose	0/1	0.494	0.500	0.626	0.484
9	Single *	Dummy of the single use	0/1	0.664	0.472	0.581	0.493
10	Coworker	Dummy accompanying colleagues	0/1	0.058	0.233	0.037	0.190
11	Family	Dummy accompanying family	0/1	0.225	0.417	0.316	0.465
12	Friend	Dummy accompanying friends	0/1	0.027	0.162	0.034	0.180
13	Partner	Dummy accompanying a partner	0/1	0.026	0.161	0.032	0.177

(continued)

Table 4. (*continued*)

No	Variable	Description	Data Type	Raw Data		Screened Data	
				Mean	SD	Mean	SD
14	Room_Double	Dummy of double rooms	0/1	0.444	0.497	0.493	0.500
15	SmokingNG	Dummy of non-smoking rooms	0/1	0.793	0.405	0.819	0.385
16	Apa *	Hotel Brand Dummy	0/1	0.161	0.367	0.159	0.365
17	Comfort		0/1	0.203	0.402	0.215	0.411
18	Daiwa_Roynet		0/1	0.067	0.249	0.056	0.231
19	Dormy_Inn		0/1	0.157	0.364	0.176	0.381
20	Keio_Presso		0/1	0.036	0.187	0.036	0.186
21	Mets		0/1	0.028	0.166	0.028	0.164
22	Richmond		0/1	0.077	0.267	0.079	0.270
23	Route_Inn		0/1	0.211	0.408	0.193	0.395
24	Sotetsu_Fresa		0/1	0.036	0.186	0.033	0.179
25	Sunroute		0/1	0.025	0.156	0.026	0.158
26	Positive	Positive word mention dummy	0/1	0.222	0.416	0.328	0.469
27	Negative	Negative word mention dummy	0/1	0.109	0.311	0.207	0.405
28	Word_Count	Word count in reviews	Number	95.589	96.283	185.366	81.465

Note: * Criteria for dummy variables; SD means standard deviation

as the objective variable, the interaction of word count in reviews (No. 28) and mean of each item rating (No. 7) were used as explanatory variables, and (No. 8–25) was used as the control variable. The verification of H3 was a similar procedure. First, Model 3 was constructed with overall rating (No. 1) as the objective variable, word count in reviews (No. 28) as the explanatory variable, and control variables (No. 2–6, 8–25). After that, H3 was verified by Model 4, which added the interaction of word count in reviews (No. 28) and positive or negative word mention dummy (No. 26–27) to Model 3. Models 2 and 4 were validated on both raw and screened data. The significance level was 5%.

4 Results

The results of Model 1 show that the overall rating is higher for leisure trips than for business and for trips with family and friends rather than trips alone. Mean_Rating—which is the mean value of individual items (location, room, meal, bath, service)—has a positive correlation with overall rating, and Word_Count—which is the number of words in a review—has a negative correlation with overall rating. In other words, the more words in the review, the lower the overall rating; hence, H1 was supported. The adjusted R-squared is 0.664, indicating high compatibility (Table 5).

Table 5. Interaction model of mean rating and word count in reviews ($***p < 0.001$; $**p < 0.01$; $*p < 0.05$).

Variable	Model 1		Model 2		Model 2'		Model 3		Model 4		Model 4'	
Intercept	0.032	**	4.111	***	4.025	***	0.229	***	0.227	***	-0.077	***
R1_Location							0.122	***	0.122	***	0.108	***
R2_Room							0.303	***	0.303	***	0.303	***
R3_Meal							0.152	***	0.152	***	0.161	***
R4_Bath							0.096	***	0.096	***	0.110	***
R5_Service							0.288	***	0.287	***	0.338	***
Leisure	0.016	***	0.013	**	0.035	***	0.016	***	0.016	***	0.034	***
Family	0.021	***	0.021	***	0.027	**	0.025	***	0.026	***	0.031	***
Friend	0.022	*	0.025	*	0.027		0.019		0.021		0.026	
Partner	0.040	***	0.044	***	0.049	*	0.038	***	0.038	***	0.043	*
Comfort	0.126	***	0.120	***	0.140	***	0.057	***	0.057	***	0.062	***
Daiwa_Roynet	0.062	***	0.060	***	0.049	**	0.022	**	0.021	**	0.013	
Keio_Presso	0.196	***	0.190	***	0.225	***	0.119	***	0.119	***	0.134	***
Mets	0.063	***	0.058	***	0.067	**	0.041	***	0.041	***	0.048	*
Richmond	0.107	***	0.103	***	0.096	***	0.048	***	0.048	***	0.030	*
Route_Inn	0.029	***	0.024	***	0.054	***	0.014	*	0.014	*	0.028	**
Sotetsu_Fresa	0.136	***	0.133	***	0.198	***	0.102	***	0.102	***	0.175	***
Sunroute	0.078	***	0.074	***	0.104	***	0.055	***	0.056	***	0.075	***
Mean_Rating	1.004	***	0.995	***	1.087	***						
Positive							0.061	***	0.053	***	0.071	***
Negative							-0.161	***	-0.154	***	-0.126	***
Word_Count	-0.056	***	-0.051	***	-0.053	***	-0.043	***	-0.045	***	-0.041	***
Mean_Rating * Word_Count			0.068	***	0.052	***						
Positive * Negative									0.071	***	0.060	***
Positive * Word_Count									0.032	***	0.020	*
Negative * Word_Count									-0.012	**	-0.025	**

(*continued*)

Table 5. (*continued*)

Variable	Model 1		Model 2		Model 2'		Model 3		Model 4		Model 4'	
Intercept	0.032	**	4.111	***	4.025	***	0.229	***	0.227	***	−0.077	***
Positive * Negative * Word_Count									−0.009		0.004	
Adjusted R-squared	0.664		0.668		0.682		0.688		0.688		0.709	

Next, based on Model 2, a significant interaction was detected between Mean_Rating and Word_Count. Similar results were confirmed for Model 2'. As shown on the left side of Fig. 3, as a result of a simple slope analysis in Model 2', a low Mean_Rating (mean - 1 standard deviation) has a greater negative impact on the overall rating than a high Mean_Rating (mean + 1 standard deviation). Therefore, H2 was supported. The variance inflation factor ranged from 1.0200 to 1.8927, with no multicollinearity, and the adjusted R-squared of 0.668 was considered good.

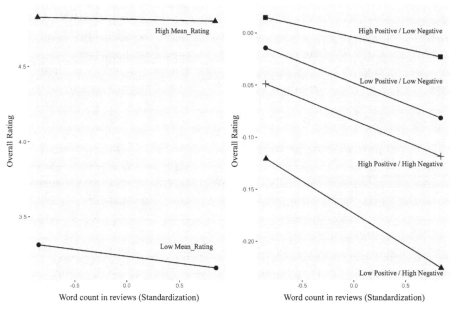

Fig. 3. Results of simple slope analysis (Left: Mean rating in Model 2', Right: Positive/negative expressions in Model 4').

Finally, positive or negative expressions were considered. Regarding the base Model 3, when positive expressions are detected, the overall rating obtains a high score; and when negative expressions are detected, the overall rating obtains a low score. Model 4, which added interaction to Model 3, detected a positive significant effect on positive expressions and Word_Count, and a negative significant effect on negative expressions and Word_Count. Similar results were obtained for Model 4'. As shown on the right side of Fig. 3, a simple slope analysis in Model 4' showed that the negative effect on the overall rating is the largest when only the negative is high. Therefore, H3 was supported.

5 Implications and Future Work

5.1 Theoretical Implications

Most existing research on customer reviews has focused on their impact on readers [11], but it is also essential to understand the underlying psychological mechanisms of those who post reviews [12, 13]. Existing research on the psychology of posters has mainly focused on their motivation for posting. The main motives for posting positive reviews are brand support [14–16] and self-enhancement [17–19], whereas the main motives for posting negative reviews are helping others [20, 21] and revenge on companies [17]. However, there is little discussion about understanding the feelings of posters based on review features. As it is difficult for customers to describe all of their experiences in writing, an approach that understands the psychology of the review characteristics is necessary. This study is among the few to adopt this approach. A negative relationship was observed between the number of review words and the overall rating on Japanese hotel reservation websites. In other words, negative emotions are stronger motivation for detailing experiences in reviews than positive emotions. Furthermore, negative content has a greater impact from the reader's perspective [5], and this study extended this finding to the poster's perspective.

5.2 Practical Implications

This study mainly provides two practical implications. First, practitioners should recognize that customers have a stronger desire to let others know about negative than positive emotions. Consumers are less likely to express their emotional attitudes through writing than speaking. This is because, in the process of writing, there is more time to ponder on things to say and less emotion [28]. A strong negative feeling is associated with posting a long sentence with considerable effort. Therefore, when a customer has a negative experience, it is necessary to take care of them immediately and make efforts to eliminate the negative feelings. Practitioners should remember that, while positive reviews are perceived as free advertising for businesses, undesirable comments can seriously damage a company's image and reputation [29]. Second, companies should include both rating numbers and review characteristics in the customer defection prediction model. Detecting positive or negative wording in reviews for prediction model is common. However, as shown in this study, understanding that negative emotions are associated with the length of review texts and efforts to incorporate this feature into the model are still insufficient. It is effective for customer understanding to identify the generation mechanism of review features that cannot be comprehended at a glance.

5.3 Limitations and Future Work

This study has several limitations. First, the results of this study are limited to Japanese hotel reservation websites. As big data of more than 80,000 cases were used, the findings have high reliability. However, it is necessary to confirm whether they are applicable to other countries and industries. Second, the present study did not consider features of customer reviews other than the length. For example, high-loyalty customers emphasize elements related to the brand concept rather than secondary elements of products or services [30]. Hence, negative feelings may be stronger when the negative part in the review is related to the concept of the target brand. These are topics for future research.

Acknowledgement. This study used "Rakuten Dataset" (https://rit.rakuten.com/data_release/) provided by Rakuten Group, Inc. Via IDR Dataset Service of National Institute of Informatics.

References

1. Wan, Y., Ma, B., Pan, Y.: Opinion evolution of online consumer reviews in the e-commerce environment. Electron. Commer. Res. **18**(2), 291–311 (2018). https://doi.org/10.1007/s10660-017-9258-7
2. Roy, G., Datta, B., Mukherjee, S.: Role of electronic word-of-mouth content and valence in influencing online purchase behavior. J. Mark. Commun. **25**(6), 661–684 (2019). https://doi.org/10.1080/13527266.2018.1497681
3. Reimer, T., Benkenstein, M.: When good WOM hurts and bad WOM gains: the effect of untrustworthy online reviews. J. Bus. Res. **69**(12), 5993–6001 (2016). https://doi.org/10.1016/j.jbusres.2016.05.014
4. Filieri, R., Mariani, M.: The role of cultural values in consumers' evaluation of online review helpfulness: a big data approach. Int. Mark. Rev. **38**(6), 1267–1288 (2021). https://doi.org/10.1108/IMR-07-2020-0172
5. Chevalier, J.A., Mayzlin, D.: The effect of word of mouth on sales: online book reviews. J. Mark. Res. **43**(3), 345–354 (2006). https://doi.org/10.1509/jmkr.43.3.345
6. Liu, Y., Pang, B.: A unified framework for detecting author spamicity by modeling review deviation. Expert Syst. Appl. **112**, 148–155 (2018). https://doi.org/10.1016/j.eswa.2018.06.028
7. Barton, B.: Ratings, reviews & ROI: how leading retailers use customer word of mouth in marketing and merchandising. J. Interact. Advert. **7**(1), 5–50 (2006). https://doi.org/10.1080/15252019.2006.10722125
8. Lin, Z.: An empirical investigation of user and system recommendations in e-commerce. Decis. Support. Syst. **68**, 111–124 (2014). https://doi.org/10.1016/j.dss.2014.10.003
9. Filieri, R.: What makes online reviews helpful? a diagnosticity-adoption framework to explain informational and normative influences in e-WOM. J. Bus. Res. **68**(6), 1261–1270 (2015). https://doi.org/10.1016/j.jbusres.2014.11.006
10. Chu, S.C., Kim, J.: The current state of knowledge on electronic word-of-mouth in advertising research. Int. J. Advert. **37**(1), 1–13 (2018). https://doi.org/10.1080/02650487.2017.1407061
11. Nam, K., Baker, J., Ahmad, N., Goo, J.: Determinants of writing positive and negative electronic word-of-mouth: empirical evidence for two types of expectation confirmation. Decis. Support. Syst. **129**, 113168 (2020). https://doi.org/10.1016/j.dss.2019.113168
12. Chen, Z., Yuan, M.: Psychology of word of mouth marketing. Curr. Opin. Psychol. **31**, 7 (2020). https://doi.org/10.1016/j.copsyc.2019.06.026

13. Fine, M.B., Gironda, J., Petrescu, M.: Prosumer motivations for electronic word-of-mouth communication behaviors. J. Hosp. Tour. Technol. **8**(2), 280–295 (2017). https://doi.org/10.1108/JHTT-09-2016-0048

14. Wolny, J., Mueller, C.: Analysis of fashion consumers' motives to engage in electronic word-of-mouth communication through social media platforms. J. Mark. Manag. **29**(5–6), 562–583 (2013). https://doi.org/10.1080/0267257X.2013.778324

15. Loureiro, S.M.C., Kaufmann, H.R.: The role of online brand community engagement on positive or negative self-expression word-of-mouth. Cogent Bus. Manag. **5**(1), 1508543 (2018). https://doi.org/10.1080/23311975.2018.1508543

16. Jeong, E., Jang, S.S.: Restaurant experiences triggering positive electronic word-of-mouth (eWOM) motivations. Int. J. Hosp. Manag. **30**(2), 356–366 (2011). https://doi.org/10.1016/j.ijhm.2010.08.005

17. Sohaib, M., Akram, U., Hui, P., Rasool, H., Razzaq, Z., Kaleem Khan, M.: Electronic word-of-mouth generation and regulatory focus. Asia Pac. J. Mark. Logist. **32**(1), 23–45 (2020). https://doi.org/10.1108/APJML-06-2018-0220

18. Ruvio, A., Bagozzi, R.P., Hult, G.T.M., Spreng, R.: Consumer arrogance and word-of-mouth. J. Acad. Mark. Sci. **48**, 1116–1137 (2020). https://doi.org/10.1007/s11747-020-00725-3

19. Packard, G., Gershoff, A.D., Wooten, D.B.: When boastful word of mouth helps versus hurts social perceptions and persuasion. J. Cons. Res. **43**(1), 26–43 (2016). https://doi.org/10.1093/jcr/ucw009

20. Nam, K., Baker, J., Ahmad, N., Goo, J.: Dissatisfaction, disconfirmation, and distrust: an empirical examination of value co-destruction through negative electronic word-of-mouth (eWOM). Inf. Syst. Front. **22**, 113–130 (2020). https://doi.org/10.1007/s10796-018-9849-4

21. Dubois, D., Bonezzi, A., De Angelis, M.: Sharing with friends versus strangers: how interpersonal closeness influences word-of-mouth valence. J. Mark. Res. **53**(5), 712–727 (2016). https://doi.org/10.1509/jmr.13.0312

22. Karabas, I., Kareklas, I., Weber, T.J., Muehling, D.D.: The impact of review valence and awareness of deceptive practices on consumers' responses to online product ratings and reviews. J. Mark. Commun. **27**(7), 685–715 (2021). https://doi.org/10.1080/13527266.2020.1759120

23. Kato, T.: Rating valence versus rating distribution: perceived helpfulness of word of mouth in e-commerce. SN Bus. Econ. **2**(11), 162, 1–24 (2022). https://doi.org/10.1007/s43546-022-00338-8

24. Pan, Y., Zhang, J.Q.: Born unequal: a study of the helpfulness of user-generated product reviews. J. Retail. **87**(4), 598–612 (2011). https://doi.org/10.1016/j.jretai.2011.05.002

25. Lo, A.S., Yao, S.S.: What makes hotel online reviews credible? an investigation of the roles of reviewer expertise, review rating consistency and review valence. Int. J. Contemp. Hosp. Manag. **31**(1), 41–60 (2019). https://doi.org/10.1108/IJCHM-10-2017-0671

26. Ismagilova, E., Dwivedi, Y.K., Slade, E.: Perceived helpfulness of eWOM: emotions, fairness and rationality. J. Retail. Consum. Serv. **53**, 101748 (2020). https://doi.org/10.1016/j.jretconser.2019.02.002

27. Rakuten Travel. Rakuten Travel. https://travel.rakuten.com/. Accessed 1 Mar 2023

28. Berger, J., Rocklage, M.D., Packard, G.: Expression modalities: how speaking versus writing shapes word of mouth. J. Cons. Res. **49**(3), 389–408 (2022). https://doi.org/10.1093/jcr/ucab076

29. Chen, Y.F., Law, R.: A review of research on electronic word-of-mouth in hospitality and tourism management. Int. J. Hosp. Tour. Adm. **17**(4), 347–372 (2016). https://doi.org/10.1080/15256480.2016.1226150

30. Kato, T.: Brand loyalty explained by concept recall: recognizing the significance of the brand concept compared to features. J. Mark. Analy. **9**(3), 185–198 (2021). https://doi.org/10.1057/s41270-021-00115-w

Extending Kryszkiewicz's Formula of Missing Values in Terms of Lipski's Approach

Michinori Nakata[1]([✉]), Norio Saito[1], Hiroshi Sakai[2], and Takeshi Fujiwara[3]

[1] Faculty of Management and Information Science, Josai International University,
1 Gumyo, Togane, Chiba 283-8555, Japan
nakatam@ieee.org, saitoh_norio@jiu.ac.jp
[2] Department of Mathematics and Computer Aided Sciences, Faculty of Engineering,
Kyushu Institute of Technology, Tobata, Kitakyushu 804-8550, Japan
sakai@mns.kyutech.ac.jp
[3] Faculty of Informatics, Tokyo University of Information Sciences, 4-1 Onaridai,
Wakaba-ku, Chiba 265-8501, Japan
fujiwara@rsch.tuis.ac.jp

Abstract. When extracting any information from a data table with incomplete information, following Lipski we only know the lower and the upper bound of the information. Methods of rough sets that are applied to data tables containing incomplete information are examined from the viewpoint of Lipski's approach based on possible world semantics. It is clarified that the formula that is first used by Kryszkiewicz, which most of the authors use, only gives the lower bound of the lower approximation and the upper bound of the upper approximation. When we focus on a value, this is because it is only considered that a missing value can be equal to that value. We extend Kryszkiewicz's formula to consider that the missing value may not be equal to the value. As a result, the extended Kryszkiewicz's formula gives the same approximations as those in terms of Lipski's approach.

Keywords: rough sets · incomplete information · missing values · lower and upper approximations · possible world semantics

1 Introduction

The framework of rough sets, proposed by Pawlak [1], is used as an effective tool in the field of data mining and related topics. In the rough sets a pair of lower and upper approximations to a target set, which correspond to inclusion and intersection operations, are derived using the indiscernibility of objects. The indiscernibility is derived from that values characterizing objects are equal.

The classical framework is constructed under data tables containing only complete information. As a matter of fact, it is well-known that real data tables usually contain incomplete information [2,3]. Lots of authors, therefore, have

K. Honda et al. (Eds.): IUKM 2023, LNAI 14376, pp. 294–305, 2024.
https://doi.org/10.1007/978-3-031-46781-3_25

dealt with data tables with incomplete information [4–14]. The method that most authors use is based on the formula proposed by Kryszkiewicz [7,15] and its extended versions are proposed [5,9,13,16–18]. These have the common characteristic of deriving only a unique pair of lower and upper approximations to a target set of objects by giving the indiscernibility between values containing missing values. However, this characteristic is incompatible with Lipski's approach based on possible world semantics. According to Lipski [19], we cannot obtain the actual answer to a query but can have nothing to obtain the lower and the upper bound of the actual answer. In other words, what we obtain is the lower and the upper bound of the lower approximation and the lower and the upper bound of the upper approximation. Therefore, we cannot say that the lower and the upper approximation obtained under Kryszkiewicz's formula are correct.

Our objective is to extend Kryszkiewicz's formula so that the results of the formula matches Lipski's approach. In this paper, we first check approximations derived from the viewpoint of methods of possible worlds. And then we extend Kryszkiewicz's formula by showing an indiscernibility relation compatible with Lipski's approach.

The paper is organized as follows. In Sect. 2, the traditional approach of Pawlak is briefly addressed under a data table with complete information. In Sect. 3, we develop Lipski's approach under a data table with incomplete information. In Sect. 4, we check approximations derived from Kryszkiewicz's formula. In Sect. 5, we clarify the relationship between Kryszkiewicz's formula and Lipski's approach. In Sect. 6, we address conclusions.

2   Pawlak's Approach

A data set is represented as a two-dimensional table, called an information table, where each row represents an object and each column represents an attribute. The information table is expressed by $(U, AT, \cup_{a \in AT} V_a)$, where U is a non-empty finite set of objects called the universe and AT is a non-empty finite set of attributes such that $\forall a \in AT : U \rightarrow V_a$. Set V_a is called the domain of attribute a. Binary relation I_a of indiscernibility for objects in U on attribute a is:

$$I_a = \{(o, o') \in U \times U \mid a(o) = a(o')\}, \tag{1}$$

where $a(o)$ is the value of attribute a for o. This relation, called an indiscernibility relation, is reflexive, symmetric, and transitive. From the indiscernibility relation, equivalence class $E(o)_a$ $(= \{o' \mid (o, o') \in I_a\})$ containing object o is obtained. This is also the set of objects that is indiscernible with object o, called the indiscernible class containing o on a. Finally, family FE_a $(= \cup_{o \in U} \{E(o)_a\})$ of equivalence classes is derived from the indiscernibility relation. Lower approximation $\underline{apr}(T)_a$ and upper approximation $\overline{apr}(T)_a$ of target set T of objects by FE_a are:

$$\underline{apr}(T)_a = \{o \in U \mid E(o)_a \subseteq T\}, \tag{2}$$

$$\overline{apr}(T)_a = \{o \in U \mid E(o)_a \cap T \neq \emptyset\}. \tag{3}$$

Example 1. Let complete information table CT be obtained as follows:

$$CT$$

O	a_1	a_2
1	z	2
2	y	1
3	x	1
4	x	1
5	y	2

In information table CT, $U = \{o_1, o_2, o_3, o_4, o_5\}$. Domains V_{a_1} and V_{a_2} of attributes a_1 and a_2 are $\{x, y, z\}$ and $\{1, 2\}$, respectively. Indiscernibility relation I_{a_1} on a_1 is:

$$I_{a_1} = \{(o_1, o_1), (o_2, o_2), (o_2, o_5), (o_3, o_3), (o_3, o_4), (o_4, o_4), (o_4, o_3), (o_5, o_5), (o_5, o_2)\}.$$

Equivalence classes containing each object on a_1 are:

$$E(o_1)_{a_1} = \{o_1\},$$
$$E(o_2)_{a_1} = E(o_5)_{a_1} = \{o_2, o_5\},$$
$$E(o_3)_{a_1} = E(o_4)_{a_1} = \{o_3, o_4\}.$$

Family FE_{a_1} of equivalence classes on a_1 is:

$$FE_{a_1} = \{\{o_1\}, \{o_2, o_5\}, \{o_3, o_4\}\}.$$

Let target set T be $\{o_2, o_3, o_4\}$. Lower approximation $\underline{apr}(T)_{a_1}$ and upper approximation $\overline{apr}(T)_{a_1}$ by FE_{a_1} are:

$$\underline{apr}(T)_{a_1} = \{o_3, o_4\},$$
$$\overline{apr}(T)_{a_1} = \{o_2, o_3, o_4, o_5\}.$$

3   Lipski's Approach Based on Possible World Semantics

Lipski [19] used the set of possible tables derived from an information table containing missing values, called an incomplete information table, by using possible world semantics. Following in Lipski's footsteps, we first obtain the set of possible tables from an incomplete information table. A possible table on an attribute is a table in which each missing value is replaced with a value in the corresponding domain. The approach addressed in the previous section is applied to each possible table. When missing values exist on attribute a in incomplete information table IT, set PT_a of possible tables on a is:

$$PT_a = \{pt_{a,1}, \ldots, pt_{a,n}\}, \tag{4}$$

where every possible table $pt_{a,i}$ has an equal possibility that it is actual, number n of possible tables is equal to $|V_a|^{m_a}$, the number of missing values is m_a on attribute a, and $|V_a|$ is the cardinality of domain V_a.

Attribute a has a value in V_a in all possible tables. Possible indiscernibility relation $PI_{a,i}$ is derived from possible table $pt_{a,i}$.

$$PI_{a,i} = \{(o, o') \in U \times U \mid a(o)_i = a(o')_i\}, \tag{5}$$

where $a(o)_i$ is the value of attribute a for o in $pt_{a,i}$. Possible equivalence class $PE(o)_{a,i}$ containing object o in $pt_{a,i}$ is:

$$PE(o)_{a,i} = \{o' \mid (o, o') \in PI_{a,i}\}. \tag{6}$$

Minimum possible equivalence class $PE(o)_{a,min}$ and maximum possible equivalence class $PE(o)_{a,max}$ containing object o on a are:

$$PE(o)_{a,min} = \cap_i PE(o)_{a,i}, \tag{7}$$
$$PE(o)_{a,max} = \cup_i PE(o)_{a,i}. \tag{8}$$

Family $FPE(o)_a$ of possible equivalence classes containing o consists of those containing o in the possible tables.

$$FPE(o)_a = \cup_i \{PE(o)_{a,i}\}. \tag{9}$$

The family has a lattice structure with the minimum and the maximum element, which are the minimum and the maximum possible equivalence class, respectively [20]. Family $FPE_{a,i}$ of equivalence classes in possible table $pt_{a,i}$ is obtained from possible indiscernibility relation $PI_{a,i}$.

$$FPE_{a,i} = \cup_{o \in U} \{PE(o)_{a,i}\}. \tag{10}$$

We have two aggregations of possible indiscernibility relations the whole and the common indiscernibility relation. Whole indiscernibility relation WPI_a is the union of $PI_{a,i}$:

$$WPI_a = \cup_i PI_{a,i}. \tag{11}$$

Common indiscernibility relation CPI_a is the intersection of $PI_{a,i}$:

$$CPI_a = \cap_i PI_{a,i}. \tag{12}$$

Family$\{CPI_a, PI_{a,1}, \cdots, WPI_a\}$ has a lattice structure with the minimum element CPI_a and the maximum element WPI_a.

Proposition 1.

$$PE(o)_{a,min} = \{o' \mid (o', o) \in CPI_a\},$$
$$PE(o)_{a,max} = \{o' \mid (o', o) \in WPI_a\}.$$

When target set T of objects is specified, lower approximation $\underline{apr}(T)_{a,i}$ and upper approximation $\overline{apr}(T)_{a,i}$ in possible table $pt_{a,i}$ are:

$$\underline{apr}(T)_{a,i} = \{o \in U \mid PE(o)_{a,i} \subseteq T\}, \tag{13}$$
$$\overline{apr}(T)_{a,i} = \{o \in U \mid PE(o)_{a,i} \cap T \neq \emptyset\}. \tag{14}$$

Minimum lower approximation $\underline{apr}(T)_{a,min}$, maximum lower approximation $\underline{apr}(T)_{a,max}$, minimum upper approximation $\overline{apr}(T)_{a,min}$, and maximum upper approximation $\overline{apr}(T)_{a,max}$ are:

$$\underline{apr}(T)_{a,min} = \cap_i \underline{apr}(T)_{a,i}, \tag{15}$$
$$\underline{apr}(T)_{a,max} = \cup_i \underline{apr}(T)_{a,i}, \tag{16}$$
$$\overline{apr}(T)_{a,min} = \cap_i \overline{apr}(T)_{a,i}, \tag{17}$$
$$\overline{apr}(T)_{a,max} = \cup_i \overline{apr}(T)_{a,i}. \tag{18}$$

Proposition 2.

$$\underline{apr}(T)_{a,min} = \{o \in U \mid PE(o)_{a,max} \subseteq T\},$$
$$\underline{apr}(T)_{a,max} = \{o \in U \mid PE(o)_{a,min} \subseteq T\},$$
$$\overline{apr}(T)_{a,min} = \{o \in U \mid PE(o)_{a,min} \cap T \neq \emptyset\},$$
$$\overline{apr}(T)_{a,max} = \{o \in U \mid PE(o)_{a,max} \cap T \neq \emptyset\}.$$

Example 2. Let incomplete information table IT be obtained as follows:

IT

O	a_1	a_2
1	z	2
2	y	1
3	x	1
4	$*$	1
5	$*$	2

In information table IT, universe U, domains V_{a_1} and V_{a_2} of attributes a_1 and a_2 are the same as CT in Example 1. We obtain nine $(= 3 \times 3)$ possible tables $pt_1, \ldots, pt_9$ from IT on a_1 because missing value $*$ on attribute a_1 of objects o_4 and o_5 is replaced by one of domain elements x, y, z of attribute a_1.

	pt_1			pt_2			pt_3			pt_4			pt_5	
O	a_1	a_2	O	a_1	a_2	O	a_1	a_2	O	a_1	a_2	O	a_1	a_2
1	z	2	1	z	2	1	z	2	1	z	2	1	z	2
2	y	1	2	y	1	2	y	1	2	y	1	2	y	1
3	x	1	3	x	1	3	x	1	3	x	1	3	x	1
4	y	1	4	y	1	4	y	1	4	x	1	4	x	1
5	y	2	5	x	2	5	z	2	5	x	2	5	y	2

pt_6			pt_7			pt_8			pt_9		
O	a_1	a_2	O	a_1	a_2	O	a_1	a_2	O	a_1	a_2
1	z	2	1	z	2	1	z	2	1	z	2
2	y	1	2	y	1	2	y	1	2	y	1
3	x	1	3	x	1	3	x	1	3	x	1
4	x	1	4	z	1	4	z	1	4	z	1
5	z	2	5	x	2	5	y	2	5	z	2

o_5 is indiscernible with o_2 and o_4 on a_1 in pt_1, whereas o_5 is discernible with these objects on a_1 in pt_2. In other words, pt_1 corresponds to the case where o_5 is indiscernible with o_2 and o_4 on a_1, whereas pt_2 does to the case where o_5 is discernible with these objects.

Each possible indiscernibility relation $PI_{a_1,i}$ on a_1 for $i = 1,9$ is:

$$PI_{a_1,1} = \{(o_1,o_1),(o_2,o_2),(o_3,o_3),(o_4,o_4),(o_5,o_5),(o_2,o_4),(o_4,o_2),(o_2,o_5),$$
$$(o_5,o_2),(o_4,o_5),(o_5,o_4)\},$$

$$PI_{a_1,2} = \{(o_1,o_1),(o_2,o_2),(o_3,o_3),(o_4,o_4),(o_5,o_5),(o_2,o_4),(o_4,o_2),(o_3,o_5),$$
$$(o_5,o_3)\},$$

$$PI_{a_1,3} = \{(o_1,o_1),(o_2,o_2),(o_3,o_3),(o_4,o_4),(o_5,o_5),(o_2,o_4),(o_4,o_2),(o_1,o_5),$$
$$(o_5,o_1)\},$$

$$PI_{a_1,4} = \{(o_1,o_1),(o_2,o_2),(o_3,o_3),(o_4,o_4),(o_5,o_5),(o_3,o_4),(o_4,o_3),(o_3,o_5),$$
$$(o_5,o_3),(o_4,o_5),(o_5,o_4)\},$$

$$PI_{a_1,5} = \{(o_1,o_1),(o_2,o_2),(o_3,o_3),(o_4,o_4),(o_5,o_5),(o_3,o_4),(o_4,o_3),(o_2,o_5),$$
$$(o_5,o_2)\},$$

$$PI_{a_1,6} = \{(o_1,o_1),(o_2,o_2),(o_3,o_3),(o_4,o_4),(o_5,o_5),(o_1,o_5),(o_5,o_1),(o_3,o_4),$$
$$(o_4,o_3)\},$$

$$PI_{a_1,7} = \{(o_1,o_1),(o_2,o_2),(o_3,o_3),(o_4,o_4),(o_5,o_5),(o_1,o_4),(o_4,o_1),(o_3,o_5),$$
$$(o_5,o_3)\},$$

$$PI_{a_1,8} = \{(o_1,o_1),(o_2,o_2),(o_3,o_3),(o_4,o_4),(o_5,o_5),(o_1,o_4),(o_4,o_1),(o_2,o_5),$$
$$(o_5,o_2)\},$$

$$PI_{a_1,9} = \{(o_1,o_1),(o_2,o_2),(o_3,o_3),(o_4,o_4),(o_5,o_5),(o_1,o_4),(o_4,o_1),(o_1,o_5),$$
$$(o_5,o_1),(o_4,o_5),(o_5,o_4)\}.$$

Possible eqivalence class $PE(o_j)_{a_1,i}$ containing object o_j in $pt_{a_1,i}$ for $i = 1,9$ and $j = 1,5$ is:

$PE(o_1)_{a_1,1} = \{o_1\}, PE(o_2)_{a_1,1} = PE(o_4)_{a_1,1} = PE(o_5)_{a_1,1} = \{o_2,o_4,o_5\}, PE(o_3)_{a_1,1} = \{o_3\},$

$PE(o_1)_{a_1,2} = \{o_1\}, PE(o_2)_{a_1,2} = PE(o_4)_{a_1,2} = \{o_2,o_4\}, PE(o_3)_{a_1,2} = PE(o_5)_{a_1,2} = \{o_3,o_5\},$

$PE(o_1)_{a_1,3} = PE(o_5)_{a_1,3} = \{o_1,o_5\}, PE(o_2)_{a_1,3} = PE(o_4)_{a_1,3} = \{o_2,o_4\}, PE(o_3)_{a_1,3} = \{o_3\},$

$PE(o_1)_{a_1,4} = \{o_1\}, PE(o_2)_{a_1,4} = \{o_2\}, PE(o_3)_{a_1,4} = PE(o_4)_{a_1,4} = PE(o_5)_{a_1,4} = \{o_3,o_4,o_5\},$

$PE(o_1)_{a_1,5} = \{o_1\}, PE(o_2)_{a_1,5} = PE(o_5)_{a_1,5} = \{o_2,o_5\}, PE(o_3)_{a_1,5} = PE(o_4)_{a_1,5} = \{o_3,o_4\},$

$PE(o_1)_{a_1,6} = PE(o_5)_{a_1,6} = \{o_1,o_5\}, PE(o_2)_{a_1,6} = \{o_2\}, PE(o_3)_{a_1,6} = PE(o_4)_{a_1,6} = \{o_3,o_4\},$

$PE(o_1)_{a_1,7} = PE(o_4)_{a_1,7} = \{o_1, o_4\}, PE(o_2)_{a_1,7} = \{o_2\}, PE(o_3)_{a_1,7} = PE(o_5)_{a_1,7} = \{o_3, o_5\},$

$PE(o_1)_{a_1,8} = PE(o_4)_{a_1,8} = \{o_1, o_4\}, PE(o_2)_{a_1,8} = PE(o_5)_{a_1,8} = \{o_2, o_5\}, PE(o_3)_{a_1,8} = \{o_3\},$

$PE(o_1)_{a_1,9} = PE(o_4)_{a_1,9} = PE(o_5)_{a_1,9} = \{o_1, o_4, o_5\}, PE(o_2)_{a_1,9} = \{o_2\}, PE(o_3)_{a_1,9} = \{o_3\}.$

Families $PFE_{a_1,i}$ of possible equivalence classes on attribute a_1 in possible tables $pt_1, \ldots, pt_9$ are:

$$FPE_{a_1,1} = \{\{o_1\}, \{o_2, o_4, o_5\}, \{o_3\}\},$$
$$FPE_{a_1,2} = \{\{o_1\}, \{o_2, o_4\}, \{o_3, o_5\}\},$$
$$FPE_{a_1,3} = \{\{o_1, o_5\}, \{o_2, o_4\}, \{o_3\}\},$$
$$FPE_{a_1,4} = \{\{o_1\}, \{o_2\}, \{o_3, o_4, o_5\}\},$$
$$FPE_{a_1,5} = \{\{o_1\}, \{o_2, o_5\}, \{o_3, o_4\}\},$$
$$FPE_{a_1,6} = \{\{o_1, o_5\}, \{o_2\}, \{o_3, o_4\}\},$$
$$FPE_{a_1,7} = \{\{o_1, o_4\}, \{o_2\}, \{o_3, o_5\}\},$$
$$FPE_{a_1,8} = \{\{o_1, o_4\}, \{o_2, o_5\}, \{o_3\}\},$$
$$FPE_{a_1,9} = \{\{o_1, o_4, o_5\}, \{o_2\}, \{o_3\}\}.$$

Minimum possible equivalence class $PE(o_j)_{a_1,min}$ and maximum possible equivalence class $PE(o_j)_{a_1,max}$ containing object o_j on a_1 for $j = 1, 5$ are:

$$PE(o_1)_{a_1,min} = \{o_1\},$$
$$PE(o_1)_{a_1,max} = \{o_1, o_4, o_5\},$$
$$PE(o_2)_{a_1,min} = \{o_2\},$$
$$PE(o_2)_{a_1,max} = \{o_2, o_4, o_5\},$$
$$PE(o_3)_{a_1,min} = \{o_3\},$$
$$PE(o_3)_{a_1,max} = \{o_3, o_4, o_5\},$$
$$PE(o_4)_{a_1,min} = \{o_4\},$$
$$PE(o_4)_{a_1,max} = \{o_1, o_2, o_3, o_4, o_5\},$$
$$PE(o_5)_{a_1,min} = \{o_5\},$$
$$PE(o_5)_{a_1,max} = \{o_1, o_2, o_3, o_4, o_5\}.$$

Common possible indiscernibility relation CPI_{a_1} and whole possible indiscernibility relation WPI_{a_1} are:

$CPI_{a_1} = \{(o_1, o_1), (o_2, o_2), (o_3, o_3), (o_4, o_4), (o_5, o_5)\},$

$WPI_{a_1} = \{(o_1, o_1), (o_1, o_4), (o_1, o_5), (o_2, o_2), (o_2, o_4), (o_2, o_5), (o_3, o_3), (o_3, o_4),$
$(o_3, o_5), (o_4, o_1), (o_4, o_2), (o_4, o_3), (o_4, o_4), (o_4, o_5), (o_5, o_1), (o_5, o_2),$
$(o_5, o_3), (o_5, o_4), (o_5, o_5)\}.$

Let target set T be $\{o_2, o_3, o_4\}$. Lower approximation $\underline{apr}(T)_{a_1,i}$ and upper approximation $\overline{apr}(T)_{a_1,i}$ in possible table pt_i for $i = 1, 9$ are:

$$\underline{apr}(T)_{a_1,1} = \{o_3\},$$

$$\overline{apr}(T)_{a_1,1} = \{o_2, o_3, o_4, o_5\},$$
$$\underline{apr}(T)_{a_1,2} = \{o_2, o_4\},$$
$$\overline{apr}(T)_{a_1,2} = \{o_2, o_3, o_4, o_5\},$$
$$\underline{apr}(T)_{a_1,3} = \{o_2, o_3, o_4\},$$
$$\overline{apr}(T)_{a_1,3} = \{o_2, o_3, o_4\},$$
$$\underline{apr}(T)_{a_1,4} = \{o_2\},$$
$$\overline{apr}(T)_{a_1,4} = \{o_2, o_3, o_4, o_5\},$$
$$\underline{apr}(T)_{a_1,5} = \{o_3, o_4\},$$
$$\overline{apr}(T)_{a_1,5} = \{o_2, o_3, o_4, o_5\},$$
$$\underline{apr}(T)_{a_1,6} = \{o_2, o_3, o_4\},$$
$$\overline{apr}(T)_{a_1,6} = \{o_2, o_3, o_4\},$$
$$\underline{apr}(T)_{a_1,7} = \{o_2\},$$
$$\overline{apr}(T)_{a_1,7} = \{o_1, o_2, o_3, o_4, o_5\},$$
$$\underline{apr}(T)_{a_1,8} = \{o_3\},$$
$$\overline{apr}(T)_{a_1,8} = \{o_1, o_2, o_3, o_4, o_5\},$$
$$\underline{apr}(T)_{a_1,9} = \{o_2, o_3\},$$
$$\overline{apr}(T)_{a_1,9} = \{o_1, o_2, o_3, o_4, o_5\}.$$

Minimum lower approximation $\underline{apr}(T)_{a_1,min}$, maximum lower approximation $\underline{apr}(T)_{a_1,max}$, minimum upper approximation $\overline{apr}(T)_{a_1,min}$, and maximum upper approximation $\overline{apr}(T)_{a_1,max}$ are:

$$\underline{apr}(T)_{a_1,min} = \emptyset,$$
$$\underline{apr}(T)_{a_1,max} = \{o_2, o_3, o_4\},$$
$$\overline{apr}(T)_{a_1,min} = \{o_2, o_3, o_4\},$$
$$\overline{apr}(T)_{a_1,max} = \{o_1, o_2, o_3, o_4, o_5\}.$$

4   Kryszkiewicz's Formula Dealing with Missing Values

Kryszkiewicz used a binary relation for object indistinguishability in information tables containing missing values. The binary relation consists of pairs of indiscernible objects and is expressed as follows:

$$I_a^{\overline{K}} = \{(o, o') \in U \times U \mid a(o) = a(o') \vee a(o) = * \vee a(o') = *\}, \qquad (19)$$

The relation is reflective, and symmetric, but not transitive. Set $E_a^{\overline{K}}(o)$ of objects that are considered as equivalent to object o is derived from the relation:

$$E_a^{\overline{K}}(o) = \{o' \mid (o, o') \in I_a^{\overline{K}}\}, \qquad (20)$$

The set is not an equivalence class. Let T be a target set of objects. By using $E_a^{\overline{K}}(o)$, lower approximation $\underline{apr}_a^{\overline{K}}(T)$ and upper approximation $\overline{apr}_a^{\overline{K}}(T)$ are:

$$\underline{apr}(T)_a^{\overline{K}} = \{o \in U \mid E(o)_a^{\overline{K}} \subseteq T\}, \qquad (21)$$

$$\overline{apr}(T)_a^{\overline{K}} = \{o \in U \mid E(o)_a^{\overline{K}} \cap T \neq \emptyset\}. \tag{22}$$

This approach contradicts Lipski's approach because he showed that the actual answer set to an inquiry cannot be obtained and what we can obtain is the lower and the upper bound of the actual answer set [19]. In addition, these formulae have the drawbacks pointed by [21], as is shown in Example 3.

Example 3. We check Kryszkiewicz's formula by using incomplete information table IT in Example 2. Let target set T be $\{o_2, o_3, o_4\}$, which is the same as in Example 2. Using formula (19) in IT, we obtain the following binary relation of indiscernibility for objects:

$$I_{a_1}^{\overline{K}} = \{(o_1, o_1), (o_1, o_4), (o_1, o_5), (o_2, o_2), (o_2, o_4), (o_2, o_5), (o_3, o_3), (o_3, o_4),$$
$$(o_3, o_5), (o_4, o_1), (o_4, o_2), (o_4, o_3), (o_4, o_4), (o_4, o_5), (o_5, o_1), (o_5, o_2),$$
$$(o_5, o_3), (o_5, o_4), (o_5, o_5)\}.$$

Class $E_{a_1}^{\overline{K}}(o_j)$ derived from the relation for $j = 1, 5$ is:

$$E_{a_1}^{\overline{K}}(o_1) = \{o_1, o_4, o_5\},$$
$$E_{a_1}^{\overline{K}}(o_2) = \{o_2, o_4, o_5\},$$
$$E_{a_1}^{\overline{K}}(o_3) = \{o_3, o_4, o_5\},$$
$$E_{a_1}^{\overline{K}}(o_4) = \{o_1, o_2, o_3, o_4, o_5\},$$
$$E_{a_1}^{\overline{K}}(o_5) = \{o_1, o_2, o_3, o_4, o_5\}.$$

Lower approximation $\underline{apr}_{a_1}^{\overline{K}}(T)$ and upper approximation $\overline{apr}_{a_1}^{\overline{K}}(T)$ from formulae (21) and (21) are:

$$\underline{apr}(T)_{a_1}^{\overline{K}} = \emptyset,$$
$$\overline{apr}(T)_{a_1}^{\overline{K}} = \{o_1, o_2, o_3, o_4, o_5\} = U.$$

Note that nothing is obtained for approximations.

Example 3 shows that we have poor results for approximation in the case of using Kryszkiewicz's formula. This is because possible classes of objects equivalent to object o_3 on a_1 are $\{o_3\}$, $\{o_3, o_4\}$, and $\{o_3, o_4, o_5\}$, but only $\{o_3, o_4, o_5\}$ is considered. In other words, only the possibility that missing value $*$ may be equal to a value is considered, but the opposite possibility is neglected.

5   Relationship Between Kryszkiewicz's Formula and Lipski's Approach

Indeed, Kryszkiewicz's formula is not compatible with Lipski's approach, but we have the following propositions.

Proposition 3.

$$I_a^{\overline{K}} = WPI_a.$$

This proposition shows that the binary relation of indiscernibility used by Kryszkiewicz is equal to the whole indiscernibility relation, the union of possible indiscernibility relations in possible tables.

Proposition 4.

$$\underline{apr}(T)_a^{\overline{K}} = \underline{apr}(T)_{a,min},$$
$$\overline{apr}(T)_a^{\overline{K}} = \overline{apr}(T)_{a,max}.$$

The proposition shows that the lower and the upper approximation by Kryszkiewicz's formula are equal to ones derived from using the whole indiscernibility relation in possible world semantics. This suggests that by adding the formula corresponding to the common indiscernibility relation, results from Kryszkiewicz's approach coincide with those from Lipski's approach. The expression corresponding to the common indiscernibility relation under the notation of Kryszkiewicz is:

$$I_a^{\underline{K}} = \{(o,o') \in U \times U \mid (o = o') \vee (a(o) = a(o') \wedge a(o) \neq * \wedge a(o') \neq *)\}. \quad (23)$$

Proposition 5.

$$I_a^{\underline{K}} = CPI_a.$$

$E_a^{\underline{K}}(o)$ of each object is derived from $I_a^{\underline{K}}$ similarly to $E_a^{\overline{K}}(o)$. By using $E_a^{\underline{K}}(o)$, lower approximation $\underline{apr}_a^{\overline{K}}(T)$ and upper approximation $\overline{apr_a}^{\underline{K}}(T)$ are:

$$\underline{apr}(T)_a^{\underline{K}} = \{o \in U \mid E(o)_a^{\underline{K}} \subseteq T\}, \quad (24)$$
$$\overline{apr}(T)_a^{\underline{K}} = \{o \in U \mid E(o)_a^{\underline{K}} \cap T \neq \emptyset\}. \quad (25)$$

Proposition 6.

$$\underline{apr}(T)_a^{\underline{K}} = \underline{apr}(T)_{a,max},$$
$$\overline{apr}(T)_a^{\underline{K}} = \overline{apr}(T)_{a,min}.$$

Extending Kryszkiewicz's formula by adding formula (23) to formula (19), we can resolve the incompatibility between Kryszkiewicz's formula and Lipski's approach.

Example 4. Using formula (23) in IT of Example 2, we obtain the following binary relation of indiscernibility:

$$I_{a_1}^{\underline{K}} = \{(o_1, o_1), (o_2, o_2), (o_3, o_3), (o_4, o_4), (o_5, o_5)\}.$$

Class $E_{a_1}^K(o_j)$ derived from the relation for $j = 1, 5$ is:

$$E_{a_1}^{\overline{K}}(o_1) = \{o_1\},$$
$$E_{a_1}^{\overline{K}}(o_2) = \{o_2\},$$
$$E_{a_1}^{\overline{K}}(o_3) = \{o_3\},$$
$$E_{a_1}^{\overline{K}}(o_4) = \{o_4\},$$
$$E_{a_1}^{\overline{K}}(o_5) = \{o_5\}.$$

Let target set T be $\{o_2, o_3, o_4\}$, which is the same as in Example 2. Lower approximation $\underline{apr}_{a_1}^K(T)$ and upper approximation $\overline{apr}_{a_1}^K(T)$ from formulae (24) and (25) are:

$$\underline{apr}(T)_{a_1}^K = \{o_2, o_3, o_4\},$$
$$\overline{apr}(T)_{a_1}^K = \{o_2, o_3, o_4\}.$$

6   Conclusions

We have described rough sets in incomplete information tables containing missing values whose values are unknown. Many authors use Kryszkiewicz's formula. The formula, however, is not compatible with Lipski's approach based on possible world semantics and creates poor results for approximations. In order to resolve this point, we have checked Kryszkiewicz's formula from the viewpoint of possible world semantics. Kryszkiewicz's formula for indiscernibility is equal to that of the whole indiscernibility relation that is the union of possible indiscernibility relations obtained from possible tables. By adding the formula corresponding to the common indiscernibility relation that is the intersection of possible indiscernibility relations, we have extended Kryszkiewicz's formula. As a result, the extended Kryszkiewicz's approach is compatible with Lipski's approach based on possible world semantics.

Acknowledgment. Part of this work is supported by JSPS KAKENHI Grant Number JP20K11954.

References

1. Pawlak, Z.: Rough Sets: Theoretical Aspects of Reasoning about Data. Kluwer Academic Publishers, Dordrecht (1991). https://doi.org/10.1007/978-94-011-3534-4
2. Parsons, S.: Current approaches to handling imperfect information in data and knowledge bases. IEEE Trans. Knowl. Data Eng. **8**, 353–372 (1996)
3. Parsons, S.: Addendum to "current approaches to handling imperfect information in data and knowledge bases". IEEE Trans. Knowl. Data Eng. **10**, 862 (1998)
4. Greco, S., Matarazzo, B., Słowinski, R.: Handling missing values in rough set analysis of multi-attribute and multi-criteria decision problems. In: Zhong, N., Skowron, A., Ohsuga, S. (eds.) RSFDGrC 1999. LNCS (LNAI), vol. 1711, pp. 146–157. Springer, Heidelberg (1999). https://doi.org/10.1007/978-3-540-48061-7_19

5. Grzymala-Busse, J.W.: Data with missing attribute values: generalization of indiscernibility relation and rule induction. Trans. Rough Sets **I**, 78–95 (2004)
6. Grzymala-Busse, J.W.: Characteristic relations for incomplete data: a generalization of the indiscernibility relation. Trans. Rough Sets **IV**, 58–68 (2005)
7. Kryszkiewicz, M.: Rules in incomplete information systems. Inf. Sci. **113**, 271–292 (1999)
8. Latkowski, R.: Flexible indiscernibility relations for missing values. Fund. Inform. **67**, 131–147 (2005)
9. Leung, Y., Li, D.: Maximum consistent techniques for rule acquisition in incomplete information systems. Inf. Sci. **153**, 85–106 (2003)
10. Nakata, M., Sakai, H.: Rough sets handling missing values probabilistically interpreted. In: Ślęzak, D., Wang, G., Szczuka, M., Düntsch, I., Yao, Y. (eds.) RSFDGrC 2005. LNCS (LNAI), vol. 3641, pp. 325–334. Springer, Heidelberg (2005). https://doi.org/10.1007/11548669_34
11. Nakata, M., Sakai, H.: Twofold rough approximations under incomplete information. Int. J. Gen. Syst. **42**, 546–571 (2013). https://doi.org/10.1080/17451000.2013.798898
12. Sakai, H.: Effective procedures for handling possible equivalence relation in nondeterministic information systems. Fund. Inform. **48**, 343–362 (2001)
13. Stefanowski, J., Tsoukiàs, A.: Incomplete information tables and rough classification. Comput. Intell. **17**, 545–566 (2001)
14. Sun, L., Wanga, L., Ding, W., Qian, Y., Xu, J.: Neighborhood multi-granulation rough sets-based attribute reduction using lebesgue and entropy measures in incomplete neighborhood decision systems. Knowl.-Based Syst. **192** (2020). Article ID 105373
15. Kryszkiewicz, M.: Rough set approach to incomplete information systems. Inf. Sci. **112**, 39–49 (1998)
16. Wang, G.: Extension of rough set under incomplete information systems. In: 2002 IEEE World Congress on Computational Intelligence. In: IEEE International Conference on Fuzzy Systems, FUZZ-IEEE 2002, pp. 1098–1103 (2002). https://doi.org/10.1109/FUZZ.2002.1006657
17. Nguyen, D.V., Yamada, K., Unehara, M.: Extended tolerance relation to define a new rough set model in incomplete information systems. Adv. Fuzzy Syst., 10 (2013). Article ID 372091. https://doi.org/10.1155/2013/372091
18. Rady, E.A., Abd El-Monsef, M.M.E., Adb El-Latif, W.A.: A modified rough sets approach to incomplete information systems. J. Appl. Math. Decis. Sci., 13 (2007). Article ID 058248
19. Lipski, W.: On semantics issues connected with incomplete information databases. ACM Trans. Database Syst. **4**, 262–296 (1979)
20. Nakata, M., Saito, N., Sakai, H, Fujiwara, T.: Structures derived from possible tables in an incomplete information table. In: Proceedings in SCIS 2022, 6 p. IEEE Press (2022). https://doi.org/10.1109/SCISISIS55246.2022.10001919
21. Nakata, M., Sakai, H.: Applying rough sets to information tables containing missing values. In: Proceedings of 39th International Symposium on Multiple-Valued Logic, pp. 286–291. IEEE Press (2009). https://doi.org/10.1109/ISMVL.2009.1

Sustainable Perishable Food Distribution: Fresh Consumption and Waste Justice

Sun Olapiriyakul(ORCID) and Warut Pannakkong[✉](ORCID)

School of Manufacturing Systems and Mechanical Engineering, Sirindhorn
International Institute of Technology, Thammasat University, Bangkok, Thailand
{suno,warut}@siit.tu.ac.th

Abstract. Food waste poses significant environmental and social challenges, impacting communities and contributing to solid waste accumulation. This research addresses the critical issue of food supply chain management, emphasizing the social sustainability impact of waste justice
and fresh food consumption. Our proposed supply chain network design
model and multi-objective optimization procedure can generate solutions
whose sustainability performance aspects, including profit, greenhouse
gas emissions, and fresh consumption of products, can be controlled to
align with managerial preferences. Through a food supply chain network
case, this research demonstrates the potential of the proposed methods
to improve residents' quality of life and promote sustainable development
in the region.

Keywords: Sustainability · Sustainable inventory · Food supply
chain · Solid waste management · Inequality · Social justice ·
Greenhouse gas emission

1 Introduction

Food is a perishable good that can spoil and lose quality over time, leading to
waste. The significance of food waste issues varies across regions and is influenced
by many factors, including the availability of fresh produce, consumer culture,
and the adequacy of logistical infrastructure. Statistically, about one-third of
food produced for human consumption is lost or wasted during distribution operations and at the end-customer level [1]. This additional waste stream not only
complicates the management of solid waste but also contributes to the accumulation of solid waste, which can pose wide-ranging negative environmental and
social impacts on communities. The well-known direct impacts of solid waste
accumulation include health hazards, an unpleasant living environment, ecosystem degradation, and economic costs. Social inequalities and waste injustice are
also significant social impacts caused by solid waste accumulation. Specifically,
the buildup of solid waste often has a greater impact on low-income and minority
neighborhoods than on others located near landfills and disposal sites [2].

© The Author(s), under exclusive license to Springer Nature Switzerland AG 2024
K. Honda et al. (Eds.): IUKM 2023, LNAI 14376, pp. 306–318, 2024.
https://doi.org/10.1007/978-3-031-46781-3_26

Mitigating waste issues in a food supply chain is a complex task requiring a multifaceted approach, which involves improving the demand forecast accuracy, storage and transportation practices, and the closed-loop nature of the supply chain. As part of a waste-minimizing strategy, implementing effective inventory planning and ordering strategies can minimize the chances of spoilage and waste before or after the products reach customers. The location of distribution and retail facilities can also be essential in reducing food waste. An emerging research challenge is to develop an approach that facilitates integrated decision-making in supply chain management, encompassing inventory planning, ordering, and facility location to optimize supply chain operations. The coordination of multiple supply chain functions while considering sustainability impacts on multiple stakeholders simultaneously is an area where understanding is still evolving within a food supply chain context.

This study contributes to the existing literature by presenting a sustainable supply chain network design (SCND) model that highlights the social sustainability impact related to the freshness of perishable food inventory and waste justice. The model integrates location, inventory planning, and ordering decisions while concurrently evaluating the three key sustainability performance aspects: profit, greenhouse gas emissions, and product freshness. The quantity of waste generated is controlled by waste justice constraints to prevent excessive waste accumulation burdens in any of the regional districts. The effort to maximize the sales of fresh products and the equitable distribution of waste accumulation burdens will bring positive benefits to the well-being of residents and support sustainable development in the region.

2   Literature Review

Our literature review starts with the previous SCND approaches for perishable products. The studies by Banasik et al. [3], and Olapiriyakul and Nguyen [4], were among the first to integrate decision-making on facility location, delivery route, and product flow. The scope of integrated decision-making expands to include inventory planning as a critical component for products with limited shelf life, as shown by Biuki et al. [5]. The literature shows that inventory planning and ordering decisions are crucial for ensuring the freshness of perishable food products Tsao et al. [6], Torabzadeh et al. [7]. Effective coordination of supply chain decisions is critical in procuring, storing, and distributing perishable food items while achieving optimal and sustainable supply chain performance.

Despite the growing interest in sustainability-related issues in SCND problems, only a few studies address social impact by comprehensively incorporating the perspectives of relevant stakeholders. Daghigh et al. [8] propose a sustainable logistics network design approach for perishable items with a social objective that aims to provide fair accessibility of products for customers and equitable job opportunities for the surrounding communities. Kalantari and Hosseininezhad [9] propose a sustainable SCND for perishable food considering global supply chain risks concerning late shipment and quality problems. The cost, carbon emissions, and job creation impacts are evaluated using multi-objective optimization techniques. The study by Tirkolaee et al. [10] and Wang et al. [11] also contributes to social sustainability by offering job creation. The social impact of the perishable goods supply chain is also related to the freshness of products due to its relevance to the improved living standard of residents. To ensure the social responsibility of a supply chain, it is essential to consider consumer satisfaction and standards with product freshness at the point of demand fulfillment [12] as well as the deterioration of product quality over time [13].

While food waste is a closely related issue to a product's freshness loss, there are very few studies evaluating the waste impact of the perishable goods supply chain. Jaigirdar et al. [14] select locations for establishing refrigerated facilities to minimize spoilage and waste to achieve a sustainable supply chain. The sustainable food supply chain model proposed by Jouzdani and Govindan [15] considers the uncertainty in product life and the impacts of refrigerated vehicle usage. It is the first time that the SDGs addressed by the model are specified, one of which is the goal of reducing food loss and waste (SDG2). After reviewing the latest literature on SCND for perishable products, the following can be concluded.

- The SCND research for perishable products should continue to consider the interdependence between inventory planning, facility location decisions, and transportation routing decisions.
- There is a significant gap in considering social impact factors and understanding their relationship with other sustainability objectives.
- The impact of municipal solid waste (MSW) generation and accumulation, critical urban issues that merit greater attention in this domain, continues to be inadequately addressed.

3   Methodology

3.1   Waste Disparity Index

The waste disparity index (WDI) is introduced to formulate waste justice constraints to prevent substantial waste generation in districts already burdened with high waste accumulated. The WDI of district i is calculated by dividing the accumulated waste in district i by the average accumulated waste amount across all districts. The allowable waste generation in terms of % of the total demand in a district is assumed in Table 1.

3.2   Optimization Model for a Sustainable Perishable Goods Distribution Network

A multi-objective mixed integer programming model is proposed in this study to facilitate inventory-location-routing decisions of a sustainable perishable goods distribution network. The model encompasses three objectives: 1) maximizing profit, 2) minimizing CO_2 emissions, and 3) maximizing freshness. These three

Table 1. Allowable waste generation for a retail center

WDI value	Allowable waste generation (% of demand)
WDI i > 2	10%
2.0 $>$ WDI i > 1.5	20%
1.5 $>$ WDI i > 1.0	30%
WDI i < 1	40%

objectives provide a comprehensive approach to address the three aspects of sustainability. It is noteworthy that the objective of maximizing freshness, along with the implementation of waste justice constraints, can positively contribute to sustainable development practices and benefits the local communities. The freshness objective value is the percentage of demand satisfied by products whose remaining shelf lives do not exceed the designated fresh period, specifically set at 3 weeks for the purpose of this illustration. By ensuring the consumption of fresh products while limiting waste generation, particularly in districts with high waste accumulation, the model promotes residents' living quality in the region. The CO_2 emissions are based on transportation activities and waste only.

3.3 Model's Notation

Sets and Indices. S is the set of suppliers, indexed by s; C is the set of retail centers (RC), indexed by c; P is the set of products, indexed by p; B is the set of ordering batch sizes, indexed by b; l is the integer representing shelf lives of the products, ranging from 1 to L; H is the set of all possible shelf lives (l) of the products; T is the set of time periods, indexed by t.

Parameters. $demand_{cpt}$ is the demand of product p at the retailer center c and period t; $price_p$ is the unit selling price of product p at the retailer centers; $prodcost_p$ is the unit cost of product p that retailer centers purchase from the suppliers; $invcost_p$ is the unit cost of product per period at the retailer centers; $weight_p$ is the unit weight (kg) of product p; $weighttruck$ is the weight (kg) of truck; $maxwaste_{cp}$ is the maximum allowable waste (ton) of product p at retailer c; $supcapa_s$ is the production capacity per period of supplier s; $dist_{sc}$ is the distance between supplier s and retail center c; $initialinv_{cpl}$ is the initial inventory of product p with shelf life l at retail center c; $ordersize_{pb}$ is the ordering size of product p at batch size b; $ordercost_{pb}$ is the ordering cost of product p at batch size b; $rcopencost_c$ is the opening cost of retail center c; $productlife_p$ is the life of product p; fr is the fresh shelf life; w is the waste shelf life; $wastecost$ is the cost of handling waste as a percentage of the product cost; CO_2tkm is the CO_2 emission equivalent per ton kilometer (kg CO_2 e per tkm) for transportation CO_2kg is the CO_2 emission equivalent per kilogram of waste (kg CO_2 e per kg of waste) $allowprofit$ is the allowable total profit decrease

(%) *allowfresh* is the allowable freshness percentage decrease (%) M is a large number.

Decision Variables. $rcopen_c$ is the binary variable, equal to 1 when retail center c is open; $order_{scptb}$ is the binary variable, equal to 1 when retail center c orders product p with batch size b from supplier s at period t; $sales_{cptl}$ is the unit sales of product p with shelf life l at retail center c and period t; pr_{cptl} is the proportion of demand of product p at retailer c and period t, which is satisfied by the product with shelf life l; inv_{cptl} is the inventory level of product c with shelf life l at retail center c and period t.

Functions

– *totalwastecost* is the function that computes the total waste cost as:

$$totalwaste = \sum_{c \in C} \sum_{p \in P} \sum_{t \in T} inv_{cpt1} \times prodcost_p \times wastecost \qquad (1)$$

– *totalprodcost* is the function that computes the total product cost as:

$$totalprodcost = \sum_{s \in S} \sum_{c \in C} \sum_{p \in P} \sum_{t \in T} \sum_{b \in B} (order_{scptb} \times ordersize_{pb} \times prodcost_p) \quad (2)$$

– *totalrcopencost* is the function that computes the total retail center opening cost as:

$$totalrcopencost = \sum_{c \in C} (rcopen_c \times rcopencost_c) \qquad (3)$$

– *totalinvcost* is the function that computes the total inventory cost as:

$$totalinvcost = \sum_{c \in C} \sum_{p \in P} \sum_{t \in T} \sum_{l=1}^{L} (inv_{cptl} \times invcost_p) \qquad (4)$$

– *totalordercost* is the function that computes the total ordering cost as:

$$totalordercost = \sum_{c \in C} \sum_{p \in P} \sum_{b \in B} \sum_{l=1}^{L} (order_{scptb} \times ordercost_{pb} \times dist_{sc}) \qquad (5)$$

– *totalsales* is the function that computes total sales as:

$$totalsales = \sum_{c \in C} \sum_{p \in P} \sum_{t \in T} \sum_{l=1}^{L} sales_{cptl} \qquad (6)$$

– *totalfreshsales* is the function that computes total sales of fresh products as:

$$totalfreshsales = \sum_{c \in C} \sum_{p \in P} \sum_{t \in T} \sum_{l=fr}^{L} sales_{cptl} \qquad (7)$$

– *totaldist* is the function that computes total transportation distance between the suppliers and retail centers as:

$$totaldist = \sum_{s \in S} \sum_{c \in C} \sum_{p \in P} \sum_{t \in T} \sum_{b \in B} (order_{scptb} \times dist_{sc}) \tag{8}$$

– $CO_2emissiontran$ is the function that computes total greenhouse gas emission in the unit of CO_2 emission equivalent (kg CO_2 e) from transportation between the suppliers and retail centers over T periods as:

$$CO_2emissiontran = \sum_{s \in S} \sum_{c \in C} \sum_{p \in P} \sum_{t \in T} \sum_{b \in B} ((order_{scptb} \times ordersize_{pb} \times dist_{sc}$$
$$\times weight_p) + weighttruck)/(1000 \times CO_2tkm) \tag{9}$$

– $CO_2emissionwaste$ is the function that computes total greenhouse gas emission in the unit of CO_2 emission equivalent (kg CO_2 e) from total waste as:

$$CO_2emissionwaste = \sum_{c \in C} \sum_{p \in P} \sum_{t \in T} (inv_{cpt1} \times CO_2kg) \tag{10}$$

– $totalCO_2emission$ is the function that computes total CO_2 emission equivalent (kg CO_2 e) as:

$$totalCO_2emission = CO_2emissiontran + CO_2emissionwaste \tag{11}$$

– *totalrev* is the function that computes total revenue as:

$$totalrev = \sum_{c \in C} \sum_{p \in P} \sum_{t \in T} \sum_{l=1}^{L} (sales_{cptl} \times price_p) \tag{12}$$

– *totalcost* is the function that computes total cost as:

$$totalcost = totalprodcost + totalrcopencost + totalinvcost + totalordercost \tag{13}$$

– *totalprofit* is the function that computes total profit as:

$$totalprofit = totalrev - totalcost \tag{14}$$

3.4   Solution Algorithms

Our solution approach addresses potentially conflicting goals in a sustainable optimization problem with a hierarchy of priority levels for the optimization goals. The first key step is to align the goals with stakeholder groups and assign priorities to goals based on their importance. According to the nature of our location-inventory case, the proposed approach classifies stakeholder groups into

1) the organization, 2) customers, and 3) the communities. These groups are arranged in order of importance, starting with internal stakeholders and progressing toward external stakeholders.

The rest of the optimization process follows the constraint method for tri-objective optimization, where maximizing the profit for the organization has the highest priority. The other two objectives, with the second and third priority, include maximizing the freshness of products for customers and minimizing CO_2 emissions for the communities. The model is solved under each objective to determine the upper and lower bounds for the tri-objective. The detail of their mathematical models is presented in the remaining parts of this section.

Maximizing Profit. This model aims to maximize the *totalprofit* function while subjects to the following constraints: demand constraints (16) - (19), supplier capacity constraint (20), retail center sale constraint (21), inventory balance constraints (22) - (24), sales and inventory at the first period constraints (25) - (26), and waste allowance constraint (27). The optimal value of *totalprofit* from this model is defined as F_p^*.

$$\text{Objective function: max } \quad totalprofit \tag{15}$$

$$\text{Subject to: } \quad sales_{cptl} \leq demand_{cpt} \qquad \forall c \in C, p \in P, t \in T, l \in H \tag{16}$$

$$\sum_{l=w+1}^{L} pr_{cptl} = 1 \qquad \forall c \in C, p \in P, t \in T \tag{17}$$

$$pr_{cptl} = 0 \qquad \forall c \in C, p \in P, t \in T, l \in H \mid l \leq w \tag{18}$$

$$\sum_{l=1}^{L} sales_{cptl} \leq demand_{cpl} \qquad \forall c \in C, p \in P, t \in T \tag{19}$$

$$\sum_{c \in C} \sum_{p \in P} \sum_{b \in B} (order_{scptb} \times ordersize_{pb} \leq supcapa_s) \qquad \forall s \in S, t \in T \tag{20}$$

$$\sum_{s \in S} \sum_{p \in P} \sum_{t \in T} \sum_{l=1}^{L} (order_{scptb} \times sales_{cptl}) \leq rcopen_c \times M \qquad \forall c \in C \tag{21}$$

$$inv_{cp(t-1)l} = initialinv_{cpl} + \sum_{s \in S} \sum_{b \in B} (order_{scptb} \times ordersize_{pb}) - sales_{cptl} \tag{22}$$
$$\forall c \in C, p \in P, t = 1, l = L$$

$$inv_{cpt(l-1)} = \sum_{s \in S} \sum_{b \in B} (order_{scptb} \times ordersize_{pb}) - sales_{cptl} \tag{23}$$
$$\forall c \in C, p \in P, t \in T \mid t \geq 1, l = L$$

$$inv_{cpt(l-1)} = inv_{cp(t-1)l} - sales_{cptl} \tag{24}$$
$$\forall c \in C, p \in P, t \in T \mid t \geq 1, l \in H \mid 2 < l \leq L - 1$$

$$sales_{cptl} = 0 \qquad \forall c \in C, p \in P, t = 1, l \in H \mid l \leq L-1 \qquad (25)$$

$$inv_{cptl} = 0 \qquad \forall c \in C, p \in P, t = 1, l \in H \mid l \leq L-2 \qquad (26)$$

$$waste_{cp} \leq \sum_{t \in T} demand_{cpl} \times maxwaste_{cp} \qquad \forall c \in C, p \in P \qquad (27)$$

$$sales_{cptl}, pr_{cptl}, inv_{cptl} \geq 0 \qquad \forall c \in C, p \in P, t \in T, l \in H \qquad (28)$$

$$rcopen_c, order_{scptb} \in \{0,1\} \, \forall s \in S, c \in C, p \in P, t \in T, b \in B \qquad (29)$$

Maximizing Freshness. This model aims to maximize freshness represented by $freshpercentage$ while taking the profit into consideration. The constraint (31) is added to account for the maximum possible value of $freshpercentage$, which is 1 or 100% when all sales are fresh products. The model maximizes $totalprofit$ while being subject to the same constraints (16)–(29) as in the profit-maximizing model. The optimal value of $totalfreshsales$ from this model is defined as F_f^*.

$$\text{Objective Function: max } totalprofit \qquad (30)$$

$$\text{Subject to: } totalfreshsales - totalsales = 0 \qquad (31)$$

$$(16) - (29)$$

Minimizing CO$_2$ Emission. This model aims to minimize the $totalCO_2$ $emission$ function. Two additional constraints, (33) and (34), are added to set minimum targets for $totalprofit$ and $freshpercentage$ with allowance factors $allowprofit$ and $allowfresh$, respectively. The remaining constraints follows the constraints (16) - (29) as in the profit maximizing model. The optimal value of $totalCO_2emission$ from this model is defined as F_c^*.

$$\text{Objective Function: min } totalCO_2emission \qquad (32)$$

$$\text{Subject to: } totalprofit \geq F_p^* \times (1 - allowprofit) \qquad (33)$$

$$totalfreshsales - totalsales \times (1 - allowfresh) \geq 0 \qquad (34)$$

$$(16) - (29)$$

Tri-Objective Optimization. This model uses a min-max technique, modified from the techniqie used by Olapiriyakul et al., [16], to obtain a set of non-dominated solutions that minimize deviation from the ideal outcomes. The normalized deviations of $totalprofit$, $totalfreshsales$, and $totalCO_2emission$ are defined as σ_p, σ_f, and σ_c, respectively.

$$\sigma_p = \frac{F_p^* - F_p}{F_p^* - F_p^{min}} \qquad (35)$$

$$\sigma_f = \frac{F_f^* - F_f}{F_f^* - F_f^{min}} \tag{36}$$

$$\sigma_c = \frac{F_c - F_c^*}{F_c^{max} - F_c^*} \tag{37}$$

where F_p^*, F_f^*, and F_c^* represent the best results obtained by solving the single objective models, while F_p^{min}, F_f^{min}, and F_c^{max} represent the worst results.

The model minimizes a new continuous decision variable z while subject to the deviation constraints (39) - (41) and the constraints (16) - (29) as in the maximizing profit model.

$$\text{Objective function: } \min z \tag{38}$$
$$\text{Subject to: } \sigma_p \leq z \tag{39}$$
$$\sigma_f \leq z \tag{40}$$
$$\sigma_c \leq z \tag{41}$$
$$(16) - (29)$$

4   A Case Study Problem

This study applies the proposed model to a two-tier food supply chain problem where a company can order food items A and B from five suppliers and deliver them to their retail centers in five districts. This case represents a general supplier-to-retailer food distribution network. Due to the paper length limitation, only the overview of parameters of the case study is given in Table 2, followed by a brief description of the analysis results.

Our results in Table 3 include both single- and tri-objective optimization outcomes obtained by using IBM ILOG CPLEX Optimization Studio in solving the proposed optimization models. The machine specifications included an Intel Core i5-7360U processor running at 3.6 GHz, and 8 GB of LPDDR3 RAM. The model is initially optimized under single-objective scenarios to gain insights into the possible range of sustainability performance, which includes profit, freshness, and CO_2 emissions. The sales amount, number of small and large batch orders, waste amount, and revenue are reported. The total cost encompasses retail center opening, ordering, products, inventory, and waste disposal. Subsequently, the total profit and satisfaction level are calculated. When maximizing profit, the number of large batch orders is the highest among all the solutions, leading to a large lot discount and a profit of \$233,920,559. However, the freshness of products offered to customers appears to be only 86%. When the freshness is maximized, the network mainly relies on small batch orders to ensure a 100% freshness level for customers. Nonetheless, this results in large CO_2 emissions due to frequent deliveries. In this scenario, the waste generated from item B in District Three is restricted by the allowable waste amount. When minimizing

Table 2. Parameters of the case study

Parameter	Value
Total period (T)	30 periods
$demand_{cpt}$	AVG: 986, SD: 806, Min: 27, Max: 3976 (unit)
$price_p$	A: 28.82; B: 57.64 (USD)
$prodcost_p$	A: 8.65; B: 14.41 (USD)
$invcost_p$	A: 0.58, B: 0.86 (USD)
$weight_p$	1 kg for all items
$weighttruck$	8000 kg
$maxwaste_{cp}$	RC1-5, A: 18, 12, 2, 4, 8; RC1-5, B: 13, 10, 1, 3, 10 (ton)
$supcapa_s$	200,000 units per period for all suppliers
$dist_{sc}$	AVG: 167; SD: 114; Min: 50; Max: 500 (km)
$initialinv_{cpl}$	1000 units with shelf life 5 period for all items and retail centers
$ordersize_{pb}$	A, Small: 5000; A, Large: 10000; B, Small: 2500; B, Large: 5000 (unit)
$ordercost_{pb}$	A, Small: 2.08; A, Large: 2.88; B, Small: 1.68; B, Large: 2.08 (USD per km)
$rcopencost_c$	RC1-5: 144079, 28816, 57632, 28816, 57632 (USD)
$productlife_p$	5 periods for all items
fr	3 periods
w	1 period
$wastecost$	10%
CO_2tkm	0.0674 kg CO_2 e per tkm
CO_2kg	2 kg CO_2 e per kg of waste
$allowprofit$	20%
$allowfresh$	20%
(1 USD = 34.7 THB)	

CO_2, the sales are minimized, just enough to achieve the sales target, resulting in CO_2 emissions of only 14,146 kg of CO_2 equivalent. Under the multi-objective scenario, our model identifies the tri-objective solution that performs relatively well in all three sustainability aspects, with the highest average satisfaction level of 66%.

The following result discussion can also be made based on the results in Table 6. Lot size decision shows to create a significant impact on all the sustainability performance. CO_2 emissions, both from transportation and waste, can be greatly affected by the sales quantity. Facility location only affects transport-related CO_2 emissions. Adding a retail center in District 3 to the solutions helps reduce the transport-related CO_2 due to its close proximity to suppliers. Less waste is created when the freshness percentage is allowed to be less, as shown when comparing the tri-objective solution to that of the freshness maximization. The tradeoffs between the objectives under single-objective optimization are evident. Maximizing profit clearly yields low freshness levels and high CO_2 emission levels. Maximizing freshness leads to significant CO_2 emissions. However, through the tri-objective technique, a well-balanced solution is achieved, with priority given to profit.

Table 3. Summary result of the case study

		Single-Objective			Tri-Objective
		Max Profit	Max Freshness	Min CO_2 Emission	
Profit	Total Product Sales (Unit)	277,054	259,339	186,283	254,905
	Opened Retail Center	1, 2, 4, and 5	1, 2, 4, and 5	All	All
	Small Batch Order Number	52	64	31	55
	Large Batch Order Number	16	10	14	11
	Waste (ton)	25.45	34.10	1.22	8.10
	Total Revenue ($)	406,849,000	388,563,000	308,180,500	384,143,500
	Retail Center Opening Cost ($)	9,000,000	9,000,000	11,000,000	11,000,000
	Ordering Cost ($)	39,873,921	42,102,033	26,981,044	37,912,719
	Product Cost ($)	113,250,000	112,750,000	76,750,000	102,750,000
	Inventory Cost ($)	9,987,040	8,777,710	6,117,816	8,106,038
	Disposal Cost ($)	817,480	1,157,730	48,563	1,880,331
	Total Cost ($)	172,928,441	173,787,473	120,897,423	161,649,087
	Total Profit ($)	233,920,559	214,775,527	187,283,077	222,494,413
	Satisfaction Level	100%	59%	0%	76%
Freshness	Fresh Product Sales (Unit)	238,508	259,339	162,094	235,844
	Total Product Sales (Unit)	277,054	259,339	186,283	254,905
	Freshness Percentage	86%	100%	87%	93%
	Satisfaction Level	0%	100%	7%	47%
CO_2 Emission	CO_2 Emission from Transportation (kg CO_2 e)	36,903	37,509	11,712	20,390
	CO_2 Emission from Waste (kg CO_2 e)	50,892	68,206	2,435	16,191
	Total CO_2 Emission (kg CO_2 e)	87,795	105,715	14,146	36,581
	Satisfaction Level	20%	0%	100%	75%
Average Satisfaction Level		*40%*	53%	36%	66%
Computational Time (Second)		19	16	60	71

5   Conclusion

This research is the first sustainable perishable product SCND that addresses social impact related to the consumption of fresh products and waste justice. The modeling and problem analysis sections of this research also illustrate the connections between perishable characteristics, SCM parameters, and the burden of waste accumulation on communities. This significantly strengthens our understanding of sustainable SCND for perishable goods, expanding the scope of the social sustainability aspect in the literature. Our analysis shows that the proposed model can be solved in single- and multi-objective scenarios to pursue specific outcome preferences. The proposed multi-objective technique employs a hierarchical approach with varying priority levels to address these conflicting goals. By setting deviations from the upper bound for each objective, the model can generate trade-off solutions and gain insights into the interrelationships between different objectives. Our results suggest that hierarchical and goal-oriented approaches effectively solve sustainable SCND problems involving multiple optimization objectives with different priorities.

Regarding practical implications, the proposed approach allows decision-makers to adjust the allowable reduction in profit and freshness as well as waste generation criteria, to obtain solutions that align with their managerial preferences. Regarding the limitations of our study, CO_2 emissions in our model are solely based on transportation and waste generation, neglecting carbon emissions resulting from inventory and facility. This omission could significantly impact the optimal solution for products requiring refrigeration or special infrastructure. Additionally, our model assumes fixed selling prices, which limits its applicability in cases where prices vary depending on the remaining shelf lives of products. The future study can explore a broader scope of supply chain decisions, particularly focusing on the disposal and closed-loop components, whose waste-induced sustainability impacts can significantly affect the SCND decisions Olapiriyakul [17]. This expansion will ensure waste reduction and promote both sustainable consumption and production practices.

References

1. World Bank: Cutting Food Loss and Waste Can Deliver Big Wins for Countries' Food Security and Environment. World Bank Press Release (9 2020), https://www.worldbank.org/en/news/press-release/2020/09/28/cutting-food-loss-and-waste-can-deliver-big-wins-for-countries-food-security-and-environment, accessed: June 1, 2023
2. Buzzelli, M., Jerrett, M., Burnett, R., Finklestein, N.: Spatiotemporal perspectives on air pollution and environmental justice in hamilton, canada, 1985–1996. Ann. Assoc. Am. Geogr. **93**(3), 557–573 (2003)
3. Banasik, A., Kanellopoulos, A., Claassen, G., Bloemhof-Ruwaard, J.M., van der Vorst, J.G.: Closing loops in agricultural supply chains using multi-objective optimization: A case study of an industrial mushroom supply chain. Int. J. Prod. Econ. **183**, 409–420 (2017)
4. Olapiriyakul, S., Nguyen, T.T.: Land use and public health impact assessment in a supply chain network design problem: A case study. J. Transp. Geogr. **75**, 70–81 (2019)
5. Biuki, M., Kazemi, A., Alinezhad, A.: An integrated location-routing-inventory model for sustainable design of a perishable products supply chain network. J. Clean. Prod. **260**, 120842 (2020)
6. Tsao, Y.C., Zhang, Q., Zhang, X., Vu, T.L.: Supply chain network design for perishable products under trade credit. J. Ind. Prod. Eng. **38**(6), 466–474 (2021)
7. Torabzadeh, S.A., Nejati, E., Aghsami, A., Rabbani, M.: A dynamic multi-objective green supply chain network design for perishable products in uncertain environments, the coffee industry case study. International Journal of Management Science and Engineering Management **17**(3), 220–237 (2022)
8. Daghigh, R., Jabalameli, M., Amiri, A., Pishvaee, M.: A multi-objective location-inventory model for 3pl providers with sustainable considerations under uncertainty. Int. J. Ind. Eng. Comput. **7**(4), 615–634 (2016)
9. Kalantari, F., Hosseininezhad, S.J.: A multi-objective cross entropy-based algorithm for sustainable global food supply chain with risk considerations: A case study. Computers & Industrial Engineering **164**, 107766 (2022)
10. Tirkolaee, E.B., Aydin, N.S.: Integrated design of sustainable supply chain and transportation network using a fuzzy bi-level decision support system for perishable products. Expert Syst. Appl. **195**, 116628 (2022)
11. Wang, C.N., Nhieu, N.L., Chung, Y.C., Pham, H.T.: Multi-objective optimization models for sustainable perishable intermodal multi-product networks with delivery time window. Mathematics **9**(4), 379 (2021)

12. Yakavenka, V., Mallidis, I., Vlachos, D., Iakovou, E., Eleni, Z.: Development of a multi-objective model for the design of sustainable supply chains: the case of perishable food products. Annals of Operations Research 294(1–2), 593–621 (2020), https://www.scopus.com, cited By :33

13. Chan, F.T., Wang, Z., Goswami, A., Singhania, A., Tiwari, M.K.: Multi-objective particle swarm optimisation based integrated production inventory routing planning for efficient perishable food logistics operations. Int. J. Prod. Res. **58**(17), 5155–5174 (2020)

14. Jaigirdar, S.M., Das, S., Chowdhury, A.R., Ahmed, S., Chakrabortty, R.K.: Multi-objective multi-echelon distribution planning for perishable goods supply chain: A case study. International Journal of Systems Science: Operations & Logistics **10**(1), 2020367 (2023)

15. Jouzdani, J., Govindan, K.: On the sustainable perishable food supply chain network design: A dairy products case to achieve sustainable development goals. J. Clean. Prod. **278**, 123060 (2021)

16. Olapiriyakul, S., Pannakkong, W., Kachapanya, W., Starita, S.: Multiobjective optimization model for sustainable waste management network design. Journal of Advanced Transportation 2019 (2019)

17. Olapiriyakul, S.: Designing a sustainable municipal solid waste management system in pathum thani, thailand. Int. J. Environ. Technol. Manage. **20**(1–2), 37–59 (2017)

Security and Privacy in Machine Learning

A Novel Privacy-Preserving Federated Learning Model Based on Secure Multi-party Computation

Anh Tu Tran[1]([✉]), The Dung Luong[1], and Xuan Sang Pham[2]

[1] Academy of Cryptography Techniques, Hanoi, Vietnam
tutran@actvn.edu.vn
[2] Hanoi University of Science and Technology, Hanoi, Vietnam

Abstract. Although supporting training deep learning models distributed without disclosing the raw privacy data, federated learning (FL) is still vulnerable to inference attacks. This paper proposes ComEnc-FL, a privacy-enhancing federated learning system that combats these vulnerabilities. ComEnc-FL uses secure multi-party computation and parameter encoding to reduce communication and computational expenses. ComEnc-FL surpasses typical secure multi-party computation systems in training time and data transfer bandwidth. ComEnc-FL matches the base FL framework and outperforms differential privacy-safe frameworks. We also show that parameter compression reduces encryption time, improving model performance over the FL.

Keywords: Privacy-preserving Deep Learning (PPDL) · Federated Learning (FL) · Privacy · Secure Multi-party Computation (SMC)

1 Introduction

Deep neural networks (DNNs) are a valuable tool for various tasks, such as Computer Vision (CV), Natural Language Processing (NLP), Recommendation Systems (RS), and Cyber Security, in the era of large data [9]. The efficacy of DNNs depends on access to vast volumes of data, which may be problematic for individual entities due to storage limits and data gathering challenges [4].

A conventional strategy is to pool data from many sources into a central server for training; however, privacy considerations require revision. Protect sensitive data like medical photos and user-specific information. The General Data Protection Regulation (GDPR) emphasizes the need to protect privacy while transferring data across businesses [3].

Federated Learning (FL) [7] is a promising answer to this issue. FL enables multiple parties to train a DNN without exchanging data. Local models are trained on each participant's data and sent to a central server for aggregation. The aggregated model, incorporating all parties' knowledge, is then returned to participants. This iterative procedure continues until the model performs well.

K. Honda et al. (Eds.): IUKM 2023, LNAI 14376, pp. 321–333, 2024.
https://doi.org/10.1007/978-3-031-46781-3_27

FL allows data privacy by letting data owners control their data, unlike standard privacy-preserving machine learning systems that send all training data to a central location. FL is a big step forward in protecting personal data. FL has several practical uses, such as loan status prediction, health evaluation, and next-word prediction [5].

Data owners in FL have exclusive rights over training data. FL's simplicity makes it susceptible to inference attacks, revealing essential training data for both sides [10,11]. Differential privacy (DP) and secure multi-party computation (SMC) are risk mitigation strategies for sensitive data breaches that would be intolerable.

DP minimizes inference attacks using the final model or model updates by injecting optimized noise into model updates using differentially private processes [2]. DP guarantees privacy, although strategies using it are commonly criticized for producing inaccurate and noisy models [14]. Privacy and model performance trade off when DP and FL are used.

In contrast, SMC approaches prevent curious or untrustworthy aggregators from examining private models without affecting accuracy. Examples of this approach include threshold Paillier, [13], functional encryption [15], pairwise masking protocols [1], and (partial) homomorphic encryption [8,12,16]. Alternative data leak prevention approaches combine DP and SMC procedures for robust differential private assurances and high-quality model performance [12,13].

Existing SMC and DP methods cannot prevent colluding party inference hazards. Most cryptography-based FL models need all parties to share keys and not collude with the server. Real-world applications cannot require all parties to use the same keys without cooperating with the server.

We present ComEnc-FL, an efficient, scalable, and secure aggregation-based privacy-preserving federated learning, to overcome the abovementioned restrictions and find a satisfactory solution to the problem of secure multi-party aggregation efficiency while counteracting potential collusion within the model. ComEnc-FL relies on an ElGamal-based encryption scheme coupled with model parameters compressing method to streamline and enhance communication and computation. It effectively manages the issue of colluding parties with the server and the communication and computational costs of SMC-based FL. Consequently, ComEnc-FL addresses the deficiencies of previous cryptography-based secure aggregation systems.

The main contributions of this paper are as follows:

- We introduce the ComEnc-FL framework that enhances privacy while addressing the limitations of previous state-of-the-art cryptographic secure aggregation systems. By leveraging the ElGamal-based encryption scheme, the proposed framework addresses the issue of colluding parties, a common problem in most cryptography-based federated learning models. By adopting a unique parameter encoding strategy, ComEnc-FL mitigates computational and communication costs commonly associated with secure aggregation in federated learning.

– We have conducted a series of experiments to critically appraise the efficacy of ComEnc-FL compared to the base-FL framework. Furthermore, we perform an exhaustive analysis concerning the efficiency of ComEnc-FL, reducing computation and communication costs and elucidate the influence of the number of decimal places on performance metrics.

The following sections of this paper provide a systematic discussion of our work. Section 2 introduces the ComEnc-FL we propose. We outline our experimental setup and analysis in Sect. 3. Ultimately, Sect. 4 contains our conclusion and our plans for future work.

2 Methodology

This section describes the proposed secure sum protocol and parameter encoding strategy. Then, we apply these techniques to the federated learning framework to obtain the ComEnc-FL framework.

2.1 Framework Description

Our proposed framework is based on federated learning algorithms, which involve a central server and a set of clients $\mathcal{P} = \{P_1, P_2, \dots P_n\}$. Each client owns a private local dataset D_i and collaboratively trains a global model without revealing their local data. The central server initializes a global model (W_0) and shares the necessary hyperparameters, such as local epoch (E) and batch size (B), with all participating clients.

In the communication round t, clients train the received model W_t on their local datasets with the local epoch E and batch size B. After the local training, clients obtain local model W_t^i and share them with the server to update the current global model. The server aggregates the new global model W_{t+1}. After that, the server broadcasts the up-to-date global model to every client. The successive communication round happens in the same manner.

The federated learning framework ensures that clients control their data, preventing direct data leakage. However, inference attacks may happen in the training phase and harm the privacy of the framework. Exchanging messages with each other, the server and clients may exploit the model updates to leak sensitive information. Hence, to ensure the privacy of the model sharing during the training phase, we propose a secure multiparty computation protocol that employs a modified ElGamal cryptosystem. Moreover, we also apply a parameter encoding strategy to the FL framework to reduce the computation cost when encrypting and the communication cost when exchanging messages.

2.2 Threat Model

In our ComEnc-FL setting, we consider following threat model:

- The central server and all clients are honest-but-curious or semi-honest. In the semi-honest context, even the server and clients correctly follow the protocol specification, they may try to inspect the model updates to learn private information.
- The central server and participants may collude, potentially exchanging messages to gain access to information that should remain confidential.

2.3 Secure Multiparty Computation Protocol

2.3.1 Protocol We propose a secure multiparty computation protocol that adopts the additively homomorphic property of a variant of ElGamal encryption. The computational difficulty of the discrete logarithm problem and the related ElGamal cryptosystem determine the privacy of our protocol. The protocol allows the server to aggregate the encrypted messages submitted by clients into the desired sum.

The proposed protocol consists of two phases illustrated as follows.

Initialization phase: To start the protocol, our protocol requires the following parameters:

- We choose a prime p and a prime q such that $p - 1$ is a multiple of q. Let $\mathbf{g}$ be a generator of a cyclic group $\mathbb{Z}_p$ that satisfies $g \neq 1$ and $\mathbf{g}^q \bmod p = 1$. In our proposed protocol, all computations are done in $\mathbb{Z}_p$ and (p, q, g) are the public parameters shared among the server and all clients.
- Each client P_i has already owned two private keys $\mathbf{x}_i, \mathbf{y}_i \in \{1, 2, \ldots, p - 1\}$ and the corresponding public keys $X_i = \mathbf{g}^{\mathbf{x}_i}, Y_i = \mathbf{g}^{\mathbf{y}_i}$. Note that the private key $\mathbf{y}_i$ is one-time use. Each client P_i submits the public keys $M_i = \{X_i, Y_i\}$ to the central server.
- The server then pre-computes

$$X = \prod_{i=1}^{n} X_i; \ Y = \prod_{i=1}^{n} Y_i$$

and broadcasts the public values $M = \{X, Y\}$ to all clients through public networks.

Secure n-clients Sum Computation Phase: Once the necessary parameters have been established, we could conduct the main phase of the protocol. This phase consists of two steps described in detail in the following.

At the first step, each client encrypts his model's secret parameters using the public value $M = \{X, Y\}$ and the private keys $\mathbf{x}_i, \mathbf{y}_i$. Denote the client P_i's models be W_i. Each client computes the encrypted messages V_i by the function $f : \mathbb{Z}_p \times \mathbb{R} \to \mathbb{R} : V_i = \frac{X^{\mathbf{y}_i}}{Y^{\mathbf{x}_i}} \mathbf{g}^{W_i}$ where $X^{\mathbf{y}_i}, Y^{\mathbf{x}_i}, \frac{X^{\mathbf{y}_i}}{Y^{\mathbf{x}_i}} \in \mathbb{Z}_p$ and $\mathbf{g}^{V_i} \in \mathbb{Z}$. When the process of computing V_i is done, each client submits V_i to the central server.

At the second step, after receiving the messages V_i from all clients, the server computes $V = \prod_{i=1}^{n} V_i$. Subsequently, the server performs Shank's algorithm to obtain the sum value S that satisfies $\mathbf{g}^S = V$.

2.3.2   Proof of Correctness

Theorem 1. *The proposed protocol for secure n−parties sum exactly computes the sum of all clients' secret values.*

Proof. We prove that if the central server finds out a value S such that the equation $\mathbf{g}^S = V$ happens, then S is the sum of all clients' secret values. Assume that $\mathbf{g}^S = V$. Then we have:

$$\mathbf{g}^S = V = \prod_{i=1}^{n} V_i = \prod_{i=1}^{n} \frac{\mathbf{g}^{W_i} X^{\mathbf{y}_i}}{Y^{\mathbf{x}_i}} = \mathbf{g}^{\sum_{i=1}^{n} W_i} \prod_{i=1}^{n} \frac{\left(\prod_{j=1}^{n} X_j\right)^{\mathbf{y}_i}}{\left(\prod_{j=1}^{n} Y_j\right)^{\mathbf{x}_i}}$$

$$= \mathbf{g}^{\sum_{i=1}^{n} W_i} \prod_{i=1}^{n} \frac{\left(\mathbf{g}^{\sum_{j=1}^{n} \mathbf{x}_j}\right)^{\mathbf{y}_i}}{\left(\mathbf{g}^{\sum_{j=1}^{n} \mathbf{y}_j}\right)^{\mathbf{x}_i}} = \mathbf{g}^{\sum_{i=1}^{n} W_i} \frac{\mathbf{g}^{\sum_{j=1}^{n} \mathbf{x}_j \sum_{i=1}^{n} \mathbf{y}_i}}{\mathbf{g}^{\sum_{j=1}^{n} \mathbf{y}_j \sum_{i=1}^{n} \mathbf{x}_i}} = \mathbf{g}^{\sum_{i=1}^{n} W_i}. \quad \square$$

2.3.3   Privacy Analysis   We have the following security definition:

Definition 1. *Assume that each user P_i has the private keys x_i, y_i and the public keys X_i, Y_i. A protocol protects private W_i against the server and t corrupted clients in the semi-honest model if, for all $I \subseteq \{1, 2, \ldots, n\}$, there exists a probabilistic polynomial-time algorithm M such that*

$$\{M(W, [W_i, \mathbf{x}_i, \mathbf{y}_i]_{i \in I}, [X_i, Y_i]_{j \notin I})\} \stackrel{c}{\equiv} \{\text{view}_{\{P_i\}_{i \in I}}([V_i, \mathbf{x}_i, \mathbf{y}_i]_{i=1}^{n})\}$$

where $\stackrel{c}{\equiv}$ is computational indistinguishability.

Intuitively, this definition means that $\|I\|$ corrupted parties can learn nothing about the private data of other parties even though they have the sum of the private message W, the corrupted clients' knowledge, and the public keys. Therefore, this definition is equivalent to the fact that the server and the corrupted clients jointly infer nothing beyond W.

Theorem 2. *The protocol for secure n-clients sum presented in Fig protects each honest client's privacy against the server and up to $(n-2)$ corrupted clients.*

Proof. Without the loss of generality, we assume that two clients P_1 and P_2 do not collude while the central server and the others $P_i \| i \in I = \{3, 4, \ldots, n\}$ collude with each other. In our secure protocol, each client only sends the encrypted value V_i and two public keys X_i, Y_i to the server. X_i, Y_i are random values because the private keys x_i, y_i are uniformly random. To prove the below theorem, we must construct a probabilistic polynomial-time algorithm that can simulate the computation for the messages W_1 and W_2 using only the final sum S, corrupted clients' knowledge $\{x_i, y_i, V_i\}$ and public keys X_1, Y_1, X_2, Y_2.

We denote the algorithm that satisfies the above assumption as M. Algorithm M uses $(u_{12}, v_{12}) = (g^{w_1} g^{x_2 y_1}, g^{x_2}), (u_{21}, v_{21}) = (g^{w_2} g^{x_1 y_2}, g^{x_1})$ as its input to simulate W_1, W_2 as follows:

$$P_1' = \frac{u_{12}.Y_1^{\sum_{i \in I} x_i}.g^{S - \sum_{i \in I} W_i}}{u_{21}.X_1^{\sum_{i \in I} y_i}}, \quad P_2' = \frac{u_{21}.Y_2^{\sum_{i \in I} x_i}.g^{S - \sum_{i \in I} W_i}}{u_{12}.X_2^{\sum_{i \in I} y_i}}.$$

According to the Definition 1, our protocol is semantically secure. □

2.4 Parameter Encoding Strategy

Deep learning neural networks can easily contain millions of trainable parameters. This enormous number of trainable parameters leads to a dramatic increase in computation and communication costs. The computation time taken for encrypting and decrypting clients' parameters may dominate the running time of our federated learning framework. In addition, setting a channel to transfer these clients' parameters is a challenge that requires a large amount of bandwidth [8]. To tackle the abovementioned problems, we propose an integer encoding strategy to reduce the time-consuming encryption and the cost during communication.

Inspired by the natural resilience of deep neural networks to low-precision fixed-point representations [6], our proposed strategy converts floating-point parameters to integers that present fewer decimal places. Our encoding technique contains three phases described in detail as follows.

At the server's site, the server already owns an exponential factor (γ) that decides the number of decimal places remaining after encoding clients' parameters. Before starting the secure protocol, the server assigns value to this factor and broadcasts it to every client. Note that the exponential factor remains constant during the protocol.

In the next phase, each client P_i encodes its float parameters into integers. After the local training, client P_i encodes its model's parameters by multiplying ten by the exponential factor to limit the precision of model weights as $W_i \leftarrow 10^\gamma W_i$. The client rounds the encoded values to integers and sends the message W_i to the central server.

In the final phase, the central server uses the value γ to decode the received parameters from client P_i.

2.5   ComEnc-FL Framework

Integerating the proposed secure multiparty sum protocol and the parameter encoding technique into the federated learning, we obtain the ComEnc-FL framework described as Framework 1.

Framework 1: The ComEnc-FL framework

Input: n clients $\mathcal{P} = \{P_1, P_2, \ldots, P_n\}$; Each client has private keys x_i, y_i correspond public keys X_i, Y_i and the number of data samples m_i.

1. Initialization phase

- The server initializes a global model W_0 and an exponential factor γ. The server then sends W_0 and γ to every clients.
- Each client informs public keys $\{X_i, Y_i\}$ and m_i to the server. The server then precomputers $X = \prod_{i=1}^{n} X_i, Y = \prod_{i=1}^{n} Y_i$, and sends back to clients.

2. Training phase

- **Client site (n clients in parallel):**
  - Client P_i trains model W_t on local data D_i to obtain W_{t+1}^i.
  - Client P_i encodes W_{t+1}^i by using γ as $W_{t+1}^i \leftarrow round(10^\gamma W_{t+1}^i)$.
  - Client P_i computes $V_i = \frac{X^{y_i}}{Y^{x_i}} g^{W_{t+1}^i}$ and sends V_i to the server.
- **Server site:**
  - Obtain V_i from client P_i.
  - Calculate $V = \prod_{i=1}^{n} V_i$ and execute the baby-step giant-step algorithm to obtain $S = \sum_{i=1}^{i} W_{t+1}^i$.
  - Aggregate the new global model as $W_{t+1} \leftarrow \sum_{i=1}^{n} \frac{m_i}{M} \frac{W_{t+1}^i}{10^\gamma}$.
  - Send W_{t+1} to every client.

2.6   Computation Cost and Communication Cost

2.6.1   Computation Cost Analysis Considering the computational demand, especially in encryption, we emphasize the computation time for $V_i = \frac{g^{w_t^i} X^{y_i}}{Y^{x_i}}$. Client P_i must carry out modular exponentiation, multiplicative inverse, and multiplication with large integers. We denote the following:

- T_e: Time for a modular exponentiation operation.
- T_m: Time for a modular multiplication operation.
- T_{in}: Time for computing a modular multiplicative inverse.

Among those operations, computing the modular exponentiation is the most expensive operation. For our proposed protocol, the total time to prepare the message V_i at the client P_i's site is $3T_e + 2T_m + T_{in}$.

We adopt an integer encoding strategy to decrease T_e for g^{W_i}. We significantly reduce computation time by representing float parameters W_i as integers. Our protocol, functioning strictly with integers, necessitates the conversion of floating-point parameters W_i into integers. Conventionally, floating-point computations maintain precision through adherence to the IEEE 754 standard, allowing for representation with 15–17 decimal places. Therefore, without limiting the number of decimal places, g^{W_i} is computed with W_i in the range of $[10^{15}, 10^{17}]$. On the other hand, by applying our encoding strategy, we could reduce the exponent W_i by choosing a smaller exponential factor γ.

For instance, by selecting a smaller exponential factor γ (such as 5), we compute g^{10^5} rather than $g^{10^{15}}$. This strategy significantly expedites the encryption phase, especially for complex neural networks with millions of parameters.

2.6.2   Communication Cost Analysis We inspect the cost during the communication of the proposed framework.

In the initialization phase, our framework needs to go through these steps:

- The server sends two hyperparameters (B and E) to n clients. In our case, a 32-bit packet is enough to transfer these hyperparameters. Hence, the total communication bandwidth is $2 \times 32 \times n$ bits.
- The server sends the initial global model to every client. This step requires $n \times ModelSize$ bits.
- Clients submit values X_i, Y_i and m_i to the server. This step costs $(2 \times ElGamalKeysize + 32) \times n$ bits of bandwidth.
- The server computes X, Y and resends to every client. The bandwidth needed to communicate is $2 \times ElGamalKeysize \times n$ bits.

We analyze the communication cost of the training phase by considering one communication round. After the local training, clients send their models to the server. Receiving local models, the server updates the global model and resends the model to all clients with a message of $ModelSize$ bits. Hence, we need $2n$ communication sessions and the total bandwidth of $2 \times ModelSize \times n$ in each communication round.

By implementing our integer encoding strategy, we can reduce communication costs. We analyze this benefit of our method by comparing it to the basic federated learning framework with the floating number transmission. In the normal condition of federated learning, to ensure the precision of floating-point number computation, we use IEEE 754 double-precision binary floating-point format to represent float numbers. This standard allows us to present a number between 10^{-308} and 10^{308} with 15–17 decimal digits precision. Performing one float number requires 64 bits. On the other hand, adopting our encoding strategy, we only need 32 bits to present an integer between 10^{-10} and 10^{10}. Therefore, we can limit the encoding precision under 10 decimal places to reduce the model's size by half. This reduction becomes significant when the federated learning framework consists of many clients or uses a complex deep learning model.

3   Experimental Results and Analysis

3.1   Experimental Setup

3.1.1   Datasets and Model Architecture   To validate the performance of the ComEnc-FL framework, we conducted the experiments based on two datasets: the MNIST and the CSIC 2010 dataset.

- The MNIST dataset, a public repository of handwritten digits as 32×32 images, consists of 60,000 training and 10,000 test samples.
- CSIC 2010 is a dataset that has been widely used for testing web intrusion detection systems. This dataset is a set of HTTP requests that contain more than 36000 benign and 25000 anomalous samples.

We only consider that the data divided among the clients follows the IID method. For each dataset, we combine all the samples and divide this dataset randomly into 80% for training and 20% for testing. We then shuffle the training set and partition it into clients so that every client has the same percentage of each label.

For the model architecture, we train two different neural network models per dataset: CNN Model for MNIST dataset and CLCNN Model for CSIC 2010 dataset.

3.2   Experimental Results

We implement four experimental setups to qualify the performance of the proposed framework and analyze the effect of several hyperparameters on the resulting models. We assume all the clients will attend the training at every communication round and use the same local batch size, 64. In the first experiment, we evaluate our framework in contrast to different baselines. In the second experiment, we vary the number of clients joining the framework to see how this factor affects the model. We examine the encoding techniques in the third experiment by limiting clients' parameters under different precision levels. Finally, in the fourth experiment, we analyze the proposed approach from the perspective of total encryption time.

3.2.1   Overall Model Performance   In this part, we conduct experiments to evaluate the quality of our proposed framework. We run 50 communication rounds on two above datasets and compare the obtained results with two baselines:

- The general Average Federated Learning training without any security setting (General Fed-Avg).
- The general Average Federated Learning training with additional differential privacy in small and large noise cases.

Table 1 presents the results of our ComEnc-FL framework, where clients' model weights are rounded to 10 decimal places. We evaluate both accuracy and F1-score, the latter encompassing precision and recall. Compared to Fed-Avg training with high noise injection, our framework performs and maintains parity with Fed-Avg training at low noise levels. However, in the absence of security measures (general Fed-Avg), a marginal decrease in ComEnc-FL's accuracy and F1-score is noted, attributed to the effect of rounding clients' parameters.

Table 1. The accuracy and F1-score comparison after 50 communication rounds.

Metric	Dataset	Model			
		ComEnc-FL	General Fed-Avg	Fed-Avg (Small noise)	Fed-Avg (Large noise)
Accuracy	MNIST	0.9834	0.9860	0.9854	0.8223
	CSIC 2010	0.9779	0.9873	0.9785	0.8412
F1-score	MNIST	0.9827	0.9860	0.9854	0.8189
	CSIC 2010	0.9757	0.9794	0.9763	0.8213

3.2.2 Impact of the Number of Clients

Running through 50 communication rounds with 5, 10, 20, 40, and 50 clients on MNIST and CSIC 2010 datasets and limiting parameters' precision level under 2 decimal places, we obtain the results summarized in Table 2. We see a trend that the more clients join the training phase, the lessen metric values our framework achieves. This trend happens since the smaller amount of data distributed to each client causes a reduction in the generality of each client model and leads to an expansion in the required training time to get the equivalent results to the cases of the small number of joint clients.

Table 2. The accuracy, F1-score, precision, and recall comparison for different number of clients after 50 communication rounds.

The number of clients	Accuracy		F1-Score		Precision		Recall	
	MNIST	CSIC 2010	MNIST	CSIC 2010	MNIST	CSIC 2010	MNIST	CSIC 2010
5	0.9381	0.9357	0.9390	0.9279	0.9459	0.9628	0.9381	0.8955
10	0.9236	0.9266	0.9251	0.9161	0.9368	0.9651	0.9236	0.8719
20	0.9041	0.8008	0.9022	0.7889	0.9168	0.9318	0.9041	0.5806
40	0.8481	0.7844	0.8418	0.7154	0.8649	0.9001	0.8481	0.5331
50	0.8308	0.7488	0.8341	0.6633	0.8673	0.8653	0.8308	0.5378

3.2.3 Impact of the Number of Decimal Places

As we explained in Sect. 2, the floating point parameters of a neural network must be represented and encoded into integers that our SMC protocol takes as inputs. We consider the effect of encoding precision on the final model's accuracy value. We run the framework with 50 communication rounds and limit the clients' parameters under 2, 3, 4, 5, 10 decimal places. As shown in Fig. 1, rounding models' parameters to 2 digits leads to significant decrease in accuracy value, while the higher encoding precision settings (3, 4, 5, 10 decimal places) have minor changes on the performance.

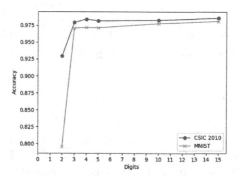

Fig. 1. The results on accuracy for different precision levels.

3.2.4   Encryption Time Evaluation In this part, we conduct an analytical investigation into the duration required for encryption of the model's parameters, explicitly focusing on the effects of decimal precision on reducing encrypting time. We examine three different scenarios. In the first two, we limit the precision level of parameters to 3 and 10 decimal places. In the third scenario, we set an upper limit of 15 decimal places for the parameters, essentially converting a float number to an integer without restrictions. The computational time required for encrypting a single parameter is assessed and correlated with varying numbers of clients.

Our results show a discernible trend: as the decimal precision of the model parameters decreases, the time required for encryption likewise diminishes, holding the number of clients constant. This empirical finding provides substantive evidence for the usefulness of our proposed compression method in easing encryption efficiency.

This reduction in encryption time bears considerable implications for scalable deployment, particularly for more extensive federated learning networks where computational efficiency is paramount (Table 3).

Table 3. The computational time required for the encryption of 1000 parameters at decimal precision levels of 3, 10, and 15.

Number of clients	Time		
	3 decimal places	10 decimal places	15 decimal places
5	1.0614	1.8689	1.8873
10	2.0987	3.6781	3.8260
20	4.2775	7.4272	6.9981
40	8.3563	13.9953	14.3905
50	11.2388	18.2978	18.5236

4   Conclusion

In this paper, we have presented ComEnc-FL, an efficient and secure framework for privacy-preserving federated learning. ComEnc-FL addresses the limitations of existing cryptographic secure aggregation systems by leveraging an ElGamal-based encryption scheme and a novel parameter encoding strategy. Our framework tackles the challenge of colluding parties, a significant concern in most cryptography-based federated learning models nowadays. The results demonstrate that ComEnc-FL enhances privacy and achieves superior training time and data transfer bandwidth compared to other secure multi-party aggregation solutions. Furthermore, our parameter compression technique significantly reduces encryption time, enabling ComEnc-FL to achieve model performance compared to the original federated learning approach.

Our work opens new possibilities for secure and efficient collaboration in federated learning scenarios. Future research can explore further enhancements to ComEnc-FL and its application in various domains to address privacy concerns while leveraging the power of distributed data in machine learning.

References

1. Bonawitz, K., et al.: Practical secure aggregation for privacy-preserving machine learning. In: Proceedings of the 2017 ACM SIGSAC Conference on Computer and Communications Security, pp. 1175–1191 (2017)
2. El Ouadrhiri, A., Abdelhadi, A.: Differential privacy for deep and federated learning: a survey. IEEE Access **10**, 22359–22380 (2022)
3. European Commission: Regulation (EU) 2016/679 of the European Parliament and of the Council of 27 April 2016 on the protection of natural persons with regard to the processing of personal data and on the free movement of such data, and repealing Directive 95/46/EC (General Data Protection Regulation) (Text with EEA relevance) (2016). https://eur-lex.europa.eu/eli/reg/2016/679/oj
4. Goodfellow, I., Bengio, Y., Courville, A.: Deep Learning. MIT Press (2016). https://www.deeplearningbook.org
5. Google: Gboard: a google keyboard (2020). Accessed 17 Apr 2023

6. Gupta, S., Agrawal, A., Gopalakrishnan, K., Narayanan, P.: Deep learning with limited numerical precision. In: International Conference on Machine Learning (2015)

7. Konečný, J., McMahan, H.B., Yu, F.X., Richtarik, P., Suresh, A.T., Bacon, D.: Federated learning: Strategies for improving communication efficiency. In: NIPS Workshop on Private Multi-Party Machine Learning (2016). https://arxiv.org/abs/1610.05492

8. Liu, C., Chakraborty, S., Verma, D.: Secure model fusion for distributed learning using partial homomorphic encryption. Policy-Based Autonomic Data Governance, pp. 154–179 (2019)

9. Moshayedi, A.J., Roy, A.S., Kolahdooz, A., Shuxin, Y.: Deep learning application pros and cons over algorithm deep learning application pros and cons over algorithm. EAI Endorsed Trans. AI Robot. 1(1) (2022)

10. Nasr, M., Shokri, R., Houmansadr, A.: Comprehensive privacy analysis of deep learning: Stand-alone and federated learning under passive and active white-box inference attacks. arXiv:abs/1812.00910 (2018)

11. Shokri, R., Stronati, M., Song, C., Shmatikov, V.: Membership inference attacks against machine learning models. In: 2017 IEEE symposium on security and privacy (SP), pp. 3–18. IEEE (2017)

12. Tran, A.T., Luong, T.D., Karnjana, J., Huynh, V.N.: An efficient approach for privacy preserving decentralized deep learning models based on secure multi-party computation. Neurocomputing **422**, 245–262 (2021)

13. Truex, S., et al.: A hybrid approach to privacy-preserving federated learning. In: Proceedings of the 12th ACM Workshop on Artificial Intelligence and Security, pp. 1–11 (2019)

14. Wei, K., et al.: Federated learning with differential privacy: algorithms and performance analysis. IEEE Trans. Inf. Forensics Secur. **15**, 3454–3469 (2020)

15. Xu, R., Baracaldo, N., Zhou, Y., Anwar, A., Ludwig, H.: Hybridalpha: an efficient approach for privacy-preserving federated learning. In: Proceedings of the 12th ACM Workshop on Artificial Intelligence and Security, pp. 13–23 (2019)

16. Zhang, C., Li, S., Xia, J., Wang, W., Yan, F., Liu, Y.: Batchcrypt: efficient homomorphic encryption for cross-silo federated learning. In: Proceedings of the 2020 USENIX Annual Technical Conference (USENIX ATC 2020) (2020)

An Elliptic Curve-Based
Privacy-Preserving Recommender System

Van Vu-Thi[✉] and Dung Luong-The

Academy of Cryptography Techniques, Hanoi, Vietnam
vanvu10101986@gmail.com

Abstract. In this research, we present a way to improve a secure multi-party computation-based recommender system solution. This solution is both efficient and protective of users' personal information. Both theoretical and empirical research demonstrate that the proposed technique protects the participants' personal information while maintaining the recommender system's accuracy. The proposed solution is also more cost-effective in terms of both communication and computing than the original ones.

Keywords: Cryptographic protocol · Elliptic Curve Cryptosystem · Homomorphic Encryption · Privacy-Preserving Recommender System · Privacy Preservation · Secure Multi-party Computation

1 Introduction

Recommender Systems (RS) are utilized in e-commerce, entertainment, education, and training. The RS are motivated by two main entities: users and items. Users give thoughts and receive recommendations and items appraised by the recommender system. The RS's output can be either a prediction or a recommendation. It is essential for reliable user data analysis. Users may not want to reveal or lie about this data because it may contain private information. This issue has raised interest in privacy-preserving recommender systems. There have been several privacy-preserving recommender system (PPRS for short) proposals.

There are two primary approaches to preserving privacy in RS. The first approach is randomization-based, in which random noise is added to the user's rating to protect user privacy for the recommendation server [13] or obfuscate the user-item connection to an untrusted server [9]. However, these methods introduce additional noise, so they have to trade-off between privacy and accuracy. The second method is cryptographic-based. In [1,3,10] offered PPRS, but these solutions require the presence of an honest seller and non-collusive parties. The ElGamal cryptosystem was used to ensure privacy when computing the cosine similarity between items and generate recommendations using collaborative filtering and content-based filtering (CBF) techniques [2]. However,

K. Honda et al. (Eds.): IUKM 2023, LNAI 14376, pp. 334–345, 2024.
https://doi.org/10.1007/978-3-031-46781-3_28

the authors in [7] found a potential drawback of this solution, and proposed a modified protocol for collaborative filtering to protect privacy to address the vulnerability identified in the solution. They used a public ledger that maintains all users' hashes and is viewable to all users. As a result, this method is extremely computationally expensive. The authors of [12] present an improved PPRS that protects rating privacy against a malicious server, as presented in [7]. They added a temporary key to cover the rating value. Although this modification improved privacy without affecting the accuracy of the proposed scheme in [2], the communication costs is comparable to the previous solution and comes with a complex calculation. In order to improve the solution of [12], the authors of [11] put forth a new PPRS. While less expensive in terms of communication and computing, this one offers the same level of security as the previous one. However, these solutions are based on the ElGamal cryptosystem, leading to low performance.

In this paper, our main goal is to develop the efficient solution for PPRS. We propose a PPRS solution to protect users' privacy during the recommendation process without affecting accuracy and improving efficiency. We optimize the original solution mentioned in [11] by using Elliptic Curve Analog of the ElGamal system. The item rate average and item-item similarity are calculated using the efficient secure multi-party computing protocols. These protocols can achieve the same accuracy as centralized item similarity computation. Our experimental results and theoretical analysis show that the performance of the proposed solution is better than that of [11,12].

The rest of this paper is structured as follows: Sect. 2 provides the theoretical foundations, and Sect. 3 introduces our PPRS based on elliptic curves, then prove its accuracy, and privacy. Section 4 analyzes the performance of the proposed protocol and compares it with the solution in [11,12]. The conclusion is in Sect. 4.

2 Background

2.1 Elliptic Curve Cryptography

In this section, we review the elliptic curve analog of the ElGamal system that is the main fundamental to construct our solution. Let $E(F_q)$ be an Elliptic curve over a finite field F_q with a point O at infinity and q be a large prime, in which Elliptic curve discrete logarithm problem is hard. In addition, G is a base point of the elliptic curve E with order q (i.e., $q.G = O$). The private key is the random number $d \in [1, q-1]$, and the corresponding public key curve point is $Q = d.G$. To encrypt the plaintext m, the sender uses the receiver's public key Q to compute the ciphertext C from the plaintext m as follows: he randomly chooses k from $[1, q-1]$ and computes the ciphertext $C = (C_1 = P_m + k.Q, C_2 = k.G)$ where P_m is a point of E with $x_{P_m} = m$. To decrypt the ciphertext C using the private key d, the receiver may compute $m = x_M$, in which $M = C_1 + (-d.C_2)$.

2.2   Recommender System

Let $U = \{u_1, u_2, ..., u_n\}$ be the set of all n users in the recommender system, and $I = \{i_1, i_2, ..., i_m\}$ be the set of items where m is the total number of items. Let R be the rating matrix where $r_{i,j}$ is the rating provided by user u_i on item i_j. The rating matrix is typically sparse due to a lack of values because it is impossible to rank all the objects submitted by all system users.

Similarity Calculation. We also used cosine similarity, as in [11,12]. The similarity between items i_j and i_k is given as:

$$s(i_j, i_k) = \frac{\sum_{i=1}^{n} r_{i,j} \cdot r_{i,k}}{\sqrt{\sum_{i=1}^{n} r_{i,j}^2} \cdot \sqrt{\sum_{i=1}^{n} r_{i,k}^2}} \tag{1}$$

Prediction Techniques. Prediction techniques are divided into two main categories [12]: Content-Based Filtering (CBF) and Collaborative Filtering (CF).

CBF-Based Recommendations The equation for predicting the recommendation using CBF is:

$$P_{i,k} = \frac{\sum_{j=1}^{m} r_{i,j} \cdot s(i_j, i_k)}{\sum_{j=1}^{m} s(i_j, i_k)} \tag{2}$$

where $P_{i,k}$ is predicted rating for user u_i on item i_k.

CF-Based Recommendations The item-based CF computes the prediction of user u_i for item i_k by

$$P_{i,k} = \frac{R_k \cdot \sum_{j=1}^{m} s(i_k, i_j) + \sum_{j=1}^{m} (r_{i,j} - R_j) \cdot s(i_k, i_j)}{\sum_{j=1}^{m} s(i_k, i_j)} \tag{3}$$

where R_j is the average rating of all users on item j. Note that the average rating of a particular item is computed by dividing the total rating by the total number of users who have rated that item.

2.3   Secure Multi-party Computation

This section reviews the secure multiparty computation framework developed by [4].

Privacy in the Semi-honest Model. In the distributed setting, let π be an n-party protocol for computing f. Let $\overline{x}$ denote $(x_1, x_2, ..., x_n)$. The view of the ith ($i \in [1, n]$) party during the execution of π on $\overline{x}$ is denoted by $view^\pi(\overline{x})$ which includes x_i, all received messages, and all internal coin flips. For every subset I of $[1, n]$, namely $I = i_1, .., i_t$, let $f_I(\overline{x})$ denote $\{y_1, .., y_t\}$ and $view_I^\pi(\overline{x}) = (I, view_{i_1}^\pi(\overline{x}), ..., view_{i_t}^\pi(\overline{x}))$. Let $OUTPUT(\overline{x})$ denote the output of all parties during the execution of π.

Definition 1. *An n-party computation protocol π for computing $f(\overline{x})$ is secure with respect to semi-honest parties if there exists a probabilistic polynomial-time algorithm denoted by S, such that for every $I \subset [1,n]$ we have*

$\{S(\overline{x}, f_I(\overline{x}), f(x))\} \overset{c}{\equiv} \{view_I^\pi(\overline{x}), output(\overline{x})\}$

where $\overset{c}{\equiv}$ denotes computational indistinguishability.

Theorem 1 *(**Composition theorem for the semi-honest model, multiparty case**). Suppose that the m-ary functionality g is privately reducible to the k-ary functionality f and that there exists a k-party protocol for privately computing f. Then there exists an m-party protocol for privately computing g.*

Detailed proof of Theorem 1 could be found in [4], and thus is omitted here.

3   An Elliptic Curve-Based Privacy-Preserving Recommender System

3.1   Set up

Let $E(Z_d)$ be an elliptic curve with a point O at infinity and d be a large prime, in which elliptic curve discrete logarithm problem is hard. In addition, G is a base point of the elliptic curve E with order d. Each user u_i keeps private values $r_{i,j}$, $j \in [1,m]$. Nobody knows these values, beyond him. Each user chooses a private key $x_i \in [1, d-1]$, after that he computes the corresponding public keys $X_i = x_i.G$. A server called a recommender server that will cooperate with users to calculate the averages as well as item-item similarities and store them in its database and assists the target user in generating the recommendation process.

Our protocol also works on the original protocol's assumption that the server is malicious and will try to obtain user's ratings anyhow but will follow the protocol such that prediction scores are not altered. Users are assumed semi-honest, and will try to obtain details of other users or the server but will not disobey the protocol. It is assumed that message (data or cipher) is shared only between involved entities.

To securely compute averages and similarities among items, all users send their public key $X_1, \cdots, X_n$ to the server. The server generates a common public key X for n users and broadcasts it to all users to encrypt their ratings. $X = \sum_{i=1}^n X_i = x.G$

3.2   Secure Multi-party Average Computation Protocol

The protocol consists of four main phases as described in Fig. 1. The value of $\sum_{i=1}^n r_{ij}$ and $\sum_{i=1}^n f_{i,j}$, is not large, therefore computing discrete logarithm is not hard.

Theorem 2. *The above-presented protocol correctly computes the average value $R_j = \frac{\sum_{i=1}^n r_{i,j}}{\sum_{i=1}^n f_{i,j}}$.*

Proof. We have $d_j^{(1)} = log_G R_j^{(1)}$, so: $d_j^{(1)}.G = R_j^{(1)}$

$= \sum_{i=1}^{n} r_{i,j}.G + \sum_{i=1}^{n} c_{i,j}^{(1)}.\sum_{i=1}^{n} x_i.G - \sum_{i=1}^{n} x_i.\sum_{i=1}^{n} c_{i,j}^{(1)}.G = \sum_{i=1}^{n} r_{i,j}.G$

Thus $d_j^{(1)}.G = \sum_{i=1}^{n} r_{i,j}.G$, and therefore $d_j^{(1)} = \sum_{i=1}^{n} r_{i,j}$. In the same way,

we can get $d_j^{(2)} = \sum_{i=1}^{n} f_{i,j}$, so $R_j = \frac{d_j^{(1)}}{d_j^{(2)}} = \frac{\sum_{i=1}^{n} r(i,j))}{\sum_{i=1}^{n} f(i,j)}$.

---

1. Phase 1: Users u_i do;
for $i \leftarrow 1$ **to** n **do**
  for $j \leftarrow 1$ **to** m **do**
    $c_{i,j}^{(1)} = Random(1, d-1); C_{i,j}^{(1)} = c_{i,j}^{(1)}.G;$
    $t_{i,j}^{(1)} = Random(1, d-1); T_{i,j}^{(1)} = t_{i,j}^{(1)}.G;$
  end
end
Sends to the server $C_{i,j}^{(1)}, T_{i,j}^{(1)}$;
2. Phase 2: The server does;
for $j \leftarrow 1$ **to** m **do**
  $C_j^{(1)} = T_j^{(1)} = O;$
  for $i \leftarrow 1$ **to** n **do**
    $C_j^{(1)} = C_j^{(1)} + C_{i,j}^{(1)}; T_j^{(1)} = T_j^{(1)} + T_{i,j}^{(1)};$
  end
  Sends to all $u_i : C_j^{(1)}, T_j^{(1)}$
end
3. Phase 3: Users u_i do;

for $i \leftarrow 1$ **to** n **do**
  for $j \leftarrow 1$ **to** m **do**
    $R_{i,j}^{(1)} = r_{i,j}.G + c_{i,j}^{(1)}.X - x_i.C_j^{(1)};$
    $F_{i,j}^{(1)} = f_{i,j}.G + t_{i,j}^{(1)}.X - x_i.T_j^{(1)};$
  end
end
Sends to the server: $R_{i,j}^{(1)}, F_{i,j}^{(1)}$;
4. Phase 4: The server does:
for $j \leftarrow 1$ **to** m **do**
  $R_j^{(1)} = F_j^{(1)} = O;$
  for $i \leftarrow 1$ **to** n **do**
    $R_j^{(1)} = R_j^{(1)} + R_{i,j}^{(1)}; F_j^{(1)} = F_j^{(1)} + F_{i,j}^{(1)}$
  end
  $d_j^{(1)} = log_G R_j^{(1)}; d_j^{(2)} = log_G F_j^{(1)};$
  $R_j = \frac{d_j^{(1)}}{d_j^{(2)}};$
end

Fig. 1. Secure Average Computation Protocol

3.3 Secure Multi-party Cosine Similarity Computation

The privacy-preserving cosine similarity computation protocol consists of four main phases described in Fig. 2.

Theorem 3. *The above-presented protocol correctly computes the similarity value between the two items i_j, i_k: $s(i_j, i_k) = \frac{\sum_{i=1}^{n} r_{i,j}.r_{i,k}}{\sqrt{\sum_{i=1}^{n} r_{i,j}^2}.\sqrt{\sum_{i=1}^{n} r_{i,k}^2}}$.*

Proof. Because $d_{j,k}^{(1)} = log_G R_{j,k}^{(2)}$, then: $d_{j,k}^{(1)}.G = \sum_{i=1}^{n} R_{i,j,k}^{(2)} = \sum_{i=1}^{n} r_{i,j}.r_{i,k}.G +$

$\sum_{i=1}^{n} c_{i,j,k}^{(2)}.\sum_{i=1}^{n} x_i.G - \sum_{i=1}^{n} x_i.\sum_{i=1}^{n} c_{i,j,k}^{(2)}.G = \sum_{i=1}^{n} r_{i,j}.r_{i,k}.G$

Thus, $d_{j,k}^{(1)}.G = \sum_{i=1}^{n} r_{i,j}.r_{i,k}$. In the same way, we can get $d_j^{(3)} = \sum_{i=1}^{n} r_{i,j}^2$,

$d_k^{(3)} = \sum_{i=1}^{n} r_{i,k}^2$, so $s(i_j, i_k) = \frac{d_{j,k}^{(1)}}{\sqrt{d_j^{(3)}}.\sqrt{d_k^{(3)}}} = \frac{\sum_{i=1}^{n} r_{i,j}.r_{i,k}}{\sqrt{\sum_{i=1}^{n} r_{i,j}^2}.\sqrt{\sum_{i=1}^{n} r_{i,k}^2}}$.

3.4 Proposed Privacy-Preserving Recommendation Generation

We assume there is just one user u_i. The user u_i holds the public key X_i to encrypt the ratings and the private key x_i to decrypt the ciphertexts. The server holds item-item similarity $s(i_j, i_k)$ and averages of items' ratings R_j.

```
1. Phase 1: Users u_i do;                    for i ← 1 to n do
for i ← 1 to n do                              for j ← 1 to m do
  for j ← 1 to m do                              F^(2)_{i,j} = r_{i,j}.r_{i,j}.G + t^(2)_{i,j}.X- x_i.T^(2)_j;
    t^(2)_{i,j} = Random(1,d-1); T^(2)_{i,j} = t^(2)_{i,j}.G;       for k ← j+1 to m do
    for k ← j+1 to m do                            R^(2)_{i,j,k} = r_{i,j}.r_{i,k}.G + c^(2)_{i,j,k}.X - x_i.C^(2)_{j,k};
      c^(2)_{i,j,k} = Random(1,d-1); C^(2)_{i,j,k} = c^(2)_{i,j,k}.G ;        end
    end                                        end
  end                                        end
end                                          Sends to the server F^(2)_{i,j}, R^(2)_{i,j,k};
Sends to the server T^(2)_{i,j}, C^(2)_{i,j,k};   4. Phase 4: The server does;
2. Phase 2: The server does;                 for j ← 1 to m do
for j ← 1 to m do                              F^(2)_j = O;
  T^(2)_j = O;                                 for i ← 1 to n do
  for i ← 1 to n do                              F^(2)_j = F^(2)_j + F^(2)_{i,j};
    T^(2)_j = T^(2)_j + T^(2)_{i,j};            end
  end                                          for k ← j+1 to m do
  for k ← j+1 to m do                            R^(2)_{j,k} = O;
    C^(2)_{j,k} = O;                             for i ← 1 to n do
    for i ← 1 to n do                              R^(2)_{j,k} = R^(2)_{j,k} + R^(2)_{i,j,k};
      C^(2)_{j,k} = C^(2)_{j,k} + C^(2)_{i,j,k};    end
    end                                          d^(3)_j = log_G F^(2)_j; d^(1)_{j,k} = log_G R^(2)_{j,k};
  end                                          end
end                                          end
Sends to all u_i : T^(2)_j, C^(2)_{j,k};      for j ← 1 to m do
3. Phase 3: Users u_i do;                      for k ← j+1 to m do
                                                 S_{j,k} = d^(1)_{j,k} / ( √(d^(3)_j) . √(d^(3)_k) );
                                               end
                                             end
```

Fig. 2. Privacy-preserving Similarity Computation Protocol

Privacy-Preserving CBF-Based Recommendation. This process are outlined in Fig. 3. Where $P_{i,k}$ denotes the predicted recommendation on item i_k,
and the item with the highest prediction is finally recommended on the user.

Theorem 4. *If all users and the server follow the protocol, then* $d^{(4)}_k = \sum_{j=1}^{m} r_{i,j}.s(i_k, i_j)$ *and* $d^{(5)} = \sum_{j=1}^{m} s(i_k, i_j)$.

Proof. Indeed, we have $d^{(4)}_k = log_G C^{(3)}_k$, so: $d^{(4)}_k.G = C^{(3)}_k = \sum_{j=1}^{m} s(i_k, i_j).(r_{i,j}.G + c^{(1)}_j.X_i) - x_i.\sum_{j=1}^{m} s(i_k, i_j).c^{(1)}_j.G = \sum_{j=1}^{m} s(i_k, i_j).r_{i,j}.G$

Thus, $d^{(4)}_k = \sum_{j=1}^{m} r_{i,j}.s(i_k, i_j)$. Similarly, we can prove $d^{(5)}_k = \sum_{j=1}^{m} s(i_k, i_j)$.

Privacy-Preserving CF-Based Recommendation. The proposed private
CF-based recommendation process also operates in three main steps as the CBF-
based method. The detailed steps are as Fig. 4. Where $P_{i,k}$ denotes the predicted
recommendation on item i_k ($k \in [1, m]$), and the item with the highest prediction
is finally recommended to the user target.

Theorem 5. *If all users and the server follow the protocol, then* $d^{(6)}_k = R_k.\sum_{j=1}^{m} s(i_k, i_j) + \sum_{j=1}^{m} (r_{i,j} - R_j).s(i_k, i_j)$ *and* $d^{(7)} = \sum_{j=1}^{m} s(i_k, i_j)$.

Proof. Because $d^{(6)}_k = log_G C^{(5)}_k$ then: $d^{(6)}_k.G = C^{(5)}_k = F_{9,k} - x_i.F_{10,k} = (R_k.\sum_{j=1}^{m} s(i_k, i_j) + \sum_{j=1}^{m} (r_{i,j} - R_j).s(i_k, i_j)).G$.

Thus: $d_k^{(6)} = R_k . \sum_{j=1}^{m} s(i_k, i_j) + \sum_{j=1}^{m} (r_{i,j} - R_j).s(i_k, i_j)$. Similarly, we can prove $d^{(7)} = \sum_{j=1}^{m} s(i_k, i_j)$.

---

1. Phase 1: The target user u_i does;
for $j \leftarrow 1$ **to** m **do**
$\quad c_j^{(1)} = Random(1, d-1);$
$\quad C_j^{(1)} = r_{i,j}.G + c_j^{(1)}.X_i; C_j^{(2)} = c_j^{(1)}.G;$
end
Sends to the server: $C_j^{(1)}, C_j^{(2)};$
2. Phase 2: The server does;
for $k \leftarrow 1$ **to** m **do**
$\quad S_k = F_{1,k} = F_{2,k} = O;$
$\quad$ **for** $k \# j \leftarrow 1$ **to** m **do**
$\quad\quad S_k = S_k + S_{k,j}; F_{1,k} = F_{1,k} + S_{k,j}.C_j^{(1)};$
$\quad\quad F_{2,k} = F_{2,k} + S_{k,j}.C_j^{(2)};$
end
$c_k^{(2)} = Random(1, d-1); F_{4,k} = c_k^{(2)}.G;;$
$F_{3,k} = S_k.G + c_k^{(2)}.X_i;$

end
Sends to the user: $F_{1,k}, F_{2,k}, F_{3,k}, F_{4,k};$
3. Phase 3: The target user u_i does:
for $k \leftarrow 1$ **to** m **do**
$\quad C_k^{(3)} = F_{1,k} - x_i.F_{2,k};$
$\quad C_k^{(4)} = F_{3,k} - x_i.F_{4,k};$
$\quad d_k^{(4)} = log_G C_k^{(3)};$
$\quad d_k^{(5)} = log_G C_k^{(4)};$
$\quad P_{j,k} = \frac{d_k^{(5)}}{d_k^{(6)}};$
end

Fig. 3. Privacy preservation of CBF-Based recommendation

Note that: The target user can submit the index of unrated items. In this case, the server does not consider those entries while calculating the ciphertexts of the recommendations. The server will do the de-decimal (say b places) by multiplying the similarities and averages by 10^b.

3.5 Proof of Privacy

Privacy Preservation of Average Computation Protocol. We prove the privacy of the protocol by the following theorem.

Theorem 6. *The protocol presented in Fig. 1 preserves each user's privacy in the semi-honest model, protects each honest user's privacy against the malicious server [7], and protects each honest user's privacy against the collusion of the server and up to $n - 2$ corrupted participants.*

Proof. Let recall that each user u_i sends $(C_{i,j}^{(1)}, R_{i,j}^{(1)})$, $(T_{i,j}^{(1)}, F_{i,j}^{(1)})$ that corresponding to two ElGamal ciphertexts using the common public key X and no one has the corresponding secret key x. Thus, the proposed protocol securely preserves honest user's privacy in the semi-honest model.

Continuously, we prove that the secure multi-party average computation protocol protects each user's privacy against malicious server acts as mentioned in [7].

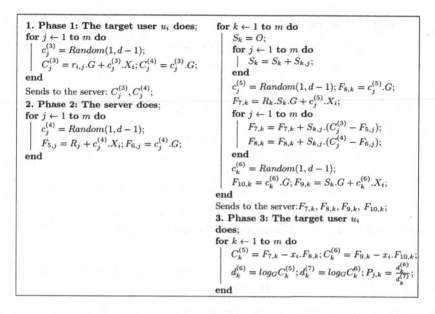

Fig. 4. Privacy preservation of CF-Based recommendation

In the proposed protocol, if a malicious server acts as mentioned in [7], the server is acting malicious and wants to get a rating of $r_{1,1}$. After, all users send their ratings and flags in encrypted form, sever broadcasts $C_j^{(1)} = C_{1,1}^{(1)} = c_{1,1}^{(1)}.G$ to all users, and they respond with $r_{i,1}.G + c_{i,1}^{(1)}.X - x_i.C_{1,1}^{(1)}$. Now server homomorphically adds these values and tries to retrieve $r_{1,1}$, for this server requires $\sum_{i=2}^{n} r_{i,1}.G$ which the server does not have. Thus, the server cannot obtain the rating $r_{1,1}$ in plaintext: $\sum_{i=1}^{n} \left(r_{i,1} \cdot G + c_{i,1}(1) \cdot X - x_i \cdot C_{1,1}(^{(1)}) \right) = r_{1,1} \cdot G + \sum_{i=2}^{n} r_{i,1} \cdot G + \sum_{i=2}^{n} c_{i,1}(^{(1)}) \cdot x_i \cdot G - \sum_{i=2}^{n} x_i \cdot c_{1,1}(^{(1)}) \cdot G$. So, the protocol protects each honest user's privacy against the malicious server [7].

Finally, to prove that the proposed protocol protects the privacy of honest users in case the server colludes up to $n-2$ users, the authors need to show a simulator M that simulates what the corrupted participants have observed during the protocol execution by a probabilistic poly-nominal time algorithm. Particularly, we need to give an algorithm that computes the joint view of the corrupted parties in polynomial time using only the corrupted parties' knowledge, the public keys, and some Elliptic curve encryptions (EC).

Without loss of generality, it is assumed that u_1 and u_2 do not collude and $I = 3, 4, ..., n$. Below is the algorithm that computes the view of the server and the corrupted users.

– M simulates $C_{1,j}{}^{(1)}, T_{1,j}{}^{(1)}, C_{2,j}{}^{(1)}, T_{2,j}{}^{(1)}$ using random EC and it computes the following values: $C'_j{}^{(1)} = \prod_{i=1}^{n} C_{i,j}^{(1)}$, $T'_j{}^{(1)} = \prod_{i=1}^{n} T_{i,j}^{(1)}$

– M using some EC $(u_{11}, v_{11}) = (r_{1,j}.G + x_2.c_{1,j}^{(1)}.G, x_2.G), (u_{12}, v_{12}) = (f_{1,j}.G + x_2.t_{1,j}^{(1)}.G, x_2.G), (u_{21}, v_{21}) = (r_{2,j}.G + x_1.c_{2,j}^{(1)}.G, x_1.G), (u_{22}, v_{22}) = (f_{2,j}.G + x_1.t_{2,j}^{(1)}.G, x_1.G)$ as its input. The algorithm of simulator M computes $(R'_{1,j}^{(1)}, F'_{1,j}^{(1)})$, and $(R'_{2,j}^{(1)}, F'_{2,j}^{(1)})$ as follows:

$$R'_{1,j}^{(1)} = u_{11} + \sum_{i \in I} x_i \cdot c_{1,j}^{(1)} + \left(d_j^{(1)} - \sum_{i \in I} r_{i,j} \right) \cdot G - \left(u_{21} + \sum_{i \in I} c_{i,j}^{(1)} \cdot X_1 \right)$$

$$R'_{2,j}^{(1)} = u_{21} + \sum_{i \in I} x_i \cdot C_{2,j}^{(1)} + \left(d_j^{(1)} - \sum_{i \in I} r_{i,j} \right) \cdot G - \left(u_{11} + \sum_{i \in I} c_{i,j}^{(1)} \cdot X_2 \right)$$

$$F'_{1,j}^{(1)} = u_{12} + \sum_{i \in I} x_i \cdot T_{1,j}^{(1)} + \left(d_j^{(2)} - \sum_{i \in I} f_{i,j} \right) \cdot G - \left(u_{22} + \sum_{i \in I} t_{i,j}^{(1)} \cdot X_1 \right)$$

$$F'_{2,j}^{(1)} = u_{22} + \sum_{i \in I} x_i \cdot T_{2,j}^{(1)} + \left(d_j^{(2)} - \sum_{i \in I} f_{i,j} \right) \cdot G - \left(u_{12} + \sum_{i \in I} t_{i,j}^{(1)} \cdot X_2 \right)$$

Thus, we have shown the simulator M that satisfies the above requirement. According to Definition 1, The protocol is semantically secure.

Privacy Preservation of Similarity Computation Protocol. We prove the privacy of the protocol by the following theorem.

Theorem 7. *The protocol in Fig. 2 preserves each user's privacy in the semi-honest model, protects each honest user's privacy against the malicious server [7], and protects each honest user's privacy against the collusion of the server and up to $n-2$ corrupted participants.*

Proof. In this protocol, All information sent by users is encrypted using the common public key X, and no one has the corresponding secret key x. Thus, the protocol securely preserves each honest user's privacy in the semihonest model.

Continuously, we prove that the secure multi-party average computation protocol protects each user's privacy against malicious server acts as mentioned in [7]. If a malicious server acts as mentioned in [7], the server cannot obtain the rating $r_{1,1}^2$ in plaintext: $\sum_{i=1}^{n} r_{i,1} \cdot r_{i,1} \cdot G + t_{i,1}^{(2)} \cdot X - x_i \cdot T_{1,1}^{(2)} = r_{1,1} \cdot r_{1,1} \cdot G + \sum_{i=2}^{n} r_{i,1} \cdot r_{i,1} \cdot G + \sum_{i=2}^{n} x_i \cdot t_{i,1}^{(2)} \cdot G - \sum_{i=2}^{n} x_i \cdot t_{1,1}^{(2)} \cdot G$.

Finally, similarly to Theorem 6, we can easily point out a simulator that matches Definition 1. Thus the theorem is proved.

Privacy Preservation of CBF-Based Recommendation. In the proposed privacy-preserving CBF-Based method, the item similarity computations are proved to be privacy-preserving and to generate recommendations in CBF, the target user encrypts item preferences using his own public key X_i and sends them to the server. The server homomorphically generates ciphertexts of recommendations leveraging the item-item similarity, which is already available to it, thereby sending the ciphertexts to the target user. While generating recommendations, the similarities among the items are encrypted using the target user's public key X_i. The ciphertexts are decrypted by the user's own secret key x_i. Therefore, during this process target user's ratings and recommendations results are not revealed and the item recommendation generations are conducted on the client side. Based on Theorem 1, we can conclude that the proposed method preserves each user's privacy in the semi-honest model, and protects each honest user's privacy against any adversary or malicious server and the collusion of the server and up to $n - 2$ corrupted participants.

Privacy Preservation of CF-Based Recommendation. The proof is similar to the privacy-preserving CBF-based recommendation, we can conclude that the proposed method preserves each user's privacy in the semi-honest model and protects each honest user's privacy against any adversary or malicious server and the collusion of the server and up to $n - 2$ corrupted participants.

3.6   Performance Evaluation

In this section, we compare the communication and computation cost in the proposed solution with the ones in two typical privacy-preserving recommendation systems: the solution of [12] and the solution of [11], (denoted as Verma's solution and Van's solution, respectively). These solutions are chosen for the comparisons since they also have a high level of privacy and the capability to ensure the generated recommendation's accuracy. We assume that: All users contribute to computing the average and similarity. Only one user (target user) participates in creating the recommendation. The user encrypts their rating and sends the ciphertext in parallel to the server, so the cost of calculation on the user side can be decreased by just computing for one user (shown in Table 1). The computation and communication costs for all users participating in the system are represented on the server side because they are dependent on all users collaborating with the server.

Note that our solution uses an Elliptic curve cryptosystem with a 192-bit private key and 384-bit public key [5], the first two solutions use an Elgamal cryptosystem with a 160-bit private key and 1024-bit public key [6] (these cryptosystems are the same level of security).

Table 1. The communication costs (in bits) comparison between solutions

	Verma's solution		Van's solution		Our solution	
Operation	User	Server	User	Server	User	Server
Average	$6144m$	$2048mn$	$4096m$	$2048mn$	$1536m$	$768mn$
Similarity	$1536m(m+1)$	$512nm(m+1)$	$1024m(m+1)$	$512nm(m+1)$	$384m(m+1)$	$192nm(m+1)$
CBF-based recommendation	$2048m$	$2048(m+1)$	$2048m$	$2048(m+1)$	$768m$	$768(m+1)$
CF-based recommendation	$2048m$	$2048(m+1)$	$2048m$	$2048(m+1)$	$768m$	$768(m+1)$

Communication Overhead. We analyze the communication costs at each phase of the compared solutions for n users and m items. The comparison is given in Table 1. Observing this table, we can see that our solution transfers a much lower number of bits than the others.

Running Time. To compare the performance, we implement the proposed solution, Verma's solution, and Van's solution in the $C\#$ language of the Visual Studio 2019 environment, using the System. Numerics namespace to compare the performance of them (i.e., communication overhead and time complexity). Our experiments carry on a laptop with a 2.60 GHz Intel Core i5 processor and 8 GB memory. The study used the same dataset used in [12], the MovieLens dataset [8] which is freely available to the public. It includes 100,000 ratings (from 1 to 5) collected from 943 users on 1982 items. In the implementation, we randomly selected 200 items from the dataset for the experiment while keeping all users. As a result, the performance analysis of the solutions includes 943 users and 200 items. After calculating the average and similarity, a user ("target user") will be assigned randomly to generate recommendations.

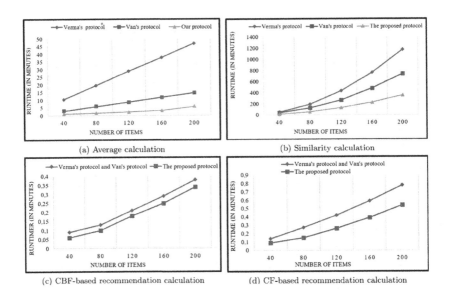

(a) Average calculation   (b) Similarity calculation

(c) CBF-based recommendation calculation   (d) CF-based recommendation calculation

Fig. 5. The computational cost comparison for all phases

We have computed the actual computation time required to generate a recommendation. The results obtained are shown in Fig. 5.a, 5.b, 5.c, and 5.d. Our results show that the modified solution runs faster than the previous solutions, especially when the number of items is large.

4   Conclusion

This paper proposed an Elliptic curve-based privacy-preserving recommendation system based on item-item similarity. It can protect user profiles and their rating history from any third parties or others. This solution even protect the privacy of honest users against up to $n-2$ corrupted users colluding with the server. Furthermore, the server can compute desired recommendations without affecting the actual rating information. The experimental results show that the proposed solution not only ensures accuracy and privacy but also has lower computation and communication costs than the previous protocols. Our future research will develop an efficient and secure recommender system for other data models.

References

1. Aïmeur, E., Brassard, G., Fernandez, J.M., Mani Onana, F.S.: ALAMBIC: a privacy-preserving recommender system for electronic commerce. Int. J. Inf. Secur. **7**(5), 307–334 (2008)
2. Badsha, S., Yi, X., Khalil, I.: A practical privacy-preserving recommender system. Data Sci. Eng. **1**, 161–177 (2016)
3. Erkin, Z., Veugen, T., Toft, T., Lagendijk, R.L.: Generating private recommendations efficiently using homomorphic encryption and data packing. IEEE Trans. Inf. Forensics Secur. **7**(3), 1053–1066 (2012)
4. Goldreich, O.: Foundations of Cryptography, vol. 2. Cambridge University Press, Cambridge (2004)
5. Kerry, C.F., Gallagher, P.D.: Digital signature standard (DSS). FIPS PUB, pp. 186–4 (2013)
6. Lepinski, M., Kent, S.: RFC 5114-additional Diffie-Hellman groups for use with IETF standards. See Section 2.1. 1024-bit MODP Group with 160-bit Prime Order Subgroup (2008)
7. Mu, E., Shao, C., Miglani, V.: Privacy preserving collaborative filtering (2017)
8. Resnick, P., Iacovou, N., Suchak, M., Bergstrom, P., Riedl, J.: GroupLens: an open architecture for collaborative filtering of netnews. In: Proceedings of the 1994 ACM Conference on Computer Supported Cooperative Work, pp. 175–186 (1994)
9. Shokri, R., Pedarsani, P., Theodorakopoulos, G., Hubaux, J.P.: Preserving privacy in collaborative filtering through distributed aggregation of offline profiles. In: Proceedings of the third ACM Conference on Recommender Systems, pp. 157–164 (2009)
10. Tang, Q., Wang, H.: Privacy-preserving hybrid recommender system. In: Proceedings of the Fifth ACM International Workshop on Security in Cloud Computing, pp. 59–66 (2017)
11. Van, V.T., Dung, L.T., Quan, H.V.: An efficient privacy-preserving recommender system. In: 2022 14th International Conference on Knowledge and Systems Engineering (KSE), pp. 1–6 (2022). https://doi.org/10.1109/KSE56063.2022.9953800
12. Verma, P., Vaishnav, H., Mathuria, A., Dasgupta, S.: An enhanced privacy-preserving recommender system. In: Nandi, S., Jinwala, D., Singh, V., Laxmi, V., Gaur, M.S., Faruki, P. (eds.) ISEA-ISAP 2019. CCIS, vol. 939, pp. 247–260. Springer, Singapore (2019). https://doi.org/10.1007/978-981-13-7561-3_18
13. Zhang, S., Ford, J., Makedon, F.: Deriving private information from randomly perturbed ratings. In: Proceedings of the 2006 SIAM International Conference on Data Mining, pp. 59–69. SIAM (2006)

A Method Against Adversarial Attacks to Enhance the Robustness of Deep Learning Models

Phi Ho Truong and Duy Trung Pham$^{(\boxtimes)}$

Academy of Cryptography Techniques, Ha Noi, Vietnam
`trungpd@actvn.edu.vn`

Abstract. Deep learning is facing a dangerous challenge because attackers are always lurking to find and exploit the model's vulnerabilities to deceive, making the model misidentify the classifier for the target model. It is dangerous if a smart device using artificial intelligence misrecognizes the object class. Attackers today often use adversarial examples, which at first glance do not differ from an image that is defined as natural when collected from sensors, or digital devices. Many studies on attacks and methods of combating these attacks have been tested by research groups and announced to be highly effective against attack or pattern recognition. Training the model with the aim of making the model able to recognize the adversarial example, a seemingly simple but effective method to make the model more robust, and capable of classification and identification. In this paper, to enhance the robustness of the model, the authors use adversarial training and experiment on the YOLOv7 model. Experiments show that this method is effective, making the model more powerful, capable of detecting and classifying adversarial examples after the model has been adversarial trained.

Keywords: Deep learning · Adversarial examples · Adversarial defense · Trained model · Object detection

1 Introduction

Since its inception, the use of artificial intelligence (AI) has brought about advancements of high importance, which have improved our daily lives and daily activities in many ways. There are more and more areas that can benefit from AI assistance from image recognition, autonomous driving [20], and natural language processing [28], to medical diagnostics, economics [25], credit risk assessment [24], intrusion detection [26], etc. Along with the explosion of internet usage, big data is increasingly valuable for related information to our lives, our shopping habits, our hobbies, or everything that concerns us. This further facilitates machine learning (ML) models to train more accurate models. By leveraging statistical methods and algorithms, machine learning models are trained using extensive datasets to perform classifications, predictions, and

K. Honda et al. (Eds.): IUKM 2023, LNAI 14376, pp. 346–357, 2024.
https://doi.org/10.1007/978-3-031-46781-3_29

uncover novel insights in data mining initiatives. These models enable the extraction of valuable information and samples from data, facilitating the discovery of new knowledge and improving predictions. This knowledge will influence decisions in applications and businesses.

However, the privacy and security issues of ML have also recently become more relevant than ever. Gartner, a renowned technology research and consulting firm recognized for its data visualization and analysis tools such as Gartner Magic Quadrants and Hype Cycle, has conducted an analysis and forecasted that AI-driven risk and security management for reliability will emerge as one of the top Ten Technology Trends in 2023 [19]. Furthermore, several surveys have highlighted the vulnerabilities associated with ML models, including the risks of model theft or decompilation, potential exposure of sensitive training data, and even recognizable facial images can be recovered from victims [17]. Moreover, recent studies have uncovered the susceptibility of ML models to adversarial examples (AEs), with the possibility of confounding being difficult to notice, which can cause the ML models to mispredict with high rate. The probability of success of adversarial examples is very disturbing, which raises security concerns for ML models. Studying and understanding adversarial attacks enhances the security of ML models.

The research community reacted strongly to this finding, proposing multiple safeguards to accurately classify adversarial examples. Osbert et al. [1] proposed new indexes and an algorithm to measure and approximate the strength of neural networks. Shixiang Gu et al. studies the structure of adversarial examples and explores the network topology, preprocessing, and training strategies to improve deep neural network (DNN) robustness and data preprocessing using a denoising auto-encoder (DAE) [11]. Ruitong Huang et al. [13] proposes reinforcement learning with a strong adversarial example (AE). Jonghoon Jin et al. introduces a new feedforward CNN model that improves durability in the presence of counter-noise (in [14]). Nicolas et al. [21] proposes a method called defensive distillation to reduce the effect of AEs on DNN. Andras Rozsa et al. focused on processing the original data to combat the adversarial example (in [22]). Since then, Shaham et al. [23] proposed improving the model to combat adversarial examples. Existing defense mechanisms have shown limited effectiveness in accurately classifying adversarial examples.

From the above challenge, recent studies have turned to trying to detect them instead. The paper examines ten detection programs proposed in many recent publications [2,6,7,10,12,15,18], these defense measures have been systematically evaluated and compared to other mitigation strategies, demonstrating a relative level of consistency in assessing their effectiveness. The emergence of new attack techniques has revealed a concerning AEs: in all instances, the defense mechanisms can be circumvented by adversarial examples specifically crafted to exploit the vulnerabilities of each respective defense. When dealing with simple datasets, the attacks exhibit a marginal increase in the bias or noise necessary to create adversarial examples. However, on more intricate datasets, these adversarial examples remain entirely indistinguishable from the original examples, posing a significant challenge for detection AEs.

The above defenses are evaluated according to three threat models. The first is a black box attack, which does not target any particular model or sensor. Second, the newly developed white-box attacks have proven to dismantle each layer of defense individually when specifically tailored to exploit a particular defense mechanism. Finally, a gray-box attack model has been introduced, leveraging the transformability property [8] to operate effectively even when the adversary lacks awareness of the specific parameters of the defensive model. With those motives, the goal of the paper is to understand the mechanism and operation of adversarial attacks and investigate some ways to prevent adversarial attacks based on previous studies. The authors propose to use the adversarial training method on a specific deep learning model, and experimentally create the adversarial example using the fast gradient sign method. Then label and train the model so that the model can recognize the adversarial example. The specific method will be presented in Sect. 3.

The remaining sections of the paper are structured as follows: First, we survey detection methods from other studies [4] in Sect. 2. Proposed method presented in Sect. 3. Section 4 presents experiment that have been implemented, and experiment results. Section 5 of the paper will present the conclusions.

2   Related Work

The survey conducted by research groups (see in [4]), evaluated detection methods. The results revealed that three of these defense measures [7,10,18], utilized SCBD (Secondary Classification Based Detection) to classify images as either natural or adversarial. Three method employed to detect statistical properties of images or network parameters is the use of Principal Component Analysis-PCA [2,12,15]. Two of the detection methods (in [5,10]), employ additional statistical tests to detect adversarial examples. Furthermore, the last two methods (in [5,15]), employ techniques such as normalization, randomization, and blurring to enhance the resilience of the models against adversarial attacks. In this study, the author presents some key points of the surveyed defense models listed as follows:

2.1   Adversarial Retraining

Grosse et al. [10] propose a novel variation on adversarial retraining that deviates from the traditional approach of attempting to correctly classify AEs. Instead, they introduce a new class, denoted as $N + 1$, specifically dedicated to adversarial examples, and train the network to identify AEs. The authors suggest the following procedure for implementing their proposed approach:

1. Using training data $\chi_0 = \chi$ to train model F_{ori}.
2. On the F_{ori} model, with $(x_i, y_i) \in \chi$, create adversarial examples χ'_i.

3. Set the label $N + 1$ as the new label for adversarial examples, with $\chi_1 = \chi_0 \cup \{x'_i, N + 1 : i \in |\chi|\}$.
4. Using training data χ_1 to train an F_{ori} secured model.

Model proposed by Gong et al. [7] follows a similar approach to model proposed by Grosse et al. but with a slight difference. Instead of employing two separate models for classifying original and adversarial examples, Gong's model combines both types of examples into a single dataset and learns to classify them collectively. This unified approach allows the model to develop a comprehensive understanding of both benign and adversarial examples, enhancing its ability to accurately classify and differentiate between the two (1):

$$\chi_1 = \{(x_i, 1) : i \in |\chi|\} \cup \{(x'_i, 0) : i \in |\chi|\} \tag{1}$$

2.2   Testing Convolutional Layers

Testing convolutional layers: Model proposed by Metzen et al. [18] focuses on the detection of a given model (ResNet), taking the output of each convolution layer for classification.

2.3   Input Image Principal Component Analysis (PCA) Detection

The model proposed by Hendrycks & Gimpel [12] employs PCA to detect the dissimilarities between adversarial examples and natural images. The essence of this approach lies in assigning higher weights to the larger principal components, which represent relevant data, and lower weights to the other components that contain unrelated data. By emphasizing the important components, the data becomes more intuitive and easier to analyze. The effectiveness of this approach can be observed in the image below, illustrating the significant impact of selecting important components and assigning them higher weights.

2.4   Hidden Layer PCA

Li et al. [15] incorporates PCA analysis on the output of each convolutional layer within the neural network (show in Fig. 1). This approach involves subjecting a benign sample to pass through all the PCA components in order to be accepted by the classifier.

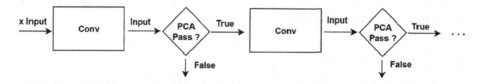

Fig. 1. Model proposed by Li et al.

2.5   Reduce Data Dimension

In [2], as described in the work of the authors, a similar approach is employed to reduce the dimensionality of the input data. The high-dimensional input, such as the 784-dimensional input for the MNIST dataset, is transformed into a smaller dimensional representation, such as 20 dimensions. Since PCA causes the model to lose its spatial location, Bhagoji et al. cannot use convolutional neural network. The model uses a fully-connected neural network and aims to make the attack models only manipulate the first K components in the hope of increasing the distortion required to generate the adversarial examples.

2.6   Maximum Mean Discrepancy

Grosse et al. [10] use the Maximum Mean Discrepancy (MMD) test as a technique for detecting differences between distributions [3,9]. Although MMD is a theoretically powerful method that can detect differences with formal proof, it is computationally expensive. Hence, Grosse et al. employed a simplified polynomial approximation that significantly reduces the computational complexity. By employing permutation test with the MMD test statistic, Grosse et al. obtained a measure of statistical significance to differentiate between distributions and detect differences between them.

2.7   Calculate the Distance to the Center of the Cluster

Feinman et al. use a Gaussian Mixture Model (GMM) [6] to model the output generated by the last hidden layer of a neural network. The underlying premise of Feinman's argument is that adversarial examples can be characterized as belonging to a distribution that differs from the distribution of natural images. For example, if an input x is classified as a particular label t, Feinman's method employs a kernel density estimate to estimate the probability of x belonging to that label. The estimation is computed using formula (2):

$$KDE(x) = \frac{1}{|X_t|} \sum_{s \in X_t} \exp \left(\frac{\left| F^{n-1}(x) - F^{n-1}(s) \right|^2}{\sigma^2} \right) \tag{2}$$

With X_t representing the set of training instances specifically labeled as t. $F^{n-1}(x)$ denotes the output of the final hidden layer when the input x is processed. A threshold τ will be chosen, if $KDE(x) < \tau$, x is adversarial, and it's natural if it returns the other value.

2.8   Dropout Randomization

In model proposed by Feinman et al., dropout is used. The objective is to ensure that a natural image consistently retains the same label regardless of the specific random values chosen during dropout. On the other hand, the adversarial examples, which possess resistance to the introduced perturbations, are expected to exhibit label variations and not consistently retain the same predicted label. By leveraging dropout in this manner, Feinman et al. aim to enhance the model's robustness against adversarial examples while maintaining label consistency for natural images. After applying a random network, with dropout enabled, the calculated uncertainty is (3):

$$U(x) = \left(\frac{1}{L} \sum_{j=1}^{L} \|F_r(x)\| \right) - \left\| \frac{1}{L} \sum_{j=1}^{L} F_r(x) \right\| \tag{3}$$

The uncertainty in model proposed by Feinman et al. is determined by computing the sum of variances across each component of the output. This calculation is performed on the output L derived from the random network. They then choose some threshold t and run several times for each selected threshold t. The sample is adversarial if the uncertainty is greater than t and vice versa.

2.9 Mean Blur

In [15], model blurs the image with a 3×3 median filter. Only then used to apply the classifier. After testing these defenses, I will now deploy a method to test resistance to adversarial attacks.

Each defense model has its strengths. However, no model is completely resistant to adversarial attacks. To contribute a part to the field of research on defense models, we propose a defensive model based on one of the methods just surveyed, which will be presented in detail in the next section.

3 Enhance the Robustness of Deep Learning Models Using Adversarial Training Method

To increase the authenticity and reliability of the presented information, the authors propose to carry out a series of experiments. Through the application of one of the surveyed machine learning methods and algorithms, the authors plan to evaluate the performance and accuracy of the models by conducting experiments on a suitable datasets. The main components of the model include:

3.1 Adversarial Attack with FGSM

Fast Gradient Signed Method (FGSM) is one of the most simple and effective adversarial attack techniques for machine learning models using DNNs. This method was introduced by Ian Goodfellow et al. in 2015 [8].

The aim of FGSM is to cause minor changes in the input data, small enough to make the machine learning model predict wrongly, but not large enough to change the meaning of the data. Since this method only uses the gradients of machine learning models, it is very fast and simple to implement. However, FGSM Attack also has certain limitations. Firstly, this method only works well for gradient-based machine learning models, and is not as suitable for non-gradient models as the decision tree-base model. Secondly, this method can be easily detected and countered if the machine learning model is trained with adversarial techniques. Expression of FGSM is presented generally and cited in many studies according to Formula (4):

$$X^{'} = X - \epsilon sign(\bigtriangledown_X J(X^{'}_N, y_t)) \tag{4}$$

With $X^{'}$ as adversarial examples image, X input as natural image, y input as the original node, ϵ to ensure low noise and J as the loss function. Ensuring that adversarial samples created using this method are identical to the real objects in their original images is not feasible. Adversarial samples are intentionally crafted to exploit vulnerabilities in machine learning models, making them appear visually similar to the original objects while causing misclassification or model manipulation. However, the goal of adversarial attacks is to deceive the model rather than accurately represent the real objects. As a result, there will always be a distinction between the original objects and their corresponding adversarial samples, as the latter are specifically designed to bypass the model's defenses and trigger unexpected behavior.

3.2   YOLOv7 Deep Learning Model and COCO Dataset

The selected deep learning model is the YOLO model version 7 (YOLOv7), the YOLOv7 model [27] was selected in this study to experiment to measure the robustness of the representative deep learning model. Through research and understanding of the authors, YOLOv7 model is a new updated version with many improvements such as increased learning speed, improved accuracy and better multitasking ability.

The COCO dataset [16], which stands for "Common Objects in Context", is a standard dataset commonly used in the field of computer vision. It is designed to study and evaluate algorithms for object detection, segmentation, and annotation. The datasets contains a large set of images representing diverse objects in various everyday situations.

One of the most powerful object detection and classification models at the moment. This is also a novelty when the author proposes an experimental model.

3.3   Proposed Model

The proposed model is shown as Fig. 2 and described in steps. Specifically, it is described through 6 steps as follows:

1. Randomly select the set S consisting of n samples from the COCO dataset, and the Test set consisting of m samples.
2. Attack the newly created image set S by FGSM attack model, obtain the set S' including noise.
3. Create the set $X = S + S'$.

4. Create a label for set X.
5. Use set X to train Yolov7 model.
6. Use the trained Yolov7 model to predict the test set S_{test}. Analyze the results obtained.

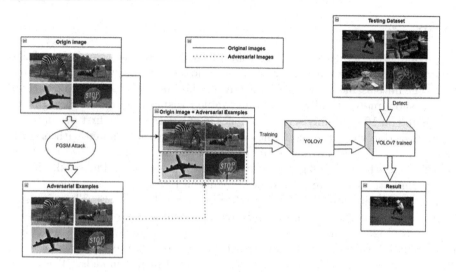

Fig. 2. Proposed model by the authors to detection adversarial example

After performing the configuration and installation steps designed according to the model observed in Fig. 2. The authors will conduct experiments according to the experimental flow in Fig. 3. Experimental process and specific results are presented in Sect. 4.

4   Experiments and Results

4.1   Experiments

Based on the proposed method and the experimental procedure show in Fig. 3. The author has experimented on a dataset of 2500 images including 1500 training images, 500 validating images, and 500 images to test after training has been completed. Adversarial examples are images taken randomly from the COCO dataset generated by the method presented in Sect. 3.1. The authors use the Makesense.ai platform to mark objects and assign new labels to these images. Next, these images are used to train the Yolov7 model. The end of the process will be the identification and evaluation of the image recognition results.

Figure 3 shows the steps in the experimental procedure concretized by the authors, showing the experimental flow diagram form.

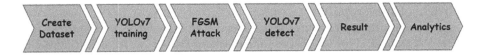

Fig. 3. Experimental flow in research

4.2 Results

After conducting experiments with 2500 images, the test set consisting of 250 AEs yielded a detection accuracy of 87.6% in identifying objects with perturbations. The misclassification rate on the adversarial samples was 30%. The authors wish to show more visually the results, Table 1 display the model Recognition rate results on 10 randomly selected images in the data set. We first evaluate the recognition rate when the model has not been adversarial trained with original (ORG) Image. We then evaluate the results of the adversarial trained model by recognizing both ORG Image and adversarial example (AE). The results in the table include the label name and the recognition rate (%).

The results are considered to be quite good by the authors. The results of Table 1 show that the model after adversarial training has recognized AE with a high rate. Figure 4 illustrates the effectiveness of the training model using two fundamental parameters: bias, and epoch (parameters are used to evaluate the effectiveness of the training process in the machine learning models). The model shows that there is good recognition ability after 500 training times and the prediction result deviation is not significant. The test run was performed on a diverse and large enough data set to ensure the effectiveness, reliability of the proposed model. After training by the built data set, the model becomes more robust and can distinguish noisy images on a random dataset.

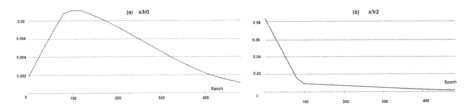

Fig. 4. The graph shows the improvement of bias (Vertical) over the number of training times (epoch - Horizontal) of the proposed model.

Experimental results have shown significant performance in object recognition and detection. The YOLOv7 model has achieved high accuracy and accurate object position detection in the image. The model has shown high performance in object recognition and detection while achieving fast processing speed and good

Table 1. Label and Recognition rate (%) on 10 random samples.

No	Image		ORG Model	Adversarial trained model	
	ORG Image	*AE*	with *ORG Image*	*ORG Image*	*AE*
1			Elephant: 93	Elephant: 85	Elep_noise: 72
2			Train: 96	Train: 92	Tr_noise: 37
3			Stop sign: 98	Stop sign:98	Ssign_noise: 88
4			Air plain: 92	Air plain: 95	Ap_noise: 75
5			Zebra: 89	Zebra: 90	Zebra_noise:70
6			Bed: 87	Bed: 75	B_noise: 93
7			Pizza: 60	Pizza: 52	Piz_noise: 75
8			Cow: 93, 94	Cow: 84, 63	Cow_noise: 61, 37
9			Toilet: 90, 90, 87	Toilet: 71, 88, 54	Toilet_noise: 80, 87, 68
10			Person: 94	Person: 87	P_noise: 95

generalization ability. The observed outcomes exemplify the potential and benefits of the proposed model within the realm of object recognition and position detection in images.

5   Conclusion

In this study, the authors investigated the popular defense models that have been studied by many groups of authors. From there, we propose a defensive model to detect adversarial examples from input data. Approaching the idea of one of the surveyed studies in model training [10]. The main contribution of the study is to use the YOLO model to perform the identification and prevention of adversarial examples. It is a seemingly simple method compared to the studies examined in the Sect. 2, but can effectively enhance the robustness of deep learning models. The experimental results indicate that the proposed model exhibits effectiveness in detecting adversarial examples under certain conditions.

The identification and detection of wrong objects are extremely dangerous for machine learning models that are applied in some areas of life such as autonomous driving, the medical field, etc. In the future, the team of authors The author hopes to perform model training with a larger amount of data in many data sets in many different fields, improving the accuracy and robustness of machine learning models and deep learning in particular.

References

1. Bastani, O., Ioannou, Y., Lampropoulos, L., Vytiniotis, D., Nori, A., Criminisi, A.: Measuring neural net robustness with constraints. In: Advances in Neural Information Processing Systems, vol. 29 (2016)
2. Bhagoji, A.N., Cullina, D., Mittal, P.: Dimensionality reduction as a defense against evasion attacks on machine learning classifiers. arXiv preprint arXiv:1704.02654, vol. 2, no. 1 (2017)
3. Borgwardt, K.M., Gretton, A., Rasch, M.J., Kriegel, H.P., Schölkopf, B., Smola, A.J.: Integrating structured biological data by kernel maximum mean discrepancy. Bioinformatics **22**(14), e49–e57 (2006)
4. Carlini, N., Wagner, D.: Adversarial examples are not easily detected: bypassing ten detection methods. In: Proceedings of the 10th ACM Workshop on Artificial Intelligence and Security, pp. 3–14 (2017)
5. Carlini, N., Wagner, D.: Towards evaluating the robustness of neural networks. In: 2017 IEEE Symposium on Security and Privacy (SP), pp. 39–57. IEEE (2017)
6. Feinman, R., Curtin, R.R., Shintre, S., Gardner, A.B.: Detecting adversarial samples from artifacts. arXiv preprint arXiv:1703.00410 (2017)
7. Gong, Z., Wang, W., Ku, W.S.: Adversarial and clean data are not twins. arXiv preprint arXiv:1704.04960 (2017)
8. Goodfellow, I.J., Shlens, J., Szegedy, C.: Explaining and harnessing adversarial examples. arXiv preprint arXiv:1412.6572 (2014)
9. Gretton, A., Borgwardt, K.M., Rasch, M.J., Schölkopf, B., Smola, A.: A kernel two-sample test. J. Mach. Learn. Res. **13**(1), 723–773 (2012)
10. Grosse, K., Manoharan, P., Papernot, N., Backes, M., McDaniel, P.: On the (statistical) detection of adversarial examples. arXiv preprint arXiv:1702.06280 (2017)
11. Gu, S., Rigazio, L.: Towards deep neural network architectures robust to adversarial examples. arXiv preprint arXiv:1412.5068 (2014)
12. Hendrycks, D., Gimpel, K.: Early methods for detecting adversarial images. arXiv preprint arXiv:1608.00530 (2016)

13. Huang, R., Xu, B., Schuurmans, D., Szepesvári, C.: Learning with a strong adversary. arXiv preprint arXiv:1511.03034 (2015)
14. Jin, J., Dundar, A., Culurciello, E.: Robust convolutional neural networks under adversarial noise. arXiv preprint arXiv:1511.06306 (2015)
15. Li, X., Li, F.: Adversarial examples detection in deep networks with convolutional filter statistics. In: Proceedings of the IEEE International Conference on Computer Vision, pp. 5764–5772 (2017)
16. Lin, T.-Y., et al.: Microsoft COCO: common objects in context. In: Fleet, D., Pajdla, T., Schiele, B., Tuytelaars, T. (eds.) ECCV 2014. LNCS, vol. 8693, pp. 740–755. Springer, Cham (2014). https://doi.org/10.1007/978-3-319-10602-1_48
17. Liu, X., et al.: Privacy and security issues in deep learning: a survey. IEEE Access 9, 4566–4593 (2020)
18. Metzen, J.H., Genewein, T., Fischer, V., Bischoff, B.: On detecting adversarial perturbations. arXiv preprint arXiv:1702.04267 (2017)
19. Nedic, B.: Gartner's top strategic technology trends (2019)
20. Ni, J., Chen, Y., Chen, Y., Zhu, J., Ali, D., Cao, W.: A survey on theories and applications for self-driving cars based on deep learning methods. Appl. Sci. 10(8), 2749 (2020)
21. Papernot, N., McDaniel, P., Wu, X., Jha, S., Swami, A.: Distillation as a defense to adversarial perturbations against deep neural networks. In: 2016 IEEE Symposium on Security and Privacy (SP), pp. 582–597. IEEE (2016)
22. Rozsa, A., Rudd, E.M., Boult, T.E.: Adversarial diversity and hard positive generation. In: Proceedings of the IEEE Conference on Computer Vision and Pattern Recognition Workshops, pp. 25–32 (2016)
23. Shaham, U., Yamada, Y., Negahban, S.: Understanding adversarial training: increasing local stability of neural nets through robust optimization. arXiv preprint arXiv:1511.05432 (2015)
24. Shen, F., Zhao, X., Kou, G., Alsaadi, F.E.: A new deep learning ensemble credit risk evaluation model with an improved synthetic minority oversampling technique. Appl. Soft Comput. 98, 106852 (2021)
25. Tran, K.A., Kondrashova, O., Bradley, A., Williams, E.D., Pearson, J.V., Waddell, N.: Deep learning in cancer diagnosis, prognosis and treatment selection. Genome Med. 13(1), 1–17 (2021)
26. Van Huong, P., Hua, T.Q., Minh, N.H., et al.: Feature generation by k-means for convolutional neural network in detecting IoT system attacks. In: 2021 IEEE International Conference on Machine Learning and Applied Network Technologies (ICMLANT), pp. 1–5. IEEE (2021)
27. Wang, C.Y., Bochkovskiy, A., Liao, H.Y.M.: Yolov7: trainable bag-of-freebies sets new state-of-the-art for real-time object detectors. In: Proceedings of the IEEE/CVF Conference on Computer Vision and Pattern Recognition, pp. 7464–7475 (2023)
28. Wang, D., Su, J., Yu, H.: Feature extraction and analysis of natural language processing for deep learning english language. IEEE Access 8, 46335–46345 (2020)

Orand - A Fast, Publicly Verifiable, Scalable Decentralized Random Number Generator Based on Distributed Verifiable Random Functions

Pham Nhat Minh[1,2,3], Chiro Hiro[3], and Khuong Nguyen-An[1,2(✉)]

[1] Department of Computer Science, Faculty of Computer Science and Engineering, Ho Chi Minh City University of Technology (HCMUT), 268 Ly Thuong Kiet Street, District 10, Ho Chi Minh City, Vietnam
{fminh.phamnminh129,nakhuong}@hcmut.edu.vn
[2] Vietnam National University Ho Chi Minh City, Linh Trung Ward, Thu Duc City, Ho Chi Minh City, Vietnam
[3] Orochi Network, Ho Chi Minh City, Vietnam
{fminh.pham,chiro}@orochi.network

Abstract. This paper introduces `Orand`, a fast, publicly verifiable, scalable decentralized random number generator designed for applications where public Proof-of-Randomness is essential. A reliable source of randomness is vital for various cryptographic applications and other applications such as decentralized gaming and blockchain proposals. Consequently, generating public randomness has attracted attention from the cryptography research community. However, attempts to generate public randomness still have limitations, such as inadequate security or high communication and computation costs. `Orand` is designed to generate public randomness in a distributed manner based on a suitable distributed verifiable random function. This approach allows `Orand` to enjoy low communication and computational costs. Moreover, `Orand` achieves the following security properties: pseudo-randomness, unbiasability, liveness, and public verifiability.

Keywords: Distributed verifiable random function · public randomness · decentralized random number generator · pseudo-randomness · blockchain

1 Introduction

Randomness has always played an important role in computational processes and decision-making in various applications in computer science, cryptography, and distributed systems. For instance, selecting the next leader to mint the next block in blockchain systems is very important and affects the entire blockchain security. If the leader's slot can be predetermined, or the randomness output can be biased, the leaders could collude to perform a double spending attack, compromising the system's security. Not only limited to distributed consensus

K. Honda et al. (Eds.): IUKM 2023, LNAI 14376, pp. 358–372, 2024.
https://doi.org/10.1007/978-3-031-46781-3_30

but various problems such as "maximal extractable value" (MEV) can only be addressed with the demand for reliable sources of randomness. By utilizing randomness to determine which transactions are included in a new block, transparency, and fairness are ensured. For application in artificial intelligence and machine learning, randomness is necessary for stochastic algorithms to improve model accuracy and avoid overfitting with small training sets. The role of randomness goes without saying for gaming since all the decisions and computations made in games rely on randomness.

However, many applications rely on local trusted parties to generate randomness and use it for conducting the application. With such a trusted randomness-generating mechanism, backdoors or biased randomness pieces may exist for some malicious actors' benefit. One of the most notorious examples is the Dual Elliptic Curve PRNG scandal[1], where a backdoor was inserted to predict the future outputs of the PRNG. Hence, creating a trustless source of randomness is crucial where no one can predict or bias the result.

Despite its significance, there has not been a successful project in creating a trustless source of randomness. To achieve feasibility, current attempts to generate such randomness require a trade-off between security and efficiency. Below, we briefly review some of the most notable attempts. Section 2 will give a more detailed survey on the constructions of Decentralized Random Number Generators in the literature.

Some blockchain oracle networks, e.g., *Chainlink*[2], combine an on-chain smart contract and off-chain server to generate random numbers [19] The smart contract listens to client requests and sends them to the server. *Chainlink*'s server employs a Verifiable Random Function in [7] to generate randomness. This guarantees that the outputs are computed correctly. The downside of this system is that it is centralized, and thus, the clients have to trust the server provider, which is unacceptable since centralized services can be easily corrupted.

Other protocols, e.g., *RANDAO* [18], *Algorand* [11] and *Ouroboros Praos* [4], generate randomness in a distributed manner. These protocols generate randomness by combining verifiable sources of partial randomness from participants. Although these protocols achieve low costs, the resulting outputs from these protocols can be biased. The last participant may refuse to reveal his partial secret, leading the protocol to abort and start over. They may do so until the protocol produces an output that benefits them.

SCRAPE [10] is one of a few protocols that achieve complete security properties. However, its communication and computation complexity is too high, making the system only be used for a few participants.

From the discussions above, we propose Orand, a decentralized system for random number generation that satisfies *pseudo-randomness*, *unbiasability*, *availability* and *publicly verifiable* and detail its construction for a use case that we have a special interest in, blockchain-based applications. Orand takes a step

---

[1] https://archive.nytimes.com/bits.blogs.nytimes.com/2013/09/10/government-announces-steps-to-restore-confidence-on-encryption-standards/.

[2] https://docs.chain.link/vrf/v2/introduction/.

further than other projects by ensuring these four mentioned properties while enhancing the system's scalability, making it more distributed and secure.

Our Contributions. We propose Orand, a *Decentralized Random Number Generator* (DRNG) for blockchain-based applications using a distributed VRF. To the best of our knowledge, our system is the first system that employs a distributed VRF for that purpose. We also apply the technique of [1] to optimize the randomness-generating process of the distributed VRF and Orand. Orand achieves the full security requirements of a DRNG while enjoying linear communication and quasi-linear computation costs in terms of participants and requires no advanced techniques such as pairings. With these properties, Orand is a scalable system suitable for a large number of participants.

The rest of the paper is organized as follows. In Sect. 2, we describe related works and compare Orand to these works. In Sect. 3, we recall basic notations and cryptographic primitives to be employed in the system, and in addition to that, we give formal definitions of a DRNG protocol and its security properties. In Sect. 4, we describe the construction and architecture of Orand with the experimental results of Orand's randomness generation time in Subsect. 4.4. We conclude the paper with Sect. 5.

2   Related Works

Using *"commit-then-reveal"* is a straightforward approach to producing public randomness. *RANDAO* [18] employs a random number generator based on the Ethereum blockchain. Each participant P_i chooses a secret value s_i and publishes a hash $\mathsf{SHA256}(s_i)$ viewed as P_i's commitment to s_i. When all parties committed their $\mathsf{SHA256}$ values, each participant P_i reveals his secret s_i. The final random number is calculated by computing the exclusive-or (XOR, $\oplus$) of the collected secret values. There is one issue with this approach. A dishonest participant can choose not to reveal his secret upon seeing the secrets of other participants, forcing the protocol to abort and restart until he obtains an output of his interest.

Using publicly verifiable secret sharing (PVSS) is another approach to constructing a DRNG. Many protocols, i.e., [9,10,13], follow this direction. Most of these protocols achieve full security properties as long as there are, at most, t dishonest participants. These protocols' limitations are that many of them have communication and computation costs exceeding n^2. *Randhound* [9] has a communication cost equal to c^2n. However, Randhound no longer achieves unbiasability due to letting the client freely separate the set of participants into different smaller groups for generating randomness.

Several protocols, namely, *Algorand* [11] and *Ouroboros Praos* [4], use *Verifiable Random Function* (VRF) to select the block producer. Participants commit to their VRF value in these protocols, determining the round's leader. The communication cost of both protocols is $\mathcal{O}(n)$. However, both protocols do not provide unbisability because the participant's contribution determines the result. An adversary who sees the VRF values of honest participants can choose which of his minions will contribute so that the resulting output would benefit him.

In [8], Dung *et al.* proposed an unpredictable number generator based on the concept of *"Flexible Proof-of-Work"* (Flexible PoW). Based on the current seed $seed_i$, each participant must find a solution to a puzzle involving the seed and a hash function for every epoch. Each solution will be hashed, and the resulting output will be the exclusive OR of these hashes. The protocol achieves $\mathcal{O}(n)$ communication cost, and its security is based on the unpredictability of the hash function. However, the protocol still faces the same problem as RANDAO since the last participant may refuse to publish his solution if the overall result does not benefit him.

In [20], Nguyen *et al.* propose to use *Homomorphic Encryption* (HE) in their DRNG system as another way to achieve linear communication and computation cost. In their protocol described in [19], each participant generates a secret, encrypts it, and publishes the ciphertext. Then, all the ciphertexts are joined. The requester with a secret key can decrypt the joined ciphertext to receive the randomness. The communication cost of [19] is $\mathcal{O}(n)$, but it requires the requester to be honest. Suppose the requester gives his secret key to a colluding participant. In that case, he can decrypt the secret of other participants before choosing his own so that the resulting secret would benefit him.

In 2019, Boneh *et al.* introduced *Verifiable Delay Function* (VDF) [5]. The main idea of a VDF is that the adversary cannot bias the output of the random beacon because it cannot calculate the result in the given time. Protocols that follow this direction are RandRunner [14] and Harmony [17]. However, VDF-based schemes do not provide pseudo-randomness; they only produce output that cannot be predicted, but these outputs may contain some bias.

Several DRNGs use a *Threshold Signature Scheme* (TSS) to provide randomness, e.g., [12,21]. Like **Orand**, these protocols require running a *Distributed Key Generation* (DKG) with a high initial setup cost, but the DKG is run only once. The main advantage of TSS-based DRNGs is that they achieve complete security properties while enjoying low communication and computation costs. However, these protocol involves complicated techniques, such as bilinear pairings.

Table 1 summarizes several aspects of the approaches mentioned above.

3   Preliminaries

In this section, we recall definitions and notions used throughout the paper, including Distributed Key Generation and Verifiable Random Function, the two main building blocks required for distributed VRF construction. We also give our formal definition and required properties of a DRNG.

Notations. Let λ be the security parameter. We say that $p(\lambda) \leq \mathsf{negl}(\lambda)$ to indicate that there exists a negligible function $\epsilon(\lambda)$ such that there exists $\lambda_0 > 0$ satisfying $p(\lambda) \leq \epsilon(\lambda)$ for all $\lambda \geq \lambda_0$. We denote by $\mathbb{G}$ a cyclic group of prime order p and let g and h be the generators of $\mathbb{G}$.

We say that an algorithm Alg is *probabilistic polynomial time* (PPT) if Alg runs within polynomial time in the size of its inputs. We denote the process

Table 1. A summary of existing DRNGs. These include: RANDAO [18], Scrape [10], HydRand [13], Algorand [11], Ouroboros Praos [4], Flexible PoW [8], Nguyen *et al.* [20], RandRunner [14], Harmony [17], drand [12] and this work

	Pseudo-randomness	Unpredictability	Unbiasability	liveness	Public Verifiability	Comm. Cost	Comp. Cost/ node	Verf. Cost / node	Honest nodes	Primitives
RANDAO	✓	✓	✗	✗	✓	$\mathcal{O}(n)$	$\mathcal{O}(1)$	$\mathcal{O}(n)$	1	Com.+ Rev
Scrape	✓	✓	✓	✓	✓	$\mathcal{O}(n^3)$	$\mathcal{O}(n^2)$	$\mathcal{O}(n^2)$	$n/2$	PVSS
HydRand	✗	✓	✓	✓	✓	$\mathcal{O}(n^2)$	$\mathcal{O}(n)$	$\mathcal{O}(n)$	$2n/3$	PVSS
Algorand	✗	✓	✗	✓	✓	$\mathcal{O}(cn)$	$\mathcal{O}(c)$	$\mathcal{O}(1)$	$2n/3$	VRF
Ouroboros Praos	✗	✓	✗	✓	✓	$\mathcal{O}(cn)$	$\mathcal{O}(c)$	$\mathcal{O}(1)$	$n/2$	VRF
Flexible PoW	✗	✓	✗	✗	✓	$\mathcal{O}(n)$	Varies	$\mathcal{O}(n)$	1	Hash func
Nguyen *et al.*	✓	✓	✗	✗	✓	$\mathcal{O}(n)$	$\mathcal{O}(n)$	$\mathcal{O}(n)$	1	HE
Randrunner	✗	✓	✓	✓	✓	$\mathcal{O}(n)$	VDF time	$\mathcal{O}(1)$	$n/2$	VDF
Harmony	✗	✓	✗	✓	✓	$\mathcal{O}(n)$	$O(1)$	$\mathcal{O}(n)$	$n/2$	VDF+VRF
drand	✗	✓	✓	✓	✓	$\mathcal{O}(n)$	$\mathcal{O}(t\log^2 t)$	$\mathcal{O}(t\log^2 t)$	$n/2$	TSS.
Orand	✓	✓	✓	✓	✓	$\mathcal{O}(n)$	$\mathcal{O}(t\log^2 t)$	$\mathcal{O}(t\log^2 t)$	$n/2$	Dist. VRF

of uniformly sampling x from a set $\mathcal{M}$ by $x \xleftarrow{\$} \mathcal{M}$. For a set $\mathcal{V}$, we define the Lagrange coefficient $\lambda_{i,\mathcal{V}}$ to be $\prod_{\substack{j \in \mathcal{V} \\ j \neq i}} \left(\frac{j}{j-i}\right)$. If $\mathcal{V} = \{v_1, v_2, \ldots, v_t\}$, denote $\mathcal{V}[a:b]$ to be the set $\{v_a, \ldots, v_b\}$.

3.1   Network and Adversarial Model

In this paper, we consider the *synchronous* network model, i.e., the time delay between messages is bounded within a time δ. It is assumed that there is a *broadcast channel*, where everyone can see a message broadcast by a participant. In addition, each pair of participants has a *private channel* that allows them to communicate with each other securely. We restrict an adversary $\mathcal{A}$ to be *static*, i.e., initially, $\mathcal{A}$ must choose and corrupt a fixed set of participants. By "corrupt", we mean that the adversary can see all the private inputs and act on behalf of the corrupted participant. In a synchronous setting, the adversary $\mathcal{A}$ is allowed to corrupt up to $n/2$ participants.

3.2   Distributed Key Generation

Distributed Key Generation A (t, n)−Distributed Key Generation (DKG) protocol allows n participants to jointly create a pair of keys (pk, sk) without having to rely on a trusted party (dealer). Each participant also receives a partial secret key and a partial public key. While the public key pk is publicly seen by everyone, the secret key sk remains secret as a (virtual) (t,n)−secret shared via a secret sharing scheme, where each share is the partial secret key of a participant [15]. No adversary can learn anything about the secret key if it does not control the required number of participants. The secret key sk will be later used in a threshold cryptosystem, e.g., to produce digital signatures or decryption.

Among existing DKG protocols, we now describe the DKG protocol of Gennaro et al. [15]. The detail of the protocol can be seen in Protocol 1.

Protocol 1. Distributed Key Generation

Generating sk :

1. Each participant P_i chooses two random polynomials $f_i(z) = a_{i0} + a_{i1}z + ... + a_{it}z^t$ and $f'_i(z) = b_{i0} + b_{i1}z + ... + b_{it}z^t$ and broadcasts $C_{ik} = g^{a_{ik}}h^{b_{ik}}$ for $k = 0, 1, ..., t$. Then P_i computes $s_{ij} = f_i(j)$, and $s'_{ij} = f'_i(j)$ and securely sends (s_{ij}, s'_{ij}) to P_j, $\forall j \neq i$.
2. Each participant P_j verifies the shares he received from P_i by checking

$$g^{s_{ij}}h^{s'_{ij}} \stackrel{?}{=} \prod_{k=0}^{t} C_{ik}^{j^k}. \tag{1}$$

If the check fails for some i, then P_j complains against P_i.
3. Each P_i who receives a complaint from P_j broadcasts (s_{ij}, s'_{ij}) that satisfy (1).
4. A participant P_i is disqualified if he receives $t + 1$ complaints or answers with value that does not satisfy (1). A set QUAL of qualified participants is then determined. For each i, the secret key sk_i of P_i is equal to $\sum_{j \in \text{QUAL}} s_{ji}$. For any set $\mathcal{V}$ of at least $t + 1$ participants, the secret key sk is equal to $\sum_{i \in \mathcal{V}} sk_i \cdot \lambda_{i,\mathcal{V}}$.

Extracting pk $= g^{sk}$:

1. Each participant P_i in the set QUAL publishes $A_{ik} = g^{a_{ik}}$ for $k = 0, 1, 2, ..., t$.
2. Each participant P_j verifies A_{ij} for each i by checking whether $g^{s_{ij}} \stackrel{?}{=} \prod_{k=0}^{t} A_{ik}^{j^k}$. If the verification fails at i, then P_j complains against P_i. If the check fails for some i, then P_j complains against P_i.
3. For each i that P_i receives at least one valid complaint, all other parties run the reconstruction phase of Pedersen VSS to reconstruct $f_i(z)$ and restore s_{i0} and A_{ij} for $j = 0, 1, ..., t$. The public key is equal to pk $= \prod_{i \in \text{QUAL}} A_{i0}$. The public key pk_i of P_i is calculated as $pk_i = g^{sk_i} = \prod_{j \in \text{QUAL}} g^{s_{ji}} = \prod_{j \in \text{QUAL}} \prod_{k=0}^{t} A_{jk}^{i^k}$.

3.3   VRF Based on Elliptic Curves (ECVRF)

Micali, Rabin, and Vadhan introduced Verifiable Random Functions in [16]. VRFs can be seen as a public key version of a pseudorandom function $F_{sk}(x)$. VRF allows a party with the secret key sk to produce pseudorandom output $F_{sk}(x)$ and a proof π certifying that the output is computed correctly. The result can be verified publicly using the corresponding public key pk.

We now describe a VRF construction based on elliptic curves in [9]. The VRF consists of three algorithms KeyGen, VRFEval and VRFVerify. The secret key holder uses VRFEval to compute the VRF output Y, while anyone can use VRFVerify to verify the correctness of Y.

Public Parameters. Let p be a prime number and $\mathbb{G}$ be a cyclic group of order p and generator g. Let H_1 be a hash function that maps a bit string to an element in $\mathbb{G}$. Let H_2 be a hash function mapping arbitrary input length to an integer. The hash functions $\mathsf{H}_1, \mathsf{H}_2$ are modeled as random oracle models. The public parameters are $p, \mathbb{G}, g, \mathsf{H}_1, \mathsf{H}_2$. The algorithms $\mathsf{VRFSetup}, \mathsf{VRFEval}$ and $\mathsf{VRFVerify}$ are described in Fig. 1.

$\mathsf{VRFSetup}(1^\lambda)$	$\mathsf{VRFEval}(\mathsf{sk}, X)$	$\mathsf{VRFVerify}(\mathsf{pk}, X, Y, \pi)$
$\mathsf{sk} \xleftarrow{\$} \mathbb{Z}_p$	Compute $h = \mathsf{H}_1(X)$ and $\gamma = h^{\mathsf{sk}}$.	Compute $h = \mathsf{H}_1(X)$.
$\mathsf{pk} := g^{\mathsf{sk}}$	Sample $k \xleftarrow{\$} \mathbb{Z}_p$	Compute $u = \mathsf{pk}^c g^s$ and $v = \gamma^c h^s$.
Return $(\mathsf{pk}, \mathsf{sk})$	Compute $c = \mathsf{H}_2(g, h, \mathsf{pk}, \gamma, g^k, h^k)$.	Check if $c = \mathsf{H}_3(g, h, \mathsf{pk}, \gamma, g^k, h^k)$
	Compute $s \equiv k - c \cdot \mathsf{sk} \pmod{p}$	Check if $Y = \gamma$
	Output $(Y, \pi) = (\gamma, (\gamma, c, s))$	

Fig. 1. Verifiable Random Function based on Elliptic Curve

3.4 DRNG Definition and Properties

This section is devoted to giving our formal description of a DRNG and the required properties of a DRNG. In the next section, we propose our protocol and system satisfying these properties based on these formal definitions.

Definition 1 (DRNG Protocol). *A (t, n)−decentralized random number generator protocol on a set of participants $\mathcal{P} = \{P_1, P_2, \ldots, P_n\}$ is an epoch-based protocol, where each epoch r consists of two interactive protocols* $\mathsf{DRNGSetup}, \mathsf{DRNGGen}$, *and an algorithm* $\mathsf{DRNGVerify}$ *and a global state* st, *working as follows*

1. $(\mathsf{st}, \mathsf{QUAL}, \mathsf{pp}, \{\mathsf{sk}_i\}_{i \in \mathsf{QUAL}}) \leftarrow \mathsf{DRNGSetup}(\lambda) \langle\{P\}_{P \in \mathcal{P}}\rangle$: *This is an interactive protocol run by all participants in $\mathcal{P}$ to determine the member list of qualified committees. At the end of the interaction, a set QUAL of qualified participants is determined, the st is initialized, and a list pp of public information is known to all participants. Each participant $P_i \in \mathsf{QUAL}$ also obtains his secret key sk_i that is only known to him.*

2. $(\mathsf{st} := \mathsf{st}', \Omega, \pi) \leftarrow \mathsf{DRNGGen}(\mathsf{st}, \mathsf{pp}) \langle\{P_i(\mathsf{sk}_i)\}_{i \in \mathsf{QUAL}}\rangle$: *This is an interactive protocol between participants in a set QUAL each holding the secret key sk_i and common inputs st, pp. At the end of the interaction, all honest participants output a value Ω and a proof π certifying the correctness of Ω made by the interaction. In addition, the state st is updated into a new state st', denoted by $\mathsf{st} := \mathsf{st}'$.*

3. $b \leftarrow \mathsf{DRNGVerify}(\mathsf{st}, \Omega, \pi, \mathsf{pp})$: *This algorithm is run by an external verifier. On input a state st, a value Ω, a proof π, a public parameter pp, this algorithm output a bit $b \in \{0, 1\}$ certifying the correctness of Ω.*

Next, we define the security properties of a DRNG abstractly, and in our forth-coming work, we will give a more formal general definition for these properties.

Definition 2 (Security of DRNG). *A secure DRNG protocol is a DRNG protocol satisfying the following properties.*

- ***Pseudo-randomness.*** *Let $\Omega_1, \Omega_2, \ldots, \Omega_r$ be outputs generated so far. We say that a $(t, n)-DRNG$ satisfies pseudo-randomness if for any future outputs Ω_j where $j > r$, for any PPT adversaries $\mathcal{A}$ who corrupts up to t participants in $\mathcal{P}$, there exists a negligible function negl such that*

$$|\Pr[\mathcal{A}(\Omega_j) = 1] - \Pr[\mathcal{A}(Y) = 1]| \leq \mathsf{negl}(\lambda),$$

where $Y \xleftarrow{\$} \{0,1\}^\lambda$ is a uniformly random element from $\{0,1\}^\lambda$.

- ***Unbiasability.*** *We say that the DRNG satisfies unbaisability if any adversary $\mathcal{A}$ who corrupts up to t participants in $\mathcal{P}$ should not be able to affect future random beacon values for his own goal.*

- ***Liveness.*** *We say that the DRNG satisfies the liveness property if, for any epoch, and for any adversary $\mathcal{A}$ who corrupts up to t participants in $\mathcal{P}$, the DRNGGen protocol is guaranteed to produce an output.*

- ***Public Verifiability.*** *We say that the DRNG satisfies publicly verifiability if given the current global state st', the public parameter pp and values $\Omega^*, \pi^* \in \{0,1\}^*$, then an external verifier can run DRNGVerify(st', pp, Ω^*, π^*) to determine if Ω^* is properly calculated from the DRNGGen protocol.*

4 Our Approach

We begin this section by describing the *Distributed VRF* (DVRF) protocol of Galindo *et al.* in [6], which will be the heart of Orand. Then, we propose Orand, our protocol, together with a system of DRNG especially designed for random-ness sources used in blockchain-based applications.

4.1 Distributed VRF

Protocol Construction. This subsection describes the DVRF protocol of Galindo *et al.* in [6]. Orand employs the DVRF protocol for generating ran-domness. The protocol is divided into epochs. In the first epoch, all participants execute the DRNGSetup protocol. From the second epoch on-wards, the value R_{r+1} of epoch r is deterministically obtained from R via the DRNGGen proto-col. The protocol details can be seen at Protocol 2. The DVRF protocol achieves *pseudo-randomness, unbiasability, liveness* and *publicly-verifiability*, assuming a synchronous network and a static adversary who corrupts up to $(n/2 - 1)$ par-ticipants. The detailed security proof of the protocol can be found in [6].

Protocol 2. Distributed VRF

In the protocol, the global state st has the form st = (st$_r$.X, st.r).

DRNGSetup(1^λ) ⟨{P}$_{P \in \mathcal{P}}$⟩ : All participants $P \in \mathcal{P}$ follow the DKG protocol of Gennaro *et al.*(see Protocol 1) to select the list QUAL of qualified participants P_i and obtain their secret and public keys (pk$_i$, sk$_i$). The public parameter pp is set to be pp := ({pk$_i$}$_{i \in \text{QUAL}}$).

DRNGGen(st, pp) ⟨{P_i(sk$_i$)}$_{i \in \text{QUAL}}$⟩ : This interactive protocol is run in each epoch to generate a pseudorandom output. It proceeds as follows:

1. The i-th participant in the set QUAL runs the algorithm VRFEval(st.X, sk$_i$) (see Figure 1) of the ECVRF to compute the value $Y_i = $ H$_1$(st.X)$^{\text{sk}_i}$ and broadcasts his value Y_i and his proof π_i;
2. For each P_i who broadcasted Y_i, other participants compute the value VRFVerify(st.X, Y_i, π_i, pk$_i$) (see Figure 1) to verify the validity of Y_i;
3. From a set $\mathcal{V}_r$ of participants with size $\geq t + 1$ who broadcasted correct outputs, participants calculate the output Ω and its proof π;
   (a) Let $\gamma = \prod_{i \in \mathcal{V}_r} Y_i^{\lambda_{i,\mathcal{V}_r}}$. The output Ω is calculated to be $\Omega = $ H$_3$(γ).
   (b) The proof π of Ω consists of {$(i, Y_i, \pi_i) \mid i \in \mathcal{V}$}.
4. Update st.r = st.r + 1 and st.X = Ω.

DRNGVerify(st, Ω, π, pp) : An external verifier, on input a state st, a value Ω, a proof $\pi = \{(i, Y_i, \pi_i) \mid i \in \mathcal{V}\}$ and a public parameter pp can verify the correctness of Ω as follow:

1. Check if VRFVerify(st.X, Y_i, π_i, pk$_i$) = 1 for all $i \in \mathcal{V}$;
2. Check if $\Omega = $ H$_3$(γ) where $\gamma = \prod_{i \in \mathcal{V}_r} Y_i^{\lambda_{i,\mathcal{V}_r}}$;
3. If all checks pass, output 1, otherwise output 0.

Optimization. In the DVRF protocol, for each epoch r, we can combine these outputs for a set $\mathcal{V}_r$ of participants who have published valid outputs to get the final result. However, the $\mathcal{V}_r$ can be different for each r. For example, some participants contributed outputs in the $(r-1)$-th epoch but suddenly became offline in the r-th epoch, and vice versa. We see that, calculate γ_r, first, we need to calculate all Lagrange coefficients $\lambda_{i,\mathcal{V}_r}$ for all $i \in \mathcal{V}_r$. Because the set $\mathcal{V}_r$ differs for each epoch r, these Lagrangian coefficients must also be recalculated for each epoch r. By doing naive Lagrange interpolation, this cost $O(t^2)$ computations. Fortunately, in [1], the authors showed that it is possible to calculate all Lagrange coefficients within $\mathcal{O}(t \log^2 t)$ steps. Hence, we apply the method of [1] to optimize the randomness-generating process for DVRF and Orand.

We assume the number of participants is a power of two. We will replace their ID from $\{1, 2, ..., n\}$ to $\{1, \omega, \omega^2, ..., \omega^{n-1}\}$, where ω is a n-th root of unity in $\mathbb{F}_p$. For a set $\mathcal{V}$, let $L_i^{\mathcal{V}}(x) = \prod_{j \in \mathcal{V}, j \neq i}(x - \omega^j)/(\omega^i - \omega^j)$. Our main interest is to calculate $L_i^{\mathcal{V}}(0)$ for all $i \in \mathcal{V}$. To do this, let $N(x) = \prod_{i \in \mathcal{V}}(x - \omega^i)$, $N_i(x) = N(x)/(x - \omega^i)$, and $D_i = \prod_{i \in \mathcal{V}}(\omega^i - \omega^j)$. Then $L_i^{\mathcal{V}}(x) = N_i(x)/D_i$. The method to calculate $L_i^{\mathcal{V}}(0)$ for all $i \in \mathcal{V}$ within $\mathcal{O}(t \log^2 t)$ steps, as described in [1] is as follow.

1. Get all coefficients of $N(x)$ recursively. This can be done within $\mathcal{O}(t \log^2 t)$ steps using Procedure 3. It can be shown in [1] that the procedure for getting all the coefficients of $N(x)$ runs in $\mathcal{O}(t \log^2 t)$ steps.

Procedure 3. GetCoefficients($N(x)$)

Input: $N(x) = \prod_{i \in \mathcal{V}} (x - \omega^i)$
Output: List of coefficients $(n_0, n_1, ..., n_t)$ satisfying $N(x) = \sum_{i=0}^{t} n_i x^i$.

1. Factor $N(x) = N_1(x) N_2(x)$ where $N_1(x) = \prod_{i \in \mathcal{V}[1:t/2]} (x - \omega^i)$ and $N_2(x) = \prod_{i \in \mathcal{V}[t/2+1:t]} (x - \omega^i)$ both have degree $t/2$.
2. Recursively execute GetCoefficient($N_1(x)$) and GetCoefficient($N_2(x)$) get the coefficients of $N_1(x)$ and $N_2(x)$. Next, calculate $N_1(\omega^i)$ and $N_2(\omega^i)$ for all $i = 0, 1, ..., n - 1$ using FFT.
3. Compute $N_1(\omega^i) N_2(\omega^i)$ for all $i = 0, 1, ..., n - 1$ and use inverse FFT to get all coefficients of $N(x)$.

---

2. Calculate $N_i(0)$ for all $i \in \mathcal{V}$. We see that $N_i(0) = N(0)/(-\omega^i)$, thus calculating all $N_i(0)$ takes $\mathcal{O}(t)$ steps.
3. Calculate the derivative $N'(x)$. If the coefficients of $N(x)$ are $a_0, a_1, ..., a_t$ then the coefficients of $N'(x)$ are $a_1, 2a_2, ..., ta_t$, thus this takes $O(t)$ steps.
4. Calculate $N'(\omega^i)$ for all $i \in \mathcal{V}$ through a single *Fast Fourier Transform* (FFT) in $\mathcal{O}(t \log t)$ steps. Finally, we see that $L_i^{\mathcal{V}}(0) = N_i(0)/N'(\omega^i)$ for all $i \in \mathcal{V}$.

4.2 Orand System Architecture

This section presents Orand, a DRNG system for blockchain-based applications. Orand employs the DVRF protocol above for randomness generation. While the DVRF protocol is used solely for generating randomness, Orand introduces additional components to realize the randomness-generating process in the blockchain environment. As a DRNG system, Orand meets all required prerequisites in [3]; therefore, it is reasonable to build Orand on a blockchain environment and deploy Orand for blockchain-based applications.

Orand has seven fundamental blocks, as seen in Fig. 2a.

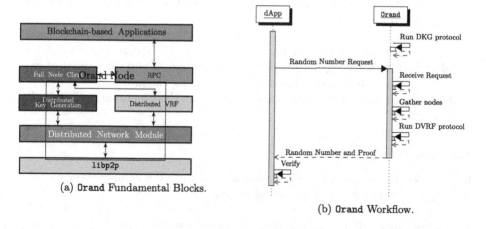

(a) Orand Fundamental Blocks.

(b) Orand Workflow.

Fig. 2. Overview of Orand.

- **Blockchain-based Application**: A client blockchain-based application that wishes to use random numbers from Orand.

- **RPC**: Remote Produce Call (RPC) enables communication between the client application and Orand.

- **Full-Node Client**: This module handles all the logical requests from the client application to Orand.

- **Distributed Key Generation**: This component is the Distributed Key Generation. The joining participants can generate public and secret keys for the Distributed VRF protocol.

- **Distributed VRF**: At the core of Orand lies the Distributed VRF protocol that allows participants to generate a pseudo-random output and its proof.

- **Distributed Network Module**: This network discovers and allows participants to connect and securely communicate with each other. It also supports node discovery, a mechanism enabling a new participant to find all other existing participants in a network.

- libp2p: This module contains various libraries that support peer-to-peer communication between participants, e.g., secure channels and messaging.

When an arbitrary decentralized application (dApp) needs to generate a random number from Orand, it follows the process in Fig. 2b. The detail is as follow.

1. Initially, the participants of Orand execute the **Distributed Key Generation** protocol to create a pair of key $(\mathsf{pk}, \mathsf{sk})$. In addition, each participant P_i also receives a partial public-secret key pair $(\mathsf{pk}_i, \mathsf{sk}_i)$.

2. When a dApp needs to receive a random number from Orand, it sends a signal to Orand to begin the generation process.

3. All qualified participants in Orand execute the DRNGGen protocol of the **Distributed VRF** to generate the random numbers and their proof of correctness. The dApp then verifies the output using the proof.

4. After the verification process, the dApp can use the generated output for its application.

4.3   Complexity Analysis

In this section, we analyze the communication cost, the computation cost for each node, and the verification cost of Orand.

Communication Cost. In the first epoch, each participant P_i has to broadcast t group elements, then send n shares s_{ij} to other participants. Hence, total

$\mathcal{O}(n^2)$ messages are sent. From the second epoch forward, each participant P_i has to send his partial output to the broadcast channel. Thus, $\mathcal{O}(n)$ messages are sent.

Computation Cost. In the first epoch, each participant P_i has to compute $(s_{ij}, s'_{ij}) = f_i(j)$. Using FFT, this requires $\mathcal{O}(n \log n)$ computation cost. From the second epoch, each participant P_i has to compute $(Y_i, \pi_i) = \mathsf{VRFEval}(X, \mathsf{sk}_i)$. Finally, when $(t + 1)$ valid outputs have been published, participants must compute γ_r. This results in $\mathcal{O}(t \log^2 t)$ computation steps.

Verification Cost. To validate the output, an external verifier first needs to check the validity of (Y_{ri}, π_{ri}) for each P_i that broadcasted his outputs, then compute γ_r and check if $R = \mathsf{H_3}(\gamma)$. This cost $\mathcal{O}(t \log^2 t)$ computations.

4.4   Experiment

We have implemented a prototype of our DRNG protocol Orand using the *Rust* programming language. Our implementation is based on the ECDSA threshold signature protocol of ING Bank. We modify the DKG protocol of ING Bank to match ours. We use the curve *secp256k1* for our implementation. We use the performance of the *libsecp256k1*[3] library for primary curve operations.

We evaluate our result by considering two aspects: *performance* and *security*. The details of these aspects can be seen below.

1. For performance, we benchmark the time required to generate a single random value in one round for different nodes. To show the efficiency of our protocol for a large number of participants, even in the worst case, especially after optimization, we experiment with a large number of participants (up to 512) and compare both the original and optimized versions. We experimented with $32, 64, 128, 256$, and 512 nodes, respectively, where each node represents a participant in the protocol. The experiments were run on a Macbook with a $2\,\mathrm{GHz}$ Quad-Core Intel Core i5 processor and 16 GB of memory.

2. For security, we use the NIST Test suite [2] to show that Orand produces outputs that satisfy the required statistical properties of a truly random output bit. This indicates that Orand has truly unbiased random output. We test our implementation with 20 numbers; each number is $100,000$ bits long.

Performance Test Result. The results of the experiments can be seen in Table 2. Regarding the table, the optimized version outperforms the original at $n = 128$. This is because the optimization cost is approximately close to $cn \log^2(n)$ for a constant c. Hence, when n is smaller than c, this cost is higher than the original cost of n^2. Nevertheless, the experiment shows that the optimized version outperforms the original when the number of nodes is quite large, i.e., $n \geq 128$. In addition, the experiments show that it only takes 11 and 31

---

[3] https://github.com/paritytech/libsecp256k1.

seconds for 256 and 512 nodes to generate a single random value, respectively. This indicates that Orand can be used for many nodes.

NIST Statistical Test Result. Following the instructions of [2, Subs. 4.2.1], for $m = 20$, the acceptable range is equal to $0.99 \pm 3\sqrt{(0.99(1 - 0.99)/20} = [0.923, 1]$. From Table 3, all the pass rates of the tests exceed 0.95, indicating that the Orand protocol passes the NIST Test Suite.

Table 2. Performance Test Result.

Value of n	32	64	128	256	512
Produce an ECVRF value (ms)	4	4	4	4	5
Verify ECVRF values (ms)	182	374	715	1409	2856
Combine (Original) (ms)	209	824	3350	14840	59301
Combine (Optimized) (ms)	401	1195	3042	10311	27650
Total (Original) (ms)	305	1202	4069	16253	62102
Total (Optimized) (ms)	587	1573	3761	11724	30511

Table 3. NIST Statistical Test Result.

Test name	p-value	Pass rate
Frequency	0.122325	20/20
Frequency in a Block	0.437274	20/20
Run	0.035174	20/20
Longest Run in a Block	0.275709	19/20
Binary Matrix Rank	0.637119	20/20
Discrete Fourier	0.350485	20/20
Non-Overlapping Template	0.964295	20/20
Overlapping Template	0.213309	20/20
Universal Statistical	0.967352	20/20
Linear Complexity	0.004301	19/20
Serial	0.739918	20/20
Approximate Entropy	0.048716	19/20
Cumulative Sums	0.275709	20/20

5   Conclusion

In this paper, we introduced Orand, a fast, publicly verifiable, scalable, decentralized random number generator for blockchain-based applications. We also found a way to optimize the protocol for faster randomness generation for each epoch. Orand has $\mathcal{O}(n)$ communication cost and $\mathcal{O}(n \log^2 n)$ computation cost after optimization while achieving full security properties. In addition, the protocol can be implemented with the curve *secp256k1*, the curve used by Ethereum and supported by many existing libraries. Thus, Orand offers great application in the context of blockchain.

In our forthcoming works, we will give complete, detailed security proofs of our DRNG and implement the full version of the system for the use of the community. Furthermore, the proof size of the protocol is $\mathcal{O}(t)$ since it consists of all $(t + 1)$ ECVRF proofs of partial outputs. We can use bilinear maps to aggregate the proofs to reduce their size. However, bilinear maps can only be used on certain curves, for instance, *BN254*. Most libraries only support the curve *secp256k1*, and this curve does not allow us to use bilinear maps. We hope to reduce the proof size of the DVRF protocol in the future.

Acknowledgement. The first author and the corresponding author acknowledge Ho Chi Minh City University of Technology (HCMUT), VNU-HCM, for supporting them in this study. The authors thank Tang Khai Hanh for valuable comments and suggestions during the manuscript preparation to help improve this work significantly.

References

1. Tomescu, A., et al.: Towards scalable threshold cryptosystems. In: 2020 IEEE SSP, pp. 877–893. IEEE (2020)
2. Rukhin, A., et al.: A statistical test suite for random and pseudorandom number generators for cryptographic applications, vol. 22, National Institute of Standards and Technology (2001)
3. Betzwieser, B., et al.: A decision model for the implementation of blockchain solutions. In: AMCIS 2019. Association for Information Systems (2019)
4. David, B., Gaži, P., Kiayias, A., Russell, A.: Ouroboros praos: an adaptively-secure, semi-synchronous proof-of-stake blockchain. In: Nielsen, J.B., Rijmen, V. (eds.) EUROCRYPT 2018. LNCS, vol. 10821, pp. 66–98. Springer, Cham (2018). https://doi.org/10.1007/978-3-319-78375-8_3
5. Boneh, D., Bonneau, J., Bünz, B., Fisch, B.: Verifiable delay functions. In: Shacham, H., Boldyreva, A. (eds.) CRYPTO 2018. LNCS, vol. 10991, pp. 757–788. Springer, Cham (2018). https://doi.org/10.1007/978-3-319-96884-1_25
6. Galindo, D., et al.: Fully distributed verifiable random functions and their application to decentralised random beacons. In: IEEE EuroS&P 2021, pp. 88–102. IEEE (2021)
7. Papadopoulos, D., et al.: Making nsec5 practical for dnssec. Cryptology ePrint Archive (2017). https://eprint.iacr.org/2017/099.pdf
8. Dung and Khuong et al.: Flexible proof-of-work: a decentralized unpredictable random number generator. In Proceedings of 10th National Conference on Fundamental and Applied IT Research (FAIR), vol. 2017, pp. 185–191. Publ. House for Sci & Tech (2017)

9. Syta, E., et al.: Scalable bias-resistant distributed randomness. In: IEEE S&P 2017, pp. 444–460. IEEE Computer Society (2017)

10. Cascudo, I., David, B.: SCRAPE: scalable randomness attested by public entities. In: Gollmann, D., Miyaji, A., Kikuchi, H. (eds.) ACNS 2017. LNCS, vol. 10355, pp. 537–556. Springer, Cham (2017). https://doi.org/10.1007/978-3-319-61204-1_27

11. Chen, J., Micali, S.: Algorand: a secure and efficient distributed ledger. Theoret. Comput. Sci. **777**, 155–183 (2019)

12. Nicolas, G., et al.: Drand - distributed randomness beacon (2020)

13. Schindler, P., et al.: Hydrand: efficient continuous distributed randomness. In: IEEE S&P 2020, pp. 73–89. IEEE (2020)

14. Schindler, P., et al.: RandRunner: distributed randomness from trapdoor VDFs with strong uniqueness. In: NDSS 2020. The Internet Society (2021)

15. Gennaro, R., Jarecki, S., Krawczyk, H., Rabin, T.: Secure distributed key generation for discrete-log based cryptosystems. In: Stern, J. (ed.) EUROCRYPT 1999. LNCS, vol. 1592, pp. 295–310. Springer, Heidelberg (1999). https://doi.org/10.1007/3-540-48910-X_21

16. Micali, S., et al.: Verifiable random functions. In: FOCS 1999, pp. 120–130. IEEE Computer Society (1999)

17. Team Harmony. Technical whitepaper - harmony (one). White Paper. https://harmony.one/whitepaper.pdf (2019)

18. Team Randao. Randao: Verifiable random number generation. White Paper. https://www.randao.org/whitepaper/Randao_v0.85_en.pdf (2017)

19. Thanh and Khuong et al.: Scalable distributed random number generation based on homomorphic encryption. In: IEEE Blockchain 2019, pp. 572–579. IEEE (2019)

20. Thanh and Khuong et al.: A system for scalable decentralized random number generation. In: 2019 IEEE 23rd International Enterprise Distributed Object Computing Workshop (EDOCW), pp. 100–103. IEEE (2019)

21. Team DFINITY. The internet computer for geeks. IACR Cryptol. ePrint Arch., p. 87 (2022)

Author Index

K. Honda et al. (Eds.): IUKM 2023, LNAI 14376, pp. 373–375, 2024.
https://doi.org/10.1007/978-3-031-46781-3

Printed in the United States
by Baker & Taylor Publisher Services

Printed in the United States
by Baker & Taylor Publisher Services